STATISTICAL METHODS FOR TECHNOLOGISTS

GENERAL ENGINEERING SERIES

General Editor

D. G. CHRISTOPHERSON, O.B.E., D.Phil., F.R.S.

Warden of the Durham Colleges,
Vice-Chancellor, University of Durham;
formerly Professor of Applied Science with reference to Engineering,
Imperial College of Science and Technology, University of London

---

Fundamental Thermodynamics for Engineers

M. H. Everdell, B.Sc., Ph.D.

Statistical Methods for Technologists

C. G. Paradine, M.A., and B. H. P. Rivett, M.Sc.

Mathematical Methods for Technologists

Arranged by Margaret Strain, M.A.

Engineering Dynamics

S. A. V. Swanson, B.Sc.(Eng.), Ph.D., D.I.C.

Matrices: Their Meaning and Manipulation

W. G. Bickley, D.Sc., A.C.G.I. (Hon.), F.R.Ae.S.
and R. S. H. G. Thompson, B.Sc.

STATISTICAL METHODS FOR TECHNOLOGISTS

C. G. PARADINE, M.A.

Formerly Senior Lecturer in Mathematics, Battersea College of Advanced Technology

B. H. P. RIVETT, M.Sc.

Professor of Operational Research, University of Sussex, Past President of the Operational Research Society

THE ENGLISH UNIVERSITIES PRESS LTD

ST PAUL'S HOUSE WARWICK LANE

LONDON E.C.4

First Printed 1953 as STATISTICS FOR TECHNOLOGISTS
Second Edition 1960
Second Impression 1960
Third Impression (with corrections) 1964
Fourth Impression 1966
Fifth Impression 1968
Sixth Impression 1970

ISBN 0 340 04901 4 Case bound Edition
ISBN 0 340 05343 7 Paper bound Edition

PRINTED AND BOUND IN ENGLAND
FOR THE ENGLISH UNIVERSITIES PRESS LTD
BY HAZELL WATSON AND VINEY LTD, AYLESBURY

EDITOR'S FOREWORD

by

D. G. CHRISTOPHERSON
O.B.E., D.Phil., F.R.S.

Warden of the Durham Colleges;
Vice-Chancellor, University of Durham;
formerly Professor of Applied Science with reference to Engineering,
Imperial College of Science and Technology, University of London

The techniques of the statistician are becoming every year of greater importance to the scientist and technologist. Not only are there completely new subjects of study, such as quality control, which are based directly on statistical methods, but the application of sound analytical methods to the great body of information obtained from experiment and research is becoming ever more widespread and more necessary.

On the other hand, although most degree courses for engineers now include some of the fundamentals of mathematical statistics, in most cases the time available for this subject is limited, and it is difficult to include all the techniques which the professional engineer is likely to find of practical value in his work. Moreover, the engineer is, in practice, very often more concerned with the details of how the computation is actually done in order to reveal the results more simply and conclusively, rather than with the fundamental underlying theory.

The present volume is intended to cover just that need. It is addressed to the practical engineer and research worker rather than the professional statistician, though statisticians may find it a useful and concise survey of all the main methods of analysis now in use. The book was most successful when first issued in 1953, and in its present form, revised and brought up to date in a number of places, it is hoped that it will prove useful to a still wider public.

PREFACE

Many of those called upon to apply statistical methods, if only from time to time, in industry or research, have to acquire their knowledge from books, after graduating in science, engineering or mathematics. More recent graduates in engineering and science will have received an introduction to probability and statistics as part of their course in mathematics, but the ground covered is likely to be very small, so that it is difficult for the student to appreciate the uses and purpose of a subject which is allotted one or two questions in an examination in mathematics.

During the last decade or two there has been a considerable raising of the standard of mathematics required of candidates for degrees or higher national certificates in science and engineering. The students are used to a mathematical presentation of the theory of their own subject. In the past many books on statistical methods have given formulae and rules of procedure without going into the underlying mathematical theory. In this book we try to steer a middle course, giving the mathematical derivation of most of the results required, with an introduction to their applications, in a sequence which it is hoped will appear logical and carry conviction. Although there can be no objection to taking the mathematics for granted if the operation of statistical methods is properly understood, the fact that statistical theory is still advancing makes it the more desirable that the basis of standard procedures should be revealed. The student who has followed the argument used in establishing a sampling distribution, for instance, is more likely to be aware of assumptions made in applying it and better equipped to read further in the subject.

There is a fairly wide choice of mathematical methods of deriving standard results of statistical theory. It has not been thought desirable to introduce generating functions for moments and cumulants, and probability generating functions are used only twice, on pages 39 and 50. Most of the more important distributions have been obtained by the use of such calculus and limits as the type of reader referred to above may be presumed to have studied. No originality can be claimed for the methods adopted, but it is thought that the use of the χ^2 table for the determination of a single sampling inspection scheme (page 144) may not have been previously noted.

Our thanks are due to the Senate of London University, to the Royal Statistical Society and to the Association of Incorporated Statisticians Ltd., for permission to use examples from examination papers. We are indebted to Professor Ronald A. Fisher, Cambridge, to Dr. Frank Yates, Rothamsted, and to Messrs. Oliver and Boyd Ltd., Edinburgh, for permission to reprint, in abridged form, Tables I to VI, IX, XI and XXXIII from their book *Statistical Tables for Biological, Agricultural and Medical Research.* For valuable criticisms and suggestions we wish to thank Dr. S. G. Hooker, O.B.E., Chief Engineer, British Aeroplane Co. Ltd., Dr. L. R. Shenton, Senior Lecturer in Mathematics, Manchester College of Science and Technology, and Mr. D. A. Wrangham, Principal of Sunderland Technical College.

Part of Professor Rivett's contribution to this book was written while he was in Government Service, and thanks are due to the Chief Scientist, Ministry of Supply, for permission to publish.

The authors would like to thank those correspondents who have drawn their attention to errors and misprints in the first edition of this book and the printers and the English Universities Press for their helpful co-operation.

C. G. PARADINE.
B. H. P. RIVETT.

CONTENTS

CHAPTER I

INTRODUCTORY

Everyone is familiar with the use of the term 'statistics' to denote sets of figures, usually tabulated in suitable form, which record facts relating to trade, production, social conditions, etc., of a community. The very rapid increase, during the last century or so, in the volume of published and unpublished statistics, led to the development of the science of Statistics, the purpose being to extract from a mass of details the most important facts and representative figures. The first of these representative figures, by means of which we replace a set of different values of a quantity by a single number, is the *average* or mean value. In using it, we are making a *generalization*, implying that for some purposes it is sufficient to replace all the particular values by their average.

A little consideration will show that averages by themselves do not tell us much and may even be misleading. The manner in which a set of figures is spread around their average can differ considerably, according to the property which the figures measure. A convenient first step in the analysis of a large set of observed values of a variable is their classification in terms of the *frequency* with which different numerical values occur. In some cases it will then be evident that values in the neighbourhood of the average are the values which occur most frequently; but there can still be considerable difference, in different cases, in the way in which the values as a whole differ from the mean. For instance, if a machine process is producing very large numbers of some component, with high precision, the whole *range* of variation in a measured dimension is small; within that range the majority of values of the dimension are in the neighbourhood of the mean value, while values near the extremes of the range are of rare occurrence. If the product of the machine process is electric lamps and the property measured is their burning life, values will be spread over a fairly wide range, though within that range the characteristic of close clustering of values about the mean is still present. On the other hand, we give presently (Chap. II, Ex. iii) an example in which the most frequently observed values are those which differ most from the mean. In order to summarize conveniently a large set of values of a variable it is necessary to devise a number of *parameters*, that is, figures which reflect the important characteristics of the whole set. Next in importance to the mean is the *standard deviation*, by which the dispersion or spread of the values is expressed. The principal parameters are considered in Chap. II.

The next problem that arises is that of *estimation*. In most cases the available data will constitute only a *sample*, sometimes large, sometimes very small, of all the possible values of the quantity measured. On the basis of the sample, we seek to estimate the true values of the parameters. In a broad sense we are all statisticians in this respect from infancy, forming opinions and making inferences as to probable future occurrences on the basis of the small sample of our individual observations. This has always been the problem of the scientist. Even the most confidently accepted scientific laws have not been proved—that is, established with certainty as true—by observation; for the laws are supposed to be valid for an infinity of possible values of the variables, and only a finite number of observations can have been made. In practice a hypothesis is formulated to explain natural phenomena and is tested by experiment; when sufficient observed values confirm the hypothesis within the limits of experimental error, it is accepted as a working hypothesis unless and until discordant facts are observed. In statistical analysis a sample of observed values is used to estimate the parameters—mean, standard deviation, etc.—which characterize the aggregate of all possible values of the variable concerned. It is important to investigate how the accuracy with which a parameter can be estimated depends on the size of the sample.

If observed data are only a sample from a larger, possibly infinite, set, there is a chance that different samples will give different results. Problems of chance are dealt with in the *Theory of Probability*, which thus enters into the theory of Statistics and indeed into every field of scientific inference. Consequently it is necessary, before dealing with samples, to give some account (Chap. III) of the Theory of Probability. There are serious philosophical difficulties to be encountered in the application of mathematical probability to problems of inference, and in the more advanced theory of Statistics there are controversial issues which have not been resolved. In a first course the reader need not worry about such difficulties. Our aim is to develop the more widely used statistical techniques, and these are generally accepted.

Besides its application to the problems of sampling, the Theory of Probability leads to certain probability 'distributions' which serve as models to which observed frequencies of values of a variable may be expected to conform. The most important probability distributions are dealt with in Chaps. III and IV, which with Chap. II are to be regarded as a necessary preliminary to the development of statistical methods. The fields in which these methods apply are surprisingly wide and numerous. For the technologist the important aspect of the subject is that it provides a method of experiment, particularly appropriate when there are many causes of variation. The method often consists of the application of a *test of significance*, by which to decide whether a set of observed values can be explained on the basis of the inherent variability of the quantity measured and the chances of sampling or, on the other hand, are so unlikely as a chance effect that they are more reasonably explained as due to some suspected cause. Here again statistical theory aims at giving precision to a method which we all use from time to time. You are not surprised if a fuse-wire is blown occasionally, but if the same fuse goes several times in a short period, you look for the cause. If one says 'it is no accident that So-and-so is seldom on hand when there's a job to be done', one is applying the principle of the significance test, that what is not easily explained as due to chance is significant of a 'real' cause—in this case deliberate dodging. The most important tests of significance are dealt with in Chaps. V and VI.

The engineer who is chiefly interested in Quality Control and Sampling Inspection can omit Chaps. V and VI, except for §§6.2.1, 6.2.2 on the mean and variance of a sample, and proceed to Chaps. VII and VIII. The chapters on Theory of Errors and the Method of Least Squares should be of interest to most technologists. Correlation and the Analysis of Variance are widely used, the latter increasingly in industrial applications, although it was originally devised by R. A. Fisher to deal with problems in biology and agriculture.

The theory of Statistics is mathematical and often of considerable difficulty, but the main results are expressed in formulae for certain quantities which are to be calculated from the data and of which the significance is then to be judged by reference to tables. Consequently the application of the theory consists essentially of arithmetical computation, together with judgment and interpretation. The computation is often arduous and best done with the aid of calculating machines. In a text-book the examples must be such as may reasonably be worked without such aids, but sufficient for the purpose of illustrating the methods. There are a number of books giving an account of the methods as applied in particular fields, without the mathematical theory, but engineers and scientists have usually a sufficient grounding in mathematics to appreciate the proofs of the formulae, which can then be applied with a clearer understanding. The validity of a formula is restricted to problems in which the conditions at least approximate to those which form the hypothesis in the mathematical proof of the formula. The reader who wishes to avoid the mathematics will find that the harder proofs have been set between asterisks; a few have been omitted, reference being made to other sources. The most complete account of the theory will be found in M. G. Kendall's *Advanced Theory of Statistics*, 2 vols. (Griffin).

It is not claimed that the engineer or scientist will find sufficient in *Statistics for Technologists* to qualify as a professional statistician. Even the statistician is usually a specialist in some field of application, but with the increasing use of statistical methods in industry, statistician and technologist need to co-operate and each know something of the other's subject.

Suggestions for further reading:

L. H. C. Tippett, *Statistics* (Oxford University Press).
G. Yule and M. G. Kendall, *Introduction to the Theory of Statistics* (Griffin).

CHAPTER II

FREQUENCY DISTRIBUTIONS. STATISTICAL PARAMETERS

Statistical data. Classification by frequency. Graphical representation. Mean, mean deviation, standard deviation. Mode, skewness. Higher moments, Sheppard's corrections. Median, quartiles, etc.

2.1 Statistical data

In their simplest form statistical data will consist of a set of values of a *variate* (variable). An observed value of the variate is obtained by measurement, using the term in the broadest sense. Measurement can be exact only when it takes the form of counting, in natural numbers (positive integers), things or minimum indivisible units—e.g. of money. In such cases the variate can take only *discrete* values. Other measurements, e.g. of length, time, weight, are on a *continuous* scale which may in theory be subdivided indefinitely. In practice there is always a limit to the precision with which the measurement can be made, and so the true value is never known. In fact, observed values are always discrete; if we stop at the third decimal place in some unit, our recorded value is an integral number of thousandths of a unit. Precision of measurement is often not of paramount importance. If we had to measure the heights of a large number of men, we should consider them finely graded if each measurement were to the nearest eighth of an inch. On the other hand, to measure the variations which occur in a mass-produced article may require the maximum precision obtainable.

Observed values must be of finite magnitude, but for purposes of statistical theory we allow that the *range* of values, whether discrete or continuous, may be conceived as either finite or infinite. It should be noted that even if the range is finite, the number of possible different values of a continuous variate is theoretically infinite.

The variability observed may be of two kinds. The first arises from differences between different specimens; e.g. different flowers (of the same species) may have different numbers of petals, different men may have different heights. A second type of variability arises from the difficulty of precise measurement. A scientific worker will often take two or three, and in some cases many, readings of the same quantity, because it is found that from causes outside our knowledge and control repeated observations of the same quantity often yield (it is to be hoped slightly) discordant results. The observer will usually adopt the arithmetic mean of the readings as the most reliable estimate of the true value.

Variability in the product of a mass-production process has features in common with both of the above kinds. Different specimens from the same machine will differ slightly in some measured characteristic, but the articles are intended to be equal in all respects and the small variations that occur must be attributed (unless a definite change in the product sets in) to unknown causes, as in the case of scientific measurement of a single quantity.

The complete aggregate of values of a variate is called a *population.* (The use of this word is due to the historical development of Statistics, originally the study of numerical data relating to the people of a country or state.) The number of individual values will usually be very large or even infinite.

A *sample* consists of a set of observed values, more or less representative of the population from which it is drawn. If a botanist counts the petals on each of a thousand specimens of a particular type of flower, his values would constitute a fairly large sample from the population of all possible values. The scientist making two or three observations of a single quantity is using a small sample from a population which must be infinite, if the quantity measured is capable of variation over a range, however small, of a continuous scale. Theoretically at least, the pressure on a thumb-screw, adjustment of a level, etc., are capable of infinitesimal graduations, and external conditions of temperature, pressure, humidity similarly contribute to the variation.

2.2 Classification by frequency

Given a population of values or a reasonably large sample therefrom, suppose the whole range of values of the variate divided into *class-intervals*, usually equal, and a count made of the *frequency* in each interval, i.e. the number of times the variate takes a value in the interval. We may now tabulate the frequency f in each class-interval of the variate x. If the variate can take only discrete values, each such value might be taken as a class-interval, or the interval might include a range of values; if the variate can take values on a continuous scale, a class-interval is bound to include an infinity of possible values. Whenever the class-interval includes more than one possible numerical value of the variate, the values are said to be *grouped.* A suitable number of groups or class-intervals may vary from, say, 6 to 20.

There are two advantages in grouping. Where the observed sample is of moderate size, from a large population, a frequency table is more likely to exhibit a rise or fall of f against x, similar to that of the parent population, if the values are grouped, than if frequencies are tabulated of each numerical value of the variate which occurs. For instance, if 20,000 candidates take the same examination and it happens that 23% of them obtain a mark in the range 50 to 59, while 2·6% of them obtain the mark 54, then in a sample of 200 candidates chosen at random we should expect much the same percentage in the range 50 to 59, but should not be greatly surprised if the mark 54 happened not to occur at all or alternatively occurred with greater relative frequency than 2·6%.

The second advantage in grouping is that in subsequent calculations much labour is saved in numerical computation by treating all individuals in a class-interval as having the value at the centre of that interval. It should be noted that for a continuous variate the limitations on precision of measurement impose an initial fine grouping. Suppose, for instance, that the lengths of a large number of mass-produced components are measured to the nearest hundredth of an inch. If we say that 113 of them have a length 3·46 in., we mean that each has a length in the range 3·455 in. to 3·465 in., but we allot the central value to all of them. If we were to measure their lengths more accurately, no doubt we should find that they were not uniformly distributed over the range, but if the grouping is fine enough the inaccuracy involved in taking the central value as the length of all 113 components is negligible. In fact, we may decide to use a broader grouping, say in class-intervals of 1/10 in., in the frequency table of the complete data. In this case it should be noted that the class-interval which includes the recorded lengths 3·40, 3·41, 3·42 . . . 3·49 in. covers the range 3·395 to 3·495 of the continuous variate and that the central value is 3·445.

If a table of frequencies is given, grouped in class-intervals of a continuous variate but with no indication of the precision of the original measurements, we must assume the central values to be at the middle of the given intervals.

2.3.1 Graphical representation

A frequency table may be represented graphically by a *frequency polygon* or by a *histogram*. To construct the former, mark off the class-intervals on a suitable horizontal scale and at the centre of each interval erect an ordinate proportional to the frequency in the interval. The straight lines joining the tops of the ordinates constitute the frequency polygon.

Ex. i

The following data show the number of accidents sustained by 313 drivers of a public utility company over a period of five years. (J.R.S.S.)

Number of accidents:	0	1	2	3	4	5	6	7	8	9	10	11
Number of drivers:	81	44	68	41	25	20	13	7	5	4	3	2

Here the number of drivers is the frequency with which a value of the variate, number of accidents, occurred. The frequency polygon is as in Fig. 1.

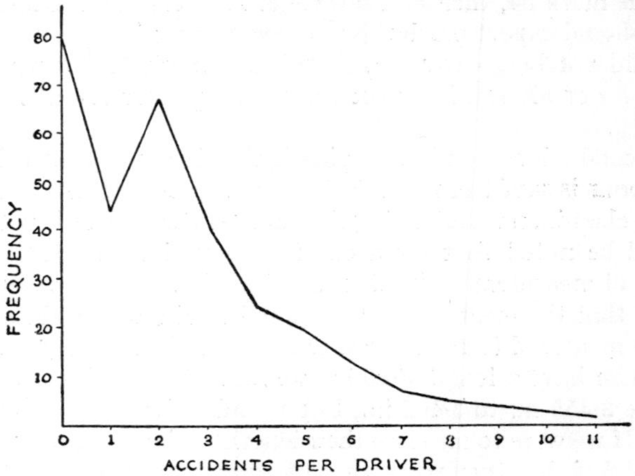
FREQUENCY
80
70
60
50
40
30
20
10
0
1
2
3
4
5
6
7
8
9
10
11
ACCIDENTS PER DRIVER

Fig. 1

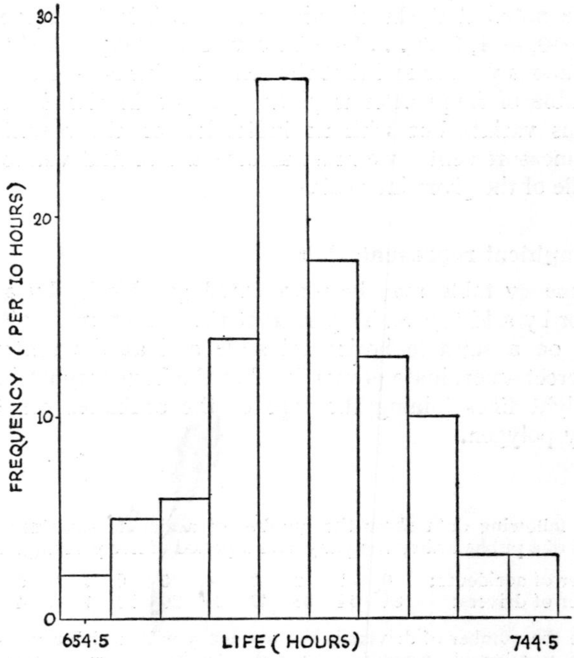
FREQUENCY (PER 10 HOURS)
30
20
10
0
654·5
LIFE (HOURS)
744·5

Fig. 2

An alternative representation, preferable for a continuous variate, is a *histogram*, which is a 'bar-graph'. On each class-interval as base, construct a rectangle whose *area* represents the frequency in the interval. Of course if, as is usually the case, the class-intervals are equal, the heights of the rectangles will be proportional to the frequencies represented. The advantage of using area as representing frequency is that we may subdivide the class-intervals indefinitely, keeping the total area constant, whereas if we maintained a constant vertical scale of frequency, the histogram would tend to disappear as the class-intervals were made indefinitely small, for the frequencies must become smaller and smaller as the total frequency is distributed over more and more intervals. Since we are much concerned, in later developments, with this indefinite subdivision of the range of the variate, it is important to realize that area is the proper representation of frequency; as the intervals become small, the decrease in frequencies in individual intervals is reflected (in the main) by decrease in the *widths*, not the heights, of the rectangles. The proper unit of the vertical scale of a histogram is a frequency *density*; e.g. if the histogram represents the frequency distribution of lengths, in class-intervals of 1/10 in., the unit of the vertical scale is frequency per 1/10 in.

Ex. ii

The life of 100 electric lamps was recorded in hours, to the nearest hour, as in the following table:

690	701	722	684	660	699	715	742	726	716
728	705	693	691	688	706	707	691	701	713
740	662	676	738	714	703	695	692	699	663
698	687	703	726	699	692	714	724	664	685
694	705	717	682	717	707	696	697	681	689
712	733	705	673	694	716	745	692	719	708
679	680	654	691	669	685	725	704	724	701
689	702	710	696	697	709	721	677	680	714
685	724	736	696	688	692	728	656	690	671
702	696	708	698	710	682	694	676	700	695

Using suitable grouping, represent the values (*a*) in a frequency table, (*b*) in a histogram.

The values range from 654 to 745 and a suitable class-interval is 10 hours. If, for example, 690- denotes recorded values from 690 to 699 inclusive, the frequency table is

Life in hours:	650-	660-	670-	680-	690-	700-	710-	720-	730-	740-
Frequency:	2	5	6	14	26	18	13	10	3	3

Ex. iii

A degree of cloudiness, recorded as x tenths,† occurred with frequency f, as follows:

x:	0	1	2	3	4	5	6	7	8	9	10
f:	320	129	74	68	45	45	55	65	90	148	676

(G. A. Pearse, *Biometrika* (1928), 20A, 336.)

† Nowadays the record would be in eighths. Moreover the occurrence of a completely cloudless sky, as opposed to slight cloud, is significant meteorologically.

Here the values of x are central values of class-intervals, except for the first and last values. Thus $x = 3$ means that, to the nearest tenth, 3-tenths of the sky was obscured by cloud so that the observer estimated the area of cloud as between 0·25 and 0·35 of the sky. But the value $x = 0$ implies that the area of cloud was less than half of one-tenth of the sky, so that the corresponding class-interval is from $x = 0$ to $x = 0{\cdot}5$ (tenths) in the continuous variate. Similarly the last interval is from $x = 9{\cdot}5$ to $x = 10$. Hence the first and last values of f must be doubled to give the frequency density.

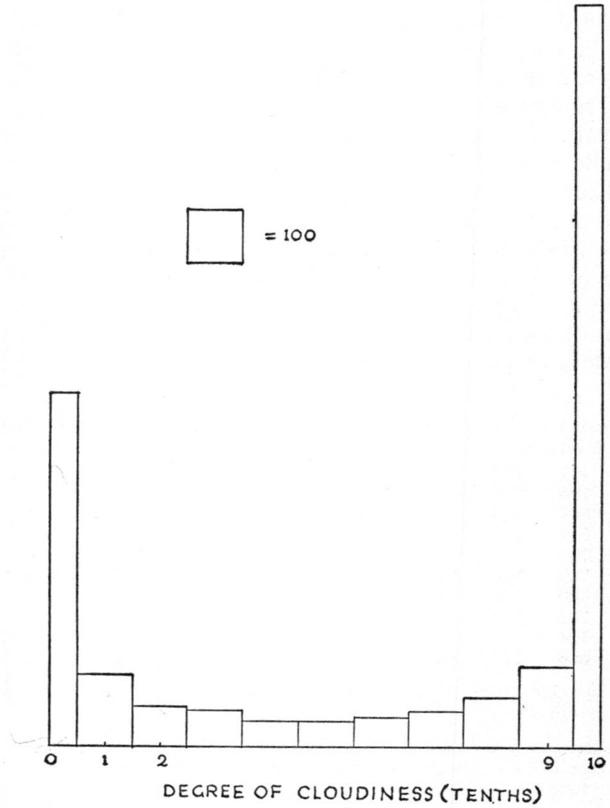

FIG. 3

2.3.2 Frequency curves

Although in practice observed frequencies must be finite, we can conceive an infinite population with a finite or infinite range of the variate. The frequency in any chosen class-interval can be expressed as a fraction of the total frequency, and this fraction is the *relative frequency* of values in the interval. A histogram of relative frequencies will have a total area of unity. If the variate is continuous the class-intervals may be imagined as subdivided indefinitely. Then the summit of the histogram, or the frequency polygon joining the tops of mid-ordinates, will tend to the form of a smooth curve as all class-intervals tend to zero (and the number of intervals to infinity). The interpretation of such a curve with equation $y = \phi(x)$, is that in any small interval δx of the variate, the relative frequency is $y\ \delta x$, the narrow strip of area under the curve, on the interval δx as base. If f denotes relative frequency

$$df/dx = y = \phi(x)$$
$$= \text{the relative frequency density of}$$

any value of x, or, in the notation of differentials,

$$df = \phi(x)\ dx.$$

The relative frequency in any chosen interval (not small) of the variate, say from $x = a$ to $x = b$, is represented by the corresponding area under the curve, namely $\int_a^b y\ dx$.

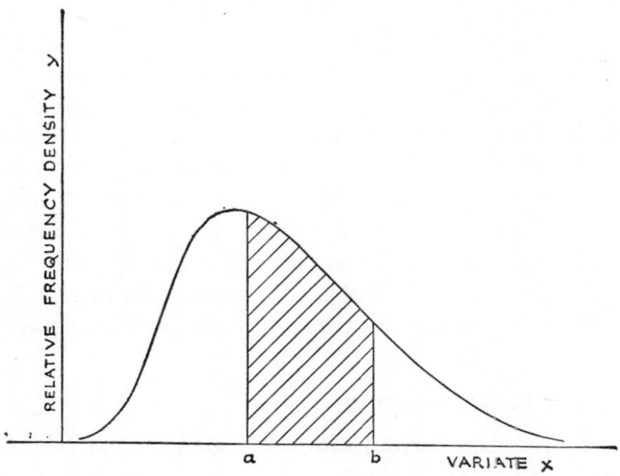

FIG. 4

The shaded area represents the relative frequency with which the variate takes a value in the range a to b. The area under the curve over the whole range of x, which may be finite or infinite, is 1. The variable y never takes negative values.

If a reasonably large sample is drawn from an infinite population, the tops of the mid-ordinates of the histogram representing the sample may be joined by a smooth curve which is likely to display, with more or less fidelity, the characteristic shape of the frequency curve of the parent population. Some common shapes may be roughly described. If the frequency curve has a single maximum with a tailing-off on both sides, it may be called 'bell-shaped'; it may be nearly symmetrical or markedly 'skew'. A histogram or resultant approximate frequency curve can give no information which is not contained in the frequency table on which it is based, but some general conclusions may be drawn from inspection of the frequency table or perhaps more vividly from the histogram. If, for instance, the variate in a bell-shaped distribution is the variable result obtained in repeated measurement of a single quantity, then a high-peaked curve reflects high precision of measurement. In this connection we may distinguish between precision and accuracy, or between 'errors' of measurement and 'mistakes'. If your watch has a seconds hand you may be able to state the time with great precision, but not necessarily with accuracy; a scientist with a personal bias in the setting or reading of an instrument may nevertheless produce results in close agreement with one another.

Other easily described shapes of frequency curve are U-shaped, J-shaped, rectangular. The histogram of degrees of cloudiness in Ex. iii is indicative of a U-shaped distribution. The extreme conditions of a completely overcast or a clear sky are much more common than any of the intermediate states. A rectangular distribution is one of constant relative frequency and can be over only a finite range of the variate. It would apply, one may assume, to errors of measurement even if perfect precision were obtained with the measuring instrument; for instance, if lengths of a certain class can be measured to the nearest thousandth of an inch and no variability can be detected with repetition of measurement, it is still true that there can be any error from $-0{\cdot}0005$ to $+0{\cdot}0005$ in. A distribution with uniform relative frequency of a discrete variate may also be described as rectangular.

2.4 Mean value. First moment

We proceed to define certain statistical parameters whose values, calculated for a given frequency distribution, measure its principal characteristics. The first of these parameters is the average or mean value of the variate.

If values $x_1, x_2 \dots x_n$ occur with frequencies $f_1, f_2 \dots f_n$, the mean value of x, denoted for the present by $\bar{x}$, is defined by

$$\bar{x} = \sum_{r=1}^{n} f_r x_r \Big/ \sum_{r=1}^{n} f_r \text{ or } \Sigma fx/\Sigma f \quad . \quad . \quad . \quad (1)$$

Alternatively, if x denotes an individual observed value, so that a value

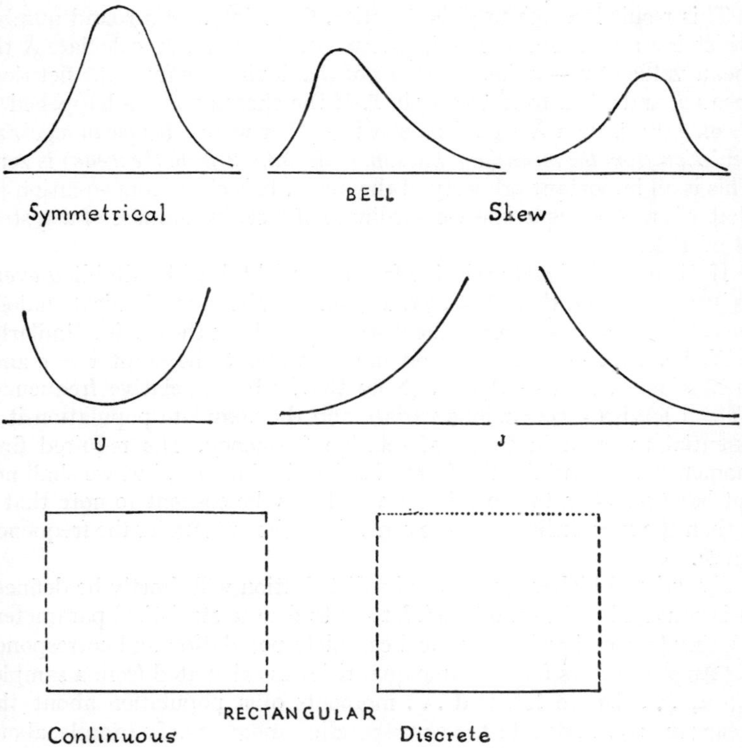

FIG. 5

occurring with frequency f occurs f times in the subsequent summation, and N, equal to Σf, is the total number of individual values, then

$$\bar{x} = \frac{1}{N} \Sigma x.$$

This is, of course, merely the arithmetic mean of all the individual values; formula (1) shortens the calculation by including in one term, fx, the sum of f equal values.

Now let $x = a + X$, where a is a constant.

$$\Sigma fx = \Sigma f(a + X)$$
$$= a\Sigma f + \Sigma fX$$

Dividing by Σf

$$\bar{x} = a + \bar{X} \quad . \quad . \quad . \quad . \quad . \quad . \quad . \quad . \quad (2)$$

where $\bar{X}$ is the mean value of X.

This result is much used in practice. Choosing as a a round number which is a rough estimate by inspection of the mean, we calculate $\bar{X}$ the mean value of $x - a$, i.e. the mean of the deviations from the fictitious mean a, and add a to $\bar{X}$ to obtain $\bar{x}$. If the chosen value a happened to be exactly $\bar{x}$, then $\bar{X}$ would be zero; in other words, *the mean algebraic deviation from the mean* (*and the sum of deviations from the mean*) is zero. This is an important property of the mean. It is clear from equation (1) that $\bar{x}$ is analogous to the x-co-ordinate of a centre of mass of a system of particles.

If f is used to denote relative frequency (obtained by dividing every frequency by N), then $\bar{x} = \Sigma fx$, and on the above mechanical analogy we call $\bar{x}$ the *first moment* of the distribution about the origin. Similarly, ΣfX, i.e. $\Sigma f(x - a)$, is the first moment about the point $x = a$ and, in particular, $\Sigma f(x - \bar{x}) = 0$. (Note that, f being relative frequency, $\Sigma f = 1$.) With a continuous variate and the complete population it is essential to work in terms of relative frequency. The required first moment will then take the form of a definite integral, but we shall not yet have occasion to use this form and may be content to note that $\bar{x}$ is then the x-co-ordinate of the centroid of the area under the frequency curve.

Second and higher moments of a distribution will shortly be defined. It is conventional to use Greek letters to denote statistical parameters relating to a real or hypothetical complete population and corresponding English letters for the same quantities as calculated from a sample. μ_1, μ_2 . . . denote 1st, 2nd . . . moments of a population about the mean; m_1, m_2 . . . denote the corresponding moments of a sample, about the mean of the sample. Moments about the origin are denoted by dashed letters—$\mu'_1, \mu'_2 \ldots; m'_1, m'_2 \ldots$.

Thus $\mu'_1 = \bar{x}$ as defined in equation (1).

$\mu_1 = 0$, always.

2.5.1 Dispersion. Mean deviation

A significant feature of a distribution is the dispersion or spread of the values about the mean. They may be evenly spread, as in the rectangular distribution, closely clustered round the mean as in a high-peaked, symmetrical, bell-shaped distribution, or it may even occur, as in the U-shaped distribution, that the bulk of values are those far from the mean.

Now, in any distribution the mean algebraic deviation from the mean is zero and so affords no measure of dispersion. To measure this characteristic we must choose a function which reckons a deviation positively, whether its value is positive or negative. One such function is the mean *absolute* deviation from the mean, called (unfortunately) the *mean deviation* and denoted by m.d.

Thus m.d. $= \Sigma f \mid x - \bar{x} \mid / \Sigma f$ . . . . (3)

and is a measure of dispersion.

Note that, if $x = a + X$, then $\bar{x} = a + \bar{X}$, so that $x - \bar{x} = X - \bar{X}$.

Ex. iv

Find (*a*) the mean life, (*b*) the mean deviation from the mean life of the electric lamps in Ex. ii.

Using the frequency table and a fictitious mean at the central value of the largest group, i.e. at 694·5, construct the following table.

$X = x - 694{\cdot}5$	f	fX		$\lvert X - \bar{X} \rvert$	$f \lvert X - \bar{X} \rvert$
− 40	2	− 80		45·2	90·4
− 30	5	− 150		35·2	176·0
− 20	6	− 120		25·2	151·2
− 10	14	− 140		15·2	212·8
0	26		0	5·2	135·2
10	18		180	4·8	86·4
20	13		260	14·8	192·4
30	10		300	24·8	248·0
40	3		120	34·8	104·4
50	3		150	44·8	134·4
	100		1010		100 | 1531·2
			− 490		15·3
			100 | 520		
			5·2		

$\bar{x} = 694{\cdot}5 + 5{\cdot}2 = 699{\cdot}7$, say 700 hours

m.d. = 15·3 hours.

2.5.2 Standard deviation. Second moment

For reasons which appear later, a more convenient measure of dispersion than the m.d. is the *root-mean-square deviation from the mean*, known as the *standard deviation* or s.d. and denoted by σ for a complete population, s for a sample.

Thus $\sigma^2 = \Sigma f (x - \bar{x})^2 / \Sigma f$ . . . . (4)

If f is relative frequency, $\sigma^2 = \Sigma f (x - \bar{x})^2$. Hence σ^2 is called the second moment about the mean and is identical with μ_2.

A great deal of statistical analysis is in terms of σ^2 rather than σ, so σ^2 is given the name *variance*. The mechanical analogue of σ is radius of gyration, for if $f_1, f_2 \ldots$ are the masses of particles situated at distances $x_1, x_2 \ldots$ from the y-axis, equation (4) defines σ^2 as the second moment of mass, about the y-axis, divided by the total mass. The well-known theorem of parallel axes for moments of inertia has its analogue, important for purposes of computation, in terms of σ^2. We have

$$\Sigma f\,(x - \bar{x})^2 = \Sigma f x^2 - 2\bar{x}\Sigma f x + \bar{x}^2 \Sigma f$$

Dividing by Σf (or, if f is relative frequency, setting $\Sigma f = 1$),

$$\begin{aligned} \sigma^2 &= \sigma_o^2 - 2\bar{x}\bar{x} + \bar{x}^2 \\ &= \sigma_o^2 - \bar{x}^2 \qquad . \quad . \quad . \quad . \quad . \qquad (5) \end{aligned}$$

where σ_o is the r.m.s. deviation from the y-axis or origin, i.e. $\sigma_o{}^2 = \mu'_2$.
More generally, let $x - a = X$, then

$$\begin{aligned} \Sigma f\,(x - \bar{x})^2 &= \Sigma f\,(X - \bar{X})^2 \\ &= \Sigma f X^2 - 2\bar{X}\Sigma f X + \bar{X}^2 \Sigma f \end{aligned}$$

and, dividing by Σf,

$$\begin{aligned} \sigma^2 &= \sigma_a^2 - 2\bar{X}\bar{X} + \bar{X}^2 \\ &= \sigma_a^2 - \bar{X}^2 \qquad . \quad . \quad . \quad . \quad . \qquad (6) \end{aligned}$$

where σ_a is the r.m.s. value of X or $(x - a)$. Hence using the same fictitious mean a as was used in calculating the mean $\bar{x}$, we may calculate σ_a^2 which is $\Sigma f\,(x - a)^2/\Sigma f$ and deduce the value of σ^2. For a worked example of calculations relating to a moderately large sample, see § 2.7.3.

Equation (6) reveals a second fundamental property of the mean. σ_a^2 is a minimum (equal to σ^2) when $\bar{X}$ is zero, i.e. when $a = \bar{x}$. Hence *mean square deviation is a minimum when deviation is measured from the mean.* (The moment of inertia, or radius of gyration, of a body is less about an axis through the centre of mass than about any parallel axis.) Given the frequency curve of a complete population with a continuous variate, σ is the radius of gyration, measured in the unit of the variate, of the area under the curve, about the ordinate at the mean value of the variate.

It is clear that the interpretation of a value of σ may involve comparison of the value with that of the mean. For instance, a standard deviation of 0·05 in. in some measured dimension of a mass-produced article represents a higher degree of precision of manufacture if the mean measurement is 25 in. than if the mean is 5 in. For this reason K. Pearson defined the ratio of the s.d. to the mean, expressed as a percentage, as the *coefficient of variation*, denoted by v. Thus

$$v = \frac{100\sigma}{\bar{x}}\ \% \qquad . \quad . \quad . \quad . \quad . \qquad (7)$$

Ex. v

The following measurements were made on a batch of rods, f being the number of rods of length l inches.

l:	30	31	32	33	34	35	36	37	38	39
f:	4	8	23	35	62	44	18	4	1	1

Calculate (*a*) the mean length, (*b*) the standard deviation from the mean.

(L.U.)

$X = l - 34$		fX		fX^2
− 4	4	− 16		64
− 3	8	− 24		72
− 2	23	− 46		92
− 1	35	− 35		35
0	62	− 121		0
1	44		44	44
2	18		36	72
3	4		12	36
4	1		4	16
5	1		5	25
	200		101	200|456
			− 121	2·28
			200|− 20	
			− 0·1	

$$
\begin{aligned}
a &= 34\\
\bar{X} &= -0{\cdot}1\\
\bar{l} &= 34 - 0{\cdot}1 = \underline{33{\cdot}9 \text{ in.}}\\
\sigma^2 &= \sigma_a{}^2 - \bar{X}^2\\
&= 2{\cdot}28 - 0{\cdot}01\\
&= 2{\cdot}27\\
\sigma &= \underline{1{\cdot}507}
\end{aligned}
$$

2.6 Mode. Skewness

If a frequency curve has one maximum ordinate, the corresponding value of the variate is called the *mode*. This must coincide with the mean if the distribution is symmetrical, the common ordinate being the axis of symmetry. An unsymmetrical distribution with a single maximum is said to be *skew* and the skewness is reflected in the displacement of the mode from the mean, but the magnitude of this displacement should be compared with the dispersion of the values of the variate. Consequently, K. Pearson defined a measure of skewness (Sk.) by the equation

$$\text{Sk.} = (\text{Mean} - \text{mode})/\sigma \quad . \quad . \quad . \quad (8)$$

By this definition skewness is positive or negative according as the mean is greater or less than the mode, as in the following sketches.

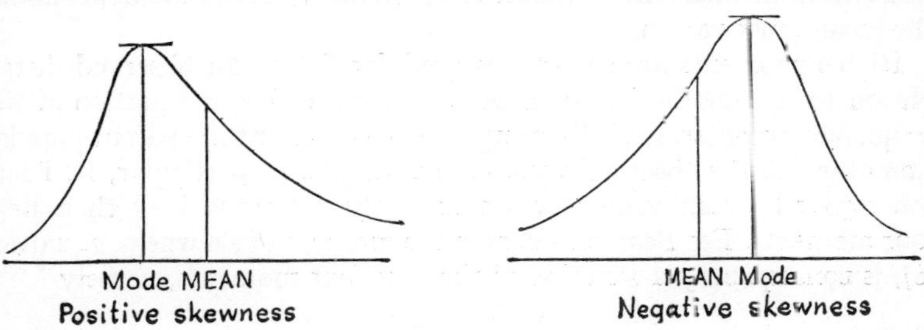

FIG. 6

The mode is not easily determined from grouped data, but may be the parameter whose value is most readily obtained for a hypothetical or theoretical distribution with a given equation of the frequency curve.

A simple example of a *bimodal* distribution is afforded by traffic flow. With time of day as variate, peak frequencies occur in the morning and evening at the times of going to and from work. In an industrial process, where high precision is desirable, the objective is a frequency distribution with small range and s.d.; in the traffic problem the ideal would be a rectangular distribution of the time variate. Cheap midday fares and staggering of hours of work are devices for reducing the peak frequencies and spreading them more evenly.

2.7.1 Higher moments

Let f denote relative frequency. The rth moment about the origin is

$$\mu'_r = \Sigma f x^r \qquad . \quad . \quad . \quad . \quad . \quad . \quad . \quad (9)$$

and the rth moment about the mean is

$$\mu_r = \Sigma f\,(x - \bar{x})^r \qquad . \quad . \quad . \quad . \quad . \quad . \quad (10)$$

As for first and second moments it is convenient to calculate higher moments about the mean from moments about a fictitious mean a.

For example
$$\begin{aligned}\mu_3 &= \Sigma f\,(x - \bar{x})^3 \\ &= \Sigma f\,(X - \bar{X})^3, \text{ where } X = x - a \\ &= \Sigma f X^3 - 3\bar{X}\Sigma f X^2 + 3\bar{X}^2\Sigma f X - \bar{X}^3\Sigma f \\ &= \mu_3^a - 3\bar{X}\mu_2^a + 3\bar{X}^3 - \bar{X}^3 \\ &= \mu_3^a - 3\bar{X}\mu_2^a + 2\bar{X}^3 \qquad . \quad . \quad . \quad . \quad . \quad (11)\end{aligned}$$

where μ_2^a, μ_3^a denote second and third moments about $x = a$.

Similarly, $\mu_4 = \mu_4^a - 4\bar{X}\mu_3^a + 6\bar{X}^2\mu_2^a - 3\bar{X}^4$ . . . . (12)

and for higher moments μ_r, the coefficients on r.h.s. are simply the binomial coefficients of $(p - q)^r$ except that the last two coefficients coalesce.

For a symmetrical distribution, since positive and negative deviations from the mean are symmetrically frequent, all *odd* moments about the mean must vanish.

Higher moments are sometimes used for fitting an observed distribution to a hypothetical distribution (with a chosen equation of the frequency curve), which is then supposed to represent the parent population of which the observed values are a sample. In particular, K. Pearson evolved a family of curves completely determined by their first four moments. For Pearson's curves the measure of skewness, equation (8), is consequently a function of the first four moments, namely

$$\frac{\sqrt{\beta_1}\,(\beta_2 + 3)}{2(5\beta_2 - 6\beta_1 - 9)}$$

where
$$\beta_1 = \mu_3^2/\mu_2^3 \quad . \quad . \quad . \quad . \quad (13)$$
$$\beta_2 = \mu_4/\mu_2^2 \quad . \quad . \quad . \quad . \quad (14)$$

These are pure numbers, invariant to change of scale of the variate, by virtue of their dimensions. Moreover, all moments about the mean are invariant to change of origin. We shall not use the above expression for skewness in terms of β_1 and β_2, but these two are worth noting for future use. For a symmetrical curve β_1 will vanish; hence the value of β_1 may itself be used as a measure of skewness, but will be positive for both types of skewness. To discriminate between the types we may adopt $\sqrt{\beta_1}$ or $-\sqrt{\beta_1}$, whichever has the sign of μ_3, as a measure of skewness. This quantity is denoted by γ_1, so that
$$\gamma_1 = \mu_3/\mu_2^{3/2} \quad . \quad . \quad . \quad . \quad (15)$$

2.7.2 Sheppard's corrections

If the distribution is continuous and $\phi(x)dx$ is the relative frequency differential, the rth moment about the origin is $\int x^r\phi(x)dx$ taken over the whole range of the variate x. In the calculation of the moment from a frequency table of grouped data, we evaluate instead $\Sigma x_k^r f_k$, where x_k is the central value of the variate in the kth group and f_k the relative frequency in the group. Without knowing $\phi(x)$ we cannot know the error resulting from the grouping, but W. F. Sheppard showed, using the Euler-Maclaurin theorem of the calculus of finite differences, that certain corrections can be applied to the grouped moments, if $\phi(x)$ satisfies certain conditions. The Euler-Maclaurin formula is an expression for the difference between a definite integral and an approximate summation by the trapezoidal rule. The difference is found to involve the derivatives of the integrand at the extremities of the range of integration. Sheppard's corrections require, when applied to moments up to the rth, that $x^r\phi^m(x) \rightarrow 0$ at the extremities of the range for all values of m up to r, where $\phi^m(x)$ denotes the mth derived function. This may be briefly expressed by saying that the frequency curve must have 'high order contact' with the x-axis at *both* extremes of the range. (The curves $y = x^2$, $y = x^3$, $y = x^4$ and so on have contact of increasingly higher order with the x-axis at the origin. Imagine such contact at the ends of the range, possibly at infinity.) The implication is that the frequency curve has a relatively long, low 'tail' at each end.

This condition is quite often satisfied and then Sheppard's results are that, for moments about the mean, μ_r denoting the corrected moment and (μ_r) the moment as calculated from the grouped data,
$$\mu_1 = (\mu_1) = 0$$
$$\mu_2 = (\mu_2) - h^2/12 \quad . \quad . \quad . \quad . \quad . \quad (16)$$
$$\mu_3 = (\mu_3)$$
$$\mu_4 = (\mu_4) - (\mu_2)h^2/2 + 7h^4/240 \quad . \quad . \quad (17)$$

and so on, where h is the class-interval. The correction $-h^2/12$ to μ_2, i.e. σ^2, is the most important.

Similar corrections are applied with m's for μ's when the moments are calculated from a table of grouped data representing a sample from a supposed infinite population with continuous variate, provided the necessary high order contact at the extremes exists. See the following worked example.

2.7.3 Calculation of the first four moments

Ex. vi

The correction x to be added to the declination of each of 450 stars as given in a certain catalogue was first given to the nearest hundredth of a second of angle. The values were then grouped in a frequency table, as follows (x = tenths of 1″):

x:	− 7 to − 6·1	− 6-	− 5-	− 4-	− 3-	− 2-	− 1-	0-	1-	2-	3-	4-
f:	2	11	12	35	82	118	88	54	32	10	5	1

Thus the highest frequency 118 is the number of stars for which the tabulated correction was in the range − 20 to − 11 (hundredths of 1″) inclusive. The centre of this class-interval is

$$-15{\cdot}5 \times 0{\cdot}01'' = -0{\cdot}155''$$

and is a suitable trial mean. To avoid unnecessary decimals and zeros, it is convenient to express X, the deviation from this trial mean, in tenths of a second. The work of computing the first four moments of the distribution may then be set out as follows:

X	f	fX	fX^2	fX^3	fX^4
− 5	2	− 10	50	− 250	1250
− 4	11	− 44	176	− 704	2816
− 3	12	− 36	108	− 324	972
− 2	35	− 70	140	− 280	560
− 1	82	− 82	82	− 82	82
		− 242		− 1640	
0	118	0	0	0	0
1	88	88	88	88	88
2	54	108	216	432	864
3	32	96	288	864	2592
4	10	40	160	640	2560
5	5	25	125	625	3125
6	1	6	36	216	1296
		363	450 | 1469	2865	450 | 16205
		242	3·264	1640	36·01
		450 | 121		450 | 1225	
		0·269		2·722	

The first two columns are the rearranged data; the table is then best completed by rows. Still working in one-tenths of a second as units,

$$\bar{x} = -1{\cdot}55 + 0{\cdot}269 = -\underline{1{\cdot}281}.$$

Uncorrected $m_2 = 3{\cdot}264 - (0{\cdot}269)^2 = 3{\cdot}192$ by equation (6)

Corrected $m_2 = 3{\cdot}192 - 1/12 = \underline{3{\cdot}109}$ „ „ (16)

$m_3 = 2{\cdot}722 - 3(0{\cdot}269)(3{\cdot}264) + 2(0{\cdot}269)^3$ „ „ (11)

$= \underline{0{\cdot}128}$

Uncorrected $m_4 = 36{\cdot}01 - 4(0{\cdot}269)\,(2{\cdot}722) + 6(0{\cdot}269)^2(3{\cdot}264) - 3(0{\cdot}269)^4$
$= 34{\cdot}46$ by equation (12)

Corrected $m_4 = 34{\cdot}46 - (3{\cdot}192)/2 + 7/240$ „ „ (17)
$= \underline{32{\cdot}89}$

$$\beta_1 = \frac{m_3^2}{m_2^3} = \frac{(0{\cdot}128)^2}{(3{\cdot}109)^3} = \underline{0{\cdot}0005.}$$

This value indicates a very slight skewness, with the sign of m_3, i.e. positive. It is, of course, evident from the frequency table that the distribution is nearly symmetrical.

$$\beta_2 = \frac{m_4}{m_2^2} = \frac{32{\cdot}89}{(3{\cdot}109)^2} = \underline{3{\cdot}40.}$$

The significance of this value will appear later (§5.3.4).

The values of β_1 and β_2 are independent of the unit of measurement; in seconds of angle, the mean correction is $-\,0{\cdot}1281''$ with a standard deviation $\sqrt{0{\cdot}03109} = 0{\cdot}1764''$. The range of values in the sample is about six times the s.d., the extreme values being about three times the s.d. above and below the mean.

2.8 Median, quartiles, etc.

Suppose the whole set of individual values of the variate to be arranged in order of magnitude. The *median* is defined as the value which divides the set into two numerically equal groups, i.e. with equal total frequencies. If the total number of values is odd, the median is the middle term; if the total number is even, we may take the median as the mean of the two middle values of the variate. In either case, of course, the median may be a value which is repeated, i.e. occurs with more than unit frequency. If the values given are grouped, the median is the value of the variate whose ordinate divides the histogram of the distribution into two equal areas; thus it is likely to fall within a class-interval and its value requires a little calculation. For example, in Ex. iii the total frequency is 1715, of which one-half is $857\frac{1}{2}$; the groups centred at 9 and 9·75 (tenths) have a total frequency 824; the required median value is

$$8{\cdot}5 - \frac{33\frac{1}{2}}{90} \text{ of } 1 = 8{\cdot}13 \text{ (tenths)}.$$

This value is quite distinct from the mean, which is just over 6 tenths, and from the mode, which is 9·75 tenths. (Here the mode is simply the most frequent value.) Given a frequency curve, the ordinate at the median will bisect the total area under the curve.

Let us consider at this point what, in terms of common sense, we learn by determining the three parameters: mean, median, mode. Statistics, like all Science, aims at investigating and giving precision to ideas first suggested by common sense. The phrases 'on the average', 'as often as not', 'most likely' are of common occurrence in discussion, and express our feeling that where a quantity is liable to more or less wide variation it is desirable to give a value which is fairly representative of all values. The above phrases are generally used vaguely, and often as if they were equivalent to one another. It is just to these three phrases that the mean, median and mode give precision. The observations tabulated in Ex. iii were made at Greenwich. On them as evidence, we infer that at Greenwich (at least over a given period) the sky was on the average about 6 tenths cloudy; that as often as not it was more than 8·13 tenths cloudy and finally that it was most often 9·75 tenths cloudy. Perhaps it is the last of these which will seem most surprising, because the phrase 'most often (or likely or probably)' is confused with 'more often than not' or with 'almost certainly'. Now, in fact, the maximum relative frequency in a distribution may not be large—it is simply larger than the others; in our example it is 676/1715, so that nearly 40% of times the sky was practically completely overcast. The next most frequent event was a practically clear sky, which was observed 18·7% of times.

It is some consolation to common sense that there are a fairly large number of types of variate conforming to a bell-shaped distribution of only moderate skewness, in which mean, median and mode will not greatly differ. There is an empirical rule for such distributions which states that

$$\text{mean} - \text{mode} = 3\,(\text{mean} - \text{median}).$$

If a distribution with a single mode is symmetrical, then mean, median and mode coincide.

There still remains the question, when they differ, as to which may be regarded as 'representative'. In statistical theory the mean is much the most important of the three; the principle of maximum likelihood (Chap. XIII) affords perhaps the only example in pure theory of preferring the mode to the mean. In special cases, however, the mean is clearly misleading if advanced as representative value. Imagine, for instance, a rural area largely inhabited by poor peasants, but including the home of a millionaire. If the average income is calculated, the inclusion of the millionaire's income may be sufficient to convey the impression of a well-to-do community, though the impression would be somewhat corrected if the coefficient of variation were also given. If we select instead the median income as representative, we count the millionaire as just another man but close our eyes to the size of his income. Finally, if we select the mode, we ignore him altogether and adopt as representative the most popular figure. There is a parallel to the case of the millionaire in the statistical theory of errors of measurement. It may happen that one of a set of observed values of a single quantity differs so widely from the others that it is reasonable to reject it as due to a mistake or a major mishap (see §9.2). Similarly in experiments to determine the average performance and variability of ammunition, an occasional 'wild' round occurs which is disregarded.

It is a property of the median that the sum, or the mean, of absolute deviations is a minimum when deviation is measured from the median. The property is not important, so we shall omit the proof.

Having divided the graded values of a variate into two groups containing equal frequencies by the median, we may further bisect these groups at two values of the variate called the *quartiles*. Thus half the total frequency will occur between the quartiles. In the case of a frequency curve for which mean, median and mode are nearly equal, the *interquartile range* of the variate is that range in which half the total relative frequency occurs and that half includes the values with the higher relative frequencies. For this reason the interquartile range is sometimes used as a rough measure of dispersion.

Other divisions of the total frequency into tenths or hundredths give values of the variate known as deciles or percentiles. If sufficient divisions are made the process amounts to reclassifying the observations in terms of equal intervals of frequency instead of equal intervals of the variate. An equivalent graphical representation consists of drawing the graph, against x, of the total relative frequency in the range from the lowest value x_1 up to any value of x. Since frequency density is never negative, the total relative frequency increases from the initial value zero to the final value 1. For bell-shaped frequency distributions the graph of the accumulated relative frequency is an ogival curve.

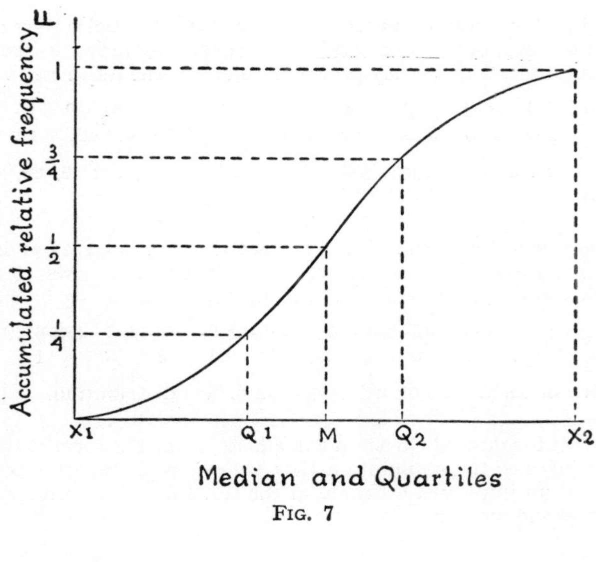

FIG. 7

In the figure it is assumed that the whole range of the variate is from x_1 to x_2. M denotes the median and Q_1 and Q_2 the quartiles.

If the range is infinite in either direction, the curve will be asymptotic to one or both of the lines F = O, F = 1.

The range of a sample must be finite. It is the simplest indication of the dispersion of the values. We have thus found four measures of dispersion: the standard deviation, the mean (absolute) deviation (from the mean), the interquartile range and the range. Of these the s.d. is the most important. The m.d. suffers from the disadvantage that its computation is not amenable to the use of a trial mean. The range of a sample, being merely the difference between the greatest and least values in the sample, is very easily found and is in consequence much used in Quality Control (Chap. VII).

Suggestions for further reading:

A. T. Doodson, *Relation of Mode, Median and Mean in Frequency Curves* (*Biometrika*, 11, 425).

W. P. Elderton, *Frequency Curves and Correlation* (Cambridge University Press).

G. Yule and M. G. Kendall, *Introduction to the Theory of Statistics*, Chapters 6, 7, 8 (Griffin).

M. G. Kendall, *Advanced Theory of Statistics*, Chapters 2, 3 (Griffin).

EXAMPLES ON CHAPTER II

1. The table shows the frequency f with which x α-particles were radiated from a disc in a given time (x takes integral values only). Represent the distribution by a frequency polygon. Calculate the mean and variance.

x:	0	1	2	3	4	5	6	7	8	9	10	11	12
f:	57	203	383	525	532	408	273	139	45	27	10	4	2

(Rutherford and Geiger, *Philosophical Magazine*, Ser. 6, 20, 1910, p. 698.)

2. The lengths of 100 steel rods are measured to the nearest hundredth of an inch and the frequencies are found to be as follows (5·4-means observed values 5·40, 5·41, 5·42 . . . up to and including 5·49).

Length:	5·4-	5·5-	5·6-	5·7-	5·8-	5·9-	6·0-	6·1-	6·2-6·29 in.
Frequency:	2	5	8	23	25	21	9	6	1

Find the mean and standard deviation of this distribution. (L.U.)

3. Represent the data of Question 2 in a histogram. The form of the distribution justifies the assumption that the use of Sheppard's correction will yield an improved estimate of the s.d. Find this corrected value. Find also the median.

4. In the following distribution the values of x are central values of the deviation, in thousandths of an inch, from a trial mean. Calculate the s.d. and the skewness, as measured by g_1, the sample value of γ_1.

x:	−5	−4	−3	−2	−1	0	1	2	3	4
f:	1	0	7	28	58	76	57	37	12	8

5. In the manufacture of a certain scientific instrument great importance is attached to the life of a particular critical component (X). This component is obtained in bulk from two sources (A and B), and in the course of inspection the lives of 1,000 of the components (X) from each source are determined and the following frequency tables are obtained:

Source A		*Source B*	
Life (hours)	*Number of Components*	*Life (hours)*	*Number of Components*
1,000–1,020	40	1,030–1,040	339
1,020–1,040	96	1,040–1,050	136
1,040–1,060	364	1,050–1,060	25
1,060–1,080	372	1,060–1,070	20
1,080–1,100	85	1,070–1,080	130
1,100–1,120	43	1,080–1,090	350

Examine the effectiveness of the measures of dispersion with which you are familiar for comparing the dispersions of the two distributions.

(R.S.S.)

6. The table given below gives the distribution of the marks of 310 candidates in an examination:

Digits

Tens	0	1	2	3	4	5	6	7	8	9	
0							1				1
10						1	1	1		1	4
20	1	2	1	3	1	4	1	1	1	4	19
30	10	8	8	9	16	13	8	3	11	4	90
40	9	5	2	8	4	10	6	2	8	10	64
50	17	9	3	3	4	7	4	8	6		61
60	11	10	2	5	3	8	5	1	3	8	56
70	5	2	1		2	3	1				14
80	1										1
	54	36	17	28	30	46	27	16	29	27	310

Thus there is one candidate with 6 marks, there are three candidates with 23 marks, and so on.

Calculate the average marks for the whole group.

Estimate the value of the mean from the table:

Range of marks:	0-	10-	20-	30-	40-	50-	60-	70-	80-	Total
Number of candidates:	1	4	19	90	64	61	56	14	1	310

To what do you attribute the difference between these results?

(R.S.S. Cert.)

7. (i) State the formula used in calculating
   (*a*) the root-mean-square value of an alternating current;
   (*b*) the radius of gyration of a plane area about a line in the plane;
   (*c*) the standard deviation of a set of measurements, and explain briefly in what way these formulae are similar.

   (ii) Prove that, if f is the frequency of a measurement x, the mean of the measurements is M and x is any number, then the standard deviation σ is given by

$$\sigma^2 = \frac{1}{N}\Sigma f(x - x_o)^2 - (x_o - M)^2$$

   where $N = \Sigma f$. (L.U.)

8. Male Population of England and Wales, 1947. Thousands.

Age Groups	*Population*	*Age Groups*	*Population*
0– 4	1,806	45–49	1,425
5– 9	1,444	50–54	1,187
10–14	1,373	55–59	1,076
15–19	1,497	60–64	933
20–24	1,567	65–69	780
25–29	1,633	70–74	580
30–34	1,631	75–79	340
35–39	1,715	80–84	149
40–44	1,629	85 and over	57

Draw a histogram to show the distribution of population by age. What can be inferred from it about the future age constitution of the population of England and Wales? Comment on any other noteworthy points in the figures. (R.S.S. Cert.)

9. Define the mean, median and mode of a frequency distribution and calculate their values for the following distribution of the number of earners in families with 6 dependants:

Number of earners:	0	1	2	3	4	5	6	7	
Number of families:	11	156	55	41	20	7	0	1	291

(R.S.S. Cert.)

10. The following data are the breaking strains, in lb.wt. to the nearest integer, at different points in a length of cord manufactured. Prepare a frequency table, grouping the data in equal class-intervals, one such interval being 1000 to 1004 inclusive. Obtain the mean breaking strain and the standard deviation, using Sheppard's correction.

992	1007	1001	1010	1001	1008	1013	1010	995	1003
1009	990	1003	1008	999	1002	992	990	998	1013
1014	992	991	1006	998	1001	991	998	1011	985
986	996	1010	1008	1008	1024	995	1023	1001	998
998	1004	1010	1000	1002	997	1013	1005	1006	1015
994	1002	999	1008	1025	998	1005	1008	992	1016
1003	981	1005	982	997	1004	1013	1018	984	1004
1009	1001	1014	1018	991	997	1009	1003	998	1007
1028	1000	988	1012	1012	1001	979	992	1002	989
1010	1017	1011	996	1006	1009	994	1011	976	1003

11. A sample of 35 values has mean 81·52 and s.d. 3·74. A second sample of 85 values from the same population has mean 82·31 and s.d. 3·62. Find the mean and s.d. of the combined sample of 120 values.

12. A discrete variate takes the values 0, 1, 2 . . . 9 with equal frequencies. Find the standard deviation.

13. A continuous variate has a rectangular frequency distribution of range a. Show that the variance is $a^2/12$. (Taking the whole area of the frequency distribution as unity, find the second moment of area about the ordinate at the mean.)

CHAPTER III

PROBABILITY. THE BINOMIAL DISTRIBUTION

Probability. Frequency in the long run. Theorems of total and compound probability. Independent events. Sampling, with and without replacement. Permutations and combinations. Bernoulli's Theorem and the binomial distribution. Distribution of proportion of successes. Continuous probability distribution.

3.1.1 Probability

In §2.8 it was hinted that the term 'likely' or 'probable' is often used of an event in relation to its relative frequency. We are used to estimating chances or probabilities, e.g. the chances of rain or of getting a seat on the next bus, and many people try to estimate the probability that a horse will win a race or a team a football match. We should be hard put to it in these cases to obtain an exact measure of the probability of the event concerned, though possibly prepared to make an intuitive attempt in the form of laying odds. In insurance of all kinds, precise calculations are made, on the basis of large masses of data concerning age at death, frequencies of fires and accidents, etc., to determine premiums and in totalizator betting the odds against the various horses are precisely determined by the proportions in which money is laid on them; these proportions reflect the public's estimates of the horses' chances of winning on the basis of their past histories. It may be necessary to take into account a factor changing with time, such as the improving form of a horse or the effect of changing social conditions on the chances of dying before a certain age.

We must now consider the problem of obtaining a numerical measure of probability.

3.1.2

The mathematical Theory of Probability, which plays a large part in the development of statistical theory, originated in the consideration of problems of games of chance. In the toss of a coin, the throw of a die or the spin of a roulette wheel, out of a number of possible events, one happens. In these examples, before the event, we can enumerate the possible events and describe them as 'equally likely' or 'equally probable'. For instance, the throw of a die may result in any of the numbers 1, 2, 3, 4, 5 and 6 as the score shown. Our reasons for regarding them as all equally likely are that the die is (almost exactly) symmetrical and the conditions of its projection are presumed not to favour any one of the numbers rather than any other. When such considerations of symmetry give understandable significance to the phrase 'equally probable' in the class of events concerned, we may make the following definition.

Definition. If an aggregate of n equally probable occurrences includes m equally probable ways in which a particular event may occur, the probability of that event is m/n.

Thus when all possible occurrences, on an assumed basis of equality, have been enumerated, the probability of a particular event is the ratio of the number of occurrences favourable to that event, to the total number of occurrences. A probability so defined may be called a *prior* probability. Simple examples are: the probability of throwing a 4 with a die is 1/6; of drawing a card higher than a 5, ace counting high, from a pack, is 36/52; of throwing 'heads' twice in succession with a coin is 1/4, the equally probable occurrences being H, H; H, T; T, H; T, T.

It is usual to denote the probability of an event by p and to call the occurrence of the event 'success'. The probability of failure, i.e. of non-occurrence of the event, is denoted by q.

Thus if $p = m/n =$ the probability of success,

$$\text{then } q = \frac{n-m}{n} = \text{the probability of failure}$$

and $p + q = 1$.

The extreme values of p are 0 and 1, corresponding, *if n is finite*, to certain failure and certain success. When a probability is expressed in the form of odds, the ratio used is of favourable to unfavourable occurrences or vice versa. To say that the odds against an event are 7 to 2 means that the probability of its occurrence is 2/9.

3.1.3 Frequency in the long run

The prior probability of an event is calculated, before the event, by classifying the 'equally probable' occurrences as favourable or unfavourable. Let us consider the relation of the prior probability to the actual events observed in a series of trials.

When a trial, the result of which may be one of several possible occurrences, is made, the observed occurrence is definite. It cannot happen a fractional number of times. We say that the probability of each of the possible scores resulting from the throw of a die is 1/6 and we certainly cannot obtain each score 1/6 of a time with a single throw. Can we then say that we shall obtain each score exactly once in six throws, or each exactly 100 times in 600 throws? The answer is clearly no, but we have a strong faith that if the die is thrown very *many times* the six possible scores will occur 'on the average' with practically equal frequencies. If this did not happen we should suspect that the die lacked symmetry or was what Damon Runyon described as 'strictly phoney'. We have, in fact, a strong underlying belief in *causality*. Symmetry, though capable in many instances of definition in a positive form, is essentially a negative fact, the fact of no known or observable difference. If the outcome of our experiment shows some consistent difference, such as one number being thrown with a die more often than the others, we believe this must be due to a cause conducive to that result and conclude that the die lacks symmetry, or that there is bias in its projection.

We shall see shortly (§3.5.4) that our intuitive belief that in a large number of trials a particular score with a die will occur with relative frequency very nearly 1/6 is paralleled in the mathematical Theory of Probability. If the die were thrown 1,000 times, the particular score could occur 0, 1, 2 . . . 999 or 1,000 times. We are able to calculate the probabilities of these results and we can show that the *proportion* of successes in n trials tends to 1/6 as n increases indefinitely. But we are still dealing in prior probabilities obtained by a classification of the 'equally probable' results of n throws. We cannot prove *mathematically* that the laws of nature are the laws of mathematical probability. The reader who has some knowledge of Statistical Mechanics will know, however, that mathematical probability has come to afford an explanation or expression of the laws of thermodynamics. Moreover, the modifications of the classical theory of Statistical Mechanics introduced in Einstein-Bose and Fermi-Dirac statistics are concerned with the classification of events or states as equally probable.

The effective equality of prior probability and relative frequency in the long run is confirmed by experience on gaming tables. By paying out, on a result on a roulette wheel, odds slightly less than those against the result given by the prior probability, the banker is assured of profit in the long run. Conversely, if an insurance company's mortality table states that for a man aged 38, the chance of dying within a year is 0·0094, this figure has been calculated from long-term observations and serves as a prior probability in calculating premiums.

3.2 Theorem of Total Probability

Consider the possible combination of two events, A and B. There are four mutually exclusive possibilities:

(i) both A and B occur,
(ii) A occurs but not B,
(iii) B occurs but not A,
(iv) neither A nor B occurs.

In the following table A, B denote the success of these events; $\bar{A}$, $\bar{B}$ denote their failure or non-occurrence. The numbers α, β, γ, δ are the frequencies of the possible resultant events (i), (ii), (iii) and (iv).

	A	$\bar{A}$
B	α	γ
$\bar{B}$	β	δ

First suppose that events A and B are themselves mutually exclusive, i.e. they cannot both occur. Then $\alpha = 0$. The total frequency is $\beta + \gamma + \delta$ and

p_1 = the probability that A occurs $= \beta/(\beta + \gamma + \delta)$,
p_2 = the probability that B occurs $= \gamma/(\beta + \gamma + \delta)$,
p = the probability that *one or other* occurs $= (\beta + \gamma)/(\beta + \gamma + \delta)$,

$$\text{so that } p = p_1 + p_2.$$

This result is almost obvious, but important. It can be extended to three or more events; for if we require the probability that one or other of three mutually exclusive events occurs we may reduce this to finding the probability that one or other of events D and C occurs, where D is the resultant event that one or other of events A and B occurs.

We have therefore the following theorem:

Theorem I. The probability that *one or other* of two or more *mutually exclusive* events shall occur is the *sum* of the probabilities of the different events.

In applying this theorem it is important to remember that the events must be mutually exclusive. To take simple examples: the probability of throwing either a 5 or a 6 with a single throw of a die is $1/6 + 1/6$; the probability of throwing a 6 with one or other of two throws of a die (or one or other of two dice thrown together) is not $1/6 + 1/6$, the events not being mutually exclusive. If the probabilities of winning a race of three horses in the race are estimated at 1/3, 1/6 and 1/18, then the probability that one of them will win is $1/3 + 1/6 + 1/18$, but this answer would be wrong if the given probabilities referred to three different races in which the horses were running. Further examples will be considered in conjunction with the next theorem.

3.3.1 Theorem of Compound Probability

We now consider the probability that both of two events A and B shall occur. They are certainly not now mutually exclusive, unless the probability of the joint event is zero.

Referring to the table in §3.2,
the probability that both occur is $\alpha/(\alpha + \beta + \gamma + \delta)$,
the probability that A occurs is $(\alpha + \beta)/(\alpha + \beta + \gamma + \delta)$,
the probability that B occurs is $(\alpha + \gamma)/(\alpha + \beta + \gamma + \delta)$.

No invariable relation connects these three probabilities. But con-

sider the probability that B will occur, *if A has already occurred*, i.e. granted that A does occur. It is $\alpha/(\alpha + \beta)$.
Thus if

p_1 = the probability that A occurs $= (\alpha + \beta)/(\alpha + \beta + \gamma + \delta)$
and p_2 = the probability that B occurs, *granted that A has* occurred $= \alpha/(\alpha + \beta)$,
then p = the probability that both A and B occur $= \alpha/(\alpha + \beta + \gamma + \delta)$ $= p_1p_2$.

Similarly if p_3 is the probability that B occurs and p_4 is the probability that A occurs, granted that B has occurred, then $p = p_3p_4$ from the identity

$$\frac{\alpha}{\alpha + \beta + \gamma + \delta} = \frac{\alpha + \gamma}{\alpha + \beta + \gamma + \delta} \cdot \frac{\alpha}{\alpha + \gamma}$$

If we require the probability that all of three events A, B, C shall occur, we can reduce this to the probability that events D, C shall both occur, where D is the joint event that both A and B occur. The probability of the triple event is the probability of the double event D multiplied by the probability of event C, granted that D has occurred, i.e. that both A and B have occurred. Similarly we may extend the theorem to four or more events.

Theorem II. The probability that events A, B, C . . . shall *all* occur is $p_1p_2p_2$ . . . where
p_1 = the probability that A occurs,
p_2 = the probability that B occurs, granted that A has occurred,
p_3 = the probability that C occurs, granted that A and B have occurred, and so on.

3.3.2 Independent events

Before considering applications of Theorem II we show that in an important special case the theorem may be stated in a modified form.

We shall say that events A and B are *independent* if the probability of B's occurrence is the same, whether A occurs or not. Referring to the table in §3.2 we see that this implies

$$\frac{\alpha}{\alpha + \beta} = \frac{\gamma}{\gamma + \delta} = \frac{\alpha + \gamma}{\alpha + \beta + \gamma + \delta}.$$

It follows that $\frac{\alpha}{\beta} = \frac{\gamma}{\delta}$ and $\frac{\alpha}{\gamma} = \frac{\beta}{\delta}$.

In the two columns of the table, the frequencies in the first and second rows are proportional; in the two rows, the frequencies in the columns are proportional.

If the numbers α, β, γ, δ in the cells are all divided by the total frequency $(\alpha + \beta + \gamma + \delta)$, the cells will contain the *probabilities* of the resultant events (i), (ii), (iii) and (iv) of §3.2.

In the case when events A, B are independent, the table of probabilities takes the special form:

	A	$\bar{A}$	Totals
B	p_1p_2	q_1p_2	p_2
$\bar{B}$	p_1q_2	q_1q_2	q_2
Totals	p_1	q_1	1

It may easily be verified algebraically that if

p_1 = the probability of event A $= (\alpha + \beta)/(\alpha + \beta + \gamma + \delta)$,
p_2 = the probability of event B $= (\alpha + \gamma)/(\alpha + \beta + \gamma + \delta)$,
$q_1 = 1 - p_1$,
$q_2 = 1 - p_2$,

and if also $\dfrac{\alpha}{\beta} = \dfrac{\gamma}{\delta}$,

then the following relations hold:

$$\frac{\alpha}{\beta} = \frac{\gamma}{\delta} = \frac{\alpha + \gamma}{\beta + \delta} = \frac{p_2}{1 - p_2} = \frac{p_2}{q_2}$$

$$\frac{\alpha}{\gamma} = \frac{\beta}{\delta} = \frac{\alpha + \beta}{\gamma + \delta} = \frac{p_1}{1 - p_1} = \frac{p_1}{q_1}$$

$$\frac{\alpha}{\alpha + \beta} = \frac{\gamma}{\gamma + \delta} = p_2$$

$$\frac{\alpha}{\alpha + \gamma} = \frac{\beta}{\beta + \delta} = p_1$$

$$\frac{\alpha}{\alpha + \beta + \gamma + \delta} = \frac{\alpha}{\alpha + \beta} \cdot \frac{\alpha + \beta}{\alpha + \beta + \gamma + \delta} = p_2p_1$$

and similarly for the probabilities in the other cells.

The events being now independent we may omit the phrase 'granted that A (or B) has occurred'. The cells may be filled in conformity with Theorem II, by multiplying together the relevant probabilities, which appear in the marginal totals. Thus the probability that B occurs but not A

= the probability of B, $\bar{A}$
= (probability of B) (probability of $\bar{A}$)
$= p_2q_1$

It is later necessary to extend the concept of independence. Instead of considering merely the two possibilities of occurrence and non-occurrence of each of two events A and B, we consider a series of possible values of each of two variates x and y.

The theorem of compound probability is so often applied to independent events that it is worth while to make a separate statement, as follows.

Theorem IIa. The probability of the occurrence of *all* of a number of *independent* events is the product of the probabilities of the events.

3.3.3 Sampling, with and without replacement

The two forms of Theorem II are well illustrated by the probabilities resulting from successive drawings, e.g. of a card from a pack, in the two circumstances

(*a*) that a card drawn is replaced (and the pack shuffled, if desired) before the next draw;

(*b*) that a card drawn is not replaced.

What, for instance, is the probability that when two players cut for deal, both cut aces? If the first card is replaced, the events (first player cutting ace, second player cutting ace) are independent, the probability being 1/13 in each case. But if the first card is not replaced the second player will have a pack of 51 cards and, if the first player drew an ace, a probability of only 3/51 or 1/17 of drawing an ace. Hence the probability that both draw aces is $1/13 \times 1/13$ if the first card drawn is replaced, but $1/13 \times 1/17$ if it is not replaced. (Note that the second factor is the probability of the second event, granted that the first has occurred.)

We have seen that a set of observed values of a variate must often be regarded as a sample drawn from a large or infinite population. Such sampling is usually without replacement, i.e. the first individual value is not, in general, returned and given a second chance of appearing in the next draw. If, however, the population is large and the proportion of individuals within a given category (e.g. values of the variate within a given class interval) not too small, the chance of drawing an individual in that category is practically not affected by replacement. In our example of cutting aces, if the pack had contained 52,000 cards of which 4,000 were aces, the factor 3/51 would have been 3,999/51,999 which is practically equal to 4,000/52,000 or 1/13, as in sampling with replacement. This point is important in relation to Sampling Inspection and Quality Control of mass production. It means, for instance, that if a large bulk of components includes a certain proportion of defectives and we draw at random a sample of moderate size, we may assume for each component drawn a constant probability of being defective, equal to the proportion of defectives in the bulk.

3.4 Permutations and Combinations. Illustrative examples

Before proceeding to illustrate the use of Theorems I and II by means of further examples, we remind the reader of some well-known formulae for numbers of permutations and combinations, which are often useful in calculating probabilities.

The number of ways of *selecting and arranging in order* r things out of n different things is called 'the number of permutations of n things r at a time' and denoted by ${}^nP_r$.

$$ {}^nP_r = n(n-1)(n-2)\ldots(n-r+1) \quad . \quad . \quad (1)$$

The maximum value of r is, of course, n and

$$ {}^nP_n = n(n-1)(n-2)\ldots3.2.1$$
$$= n! \text{ or 'factorial n'} \quad . \quad . \quad . \quad . \quad . \quad (2)$$

Thus $n!$ is the number of ways of arranging n things in order; there is no longer any question of selection, since all n things are used.

If the n things are not all different but consist of a of one kind, b of another, c of another, etc., so that $a + b + c + \ldots = n$, then the number of different orders in which all n things can be placed is

$$\frac{n!}{a!\,b!\,c!\ldots} \quad . \quad . \quad . \quad . \quad (3)$$

The number of ways of *selecting* r things from n different things is called 'the number of combinations of n things r at a time' and denoted by ${}^nC_r$.

$$ {}^nC_r = {}^nP_r/r!$$

$$= \frac{n!}{r!\,(n-r)!} \quad . \quad . \quad . \quad . \quad (4)$$

It is usual to derive the formula for ${}^nC_r$ from that for ${}^nP_r$ by the argument that the group of r things constituting any one selection from n things can, by formula (2), be arranged in $r!$ different orders. When all selections are so arranged, all permutations of the n things r at a time are obtained, so that ${}^nC_r \times r! = {}^nP_r$. It may be noted, however, that formula (4) is only a particular case of (3), with $a = r$ and $b = n - r$. In fact, the problem of selecting r things out of n is equivalent to the problem of distributing or arranging n labels, r of which are of one kind, marked 'chosen' and $n - r$ of another, marked 'rejected', amongst the n things. The exponents of permutations for football pools never use the term combinations. They are right in that the problem of the alternative ways of filling in a coupon may be regarded as one of permuting or arranging in order the labels 1, 2, x after deciding how many of each kind are to occur. But the filling of the coupon may equally be regarded as a problem of combinations, viz. of selecting the matches which are to be marked with each of the three symbols.

The most important property of combinations is

$$^{n}C_{r} = {}^{n}C_{n-r} \quad . \quad . \quad . \quad . \quad (5)$$

which is evident from (4) but also from the fact that to every selection of r things out of n must correspond a selection of $n - r$ things which are left.

Ex. i

If two cards are drawn from a pack, what is the probability that both are aces?

This is the problem of §3.3.3(*b*). Instead of the solution there given we may say:
there are $^{52}C_2$ possible pairs which can be drawn and of these $^{4}C_2$ pairs are pairs of aces.

The required probability $= {}^{4}C_2/{}^{52}C_2 = 1/221$.

Ex. ii

Six football matches are selected in each of which the chance of an away win is 3/5. Investigate the chances of one or more correct forecasts of four away wins, using combinations from the six matches.

If only a single forecast is submitted it consists of 4 matches chosen at random. By Theorem II the probability of a correct forecast, i.e. that all 4 matches result in an away win is $(3/5)^4$ or 0·13 approximately.

The number of forecasts possible is $^{6}C_4 = {}^{6}C_2 = 15$. Each has the above chance of being correct and the average number of correct forecasts to be expected is $15 \times (3/5)^4$ or 1·94 approximately. The bettor has merely enlarged the scale of his betting. The mean number of correct forecasts may be otherwise found as follows.

If there are 4 away wins, 1 forecast is correct. This can happen in $^{6}C_4$, i.e. 15 ways and for each the probability is $(3/5)^4(2/5)^2$. The probability of 1 correct forecast $= p_1 = 15(3/5)^4(2/5)^2$.

If there are 5 away wins, $^{5}C_4$, i.e. 5 forecasts are correct. This can happen in $^{6}C_5$, i.e. 6 ways and for each the probability is $(3/5)^5(2/5)$. The probability of 5 correct forecasts $= p_2 = 6(3/5)^5(2/5)$.

If there are 6 away wins, all 15 forecasts are correct. This can happen in only one way. The probability of 15 correct forecasts $= p_3 = (3/5)^6$.

There is a residual probability $p_4 = 1 - p_1 - p_2 - p_3$ that fewer than 4 of the 6 matches result in away wins and so no correct forecast is made. Remembering that these probabilities are theoretical relative frequencies in the long run, the mean number of correct forecasts is $0p_4 + 1p_1 + 5p_2 + 15p_3$. The reader may verify that this reduces to $15(3/5)^4$.

Ex. iii

The probabilities of two independent events are p_1 and p_2 respectively. What is the probability that one at least will occur?

The phrase 'one at least' is not to be confused with the phrase 'one or other' of Theorem I, which relates to mutually exclusive events. Such events are certainly not independent, for if one happens the other cannot.

The simplest way of answering the question is to note that the probability that *neither* event occurs is q_1q_2. In any other case one at least occurs, so that the required probability is $1 - q_1q_2$.
To check this, call the events A and B.

The probability (i) that both A and B occur is p_1p_2,
(ii) that A occurs but not B is p_1q_2,
(iii) that B occurs but not A is q_1p_2,
(iv) that neither occurs is q_1q_2.

The sum of these $= (p_1 + q_1)(p_2 + q_2) = 1$. We require the sum of the first three, which is therefore $1 - q_1q_2$.

Suppose the chances of winning of two racehorses are 1/3 and 1/6 respectively. If the horses are running in *different* races the chance that *one at least* will win is $1 - (2/3)(5/6) = 4/9$. If they are running in the *same* race the chance that *one or other* will win is $1/3 + 1/6 = 1/2$.

Ex. iv

If the chance that an aircraft fails to return from any single operational flight is 5%, what is the chance that it will survive 30 operational flights?

It must survive 30 times in succession and the chance of survival is 95% each time, so that the required probability is $(0{\cdot}95)^{30} = 21{\cdot}5\%$.

It follows that the probability of failing to survive 30 flights is $1 - 0{\cdot}215 = 78{\cdot}5\%$. Note that if we had set out to calculate this latter probability in terms of failing to return, the chance for each operation depends on having survived all earlier flights. The total probability of failing to return, at some stage or other, is

$$\tfrac{1}{20} + \tfrac{1}{20}\cdot\tfrac{19}{20} + \tfrac{1}{20}\cdot\left(\tfrac{19}{20}\right)^2 + \ldots \text{ in G.P. to 30 terms.}$$

The terms of the G.P. are the probabilities of the mutually exclusive events:
(i) failing to return from first flight,
(ii) surviving first flight but failing on second,
(iii) surviving two flights but failing on third, and so on.

It is easily verified that the sum of the G.P. is $1 - (19/20)^{30}$.

Ex. v

A batch of 50 components includes 3 defectives. If the components are tested in succession and the first defective found is the rth component tested, what is (i) the most probable value of r, (ii) the mean value of r?

The probability that $r = 1$ is $\frac{3}{50}$

$r = 2$ $\quad \frac{47}{50}\cdot\frac{3}{49}$

$r = 3$ $\quad \frac{47}{50}\cdot\frac{46}{49}\cdot\frac{3}{48}$

$r = 4$ $\quad \frac{47}{50}\cdot\frac{46}{49}\cdot\frac{45}{48}\cdot\frac{3}{47}$

$r = 5$ $\quad \frac{47}{50}\cdot\frac{46}{49}\cdot\frac{45}{48}\cdot\frac{44}{47}\cdot\frac{3}{46}$.

and so on. It will be noticed that from $r = 4$ onwards cancellation of identical factors is possible; in fact, for *all* values of r from 1 to 48 the probability is

$$\frac{3(50 - r)(49 - r)}{(50)(49)(48)}.$$

Each successive probability is obtained from the preceding one by multiplying by a factor less than 1. Hence the *most probable value of r is* 1.

Denoting the probability, or theoretical relative frequency, of any value of r by f,

the mean value of $r = \Sigma fr$

$$= \sum_{r=1}^{48} \frac{r(50 - r)(49 - r)}{39200}.$$

The reader familiar with a method of summing a series whose r^{th} term is a polynomial in r will be able to verify that the mean value is $12\frac{3}{4}$. It is interesting to note that this value may be obtained by observing that the 3 defectives distributed at random in the 50 components divide the remaining components into 4 groups; for all possible placings of the defectives the average number in any of the 4 groups is $(50 - 3)/4 = 11\frac{3}{4}$. The average number of articles tested, up to and including the first defective, is $11\frac{3}{4} + 1 = 12\frac{3}{4}$.

The reader may now with advantage work numbers 1 to 10 of the examples at the end of the chapter.

3.5.1 Bernoulli's Theorem and the Binomial Distribution

Theorem III. If the probability of success in each trial is p, then the probability of r successes in n trials is ${}^nC_r p^r q^{n-r}$.

Here $q = 1 - p$ = the probability of failure in each trial. There are ${}^nC_r$ ways of choosing the r trials in which the successes occur. For each, the compound probability of the r successes and $n - r$ failures is $p^r q^{n-r}$. Hence the total probability of r successes and $n - r$ failures is ${}^nC_r p^r q^{n-r}$.

The reader should compare the probabilities p_1, p_2, p_3 obtained in Ex. ii above. We could now say at once that the probability of 4 away wins in 6 matches, in each of which the probability of an away win is 3/5, i.e. the probability of 4 successes in 6 trials when the probability of success in a single trial is 3/5, is ${}^6C_4(3/5)^4(2/5)^2$.

The Binomial Distribution

If the r of Bernoulli's Theorem is regarded as a variate capable of taking the $n + 1$ discrete values 0, 1, 2 . . . n, the theorem gives the probabilities or theoretical relative frequencies of these values of the variate.

Suppose, for instance, that of a large bulk of mass-produced articles a fraction p are defective. If a sample of 10 articles is drawn at random, then on account of the largeness of the bulk we may assume that as each article is drawn there is a constant probability p that it is defective. The whole sample of 10 may contain

0, 1, 2 . . . 10 defectives with probabilities

$$q^{10},\ 10q^9p,\ {}^{10}C_2q^8p^2 \ldots p^{10} \text{ respectively.}$$

These probabilities are the terms of the expansion of $(q + p)^{10}$ in ascending powers of p.

Since probabilities are *relative frequencies* in the long run, we could now say that if a large number N of such samples were drawn the expected *frequencies* of samples containing 0, 1, 2 . . . 10 defectives are the terms of $N(q + p)^{10}$.

In Ex. ii above, the probabilities of 0, 1, 2 . . . 6 away wins in the 6 matches are the terms of $(2/5 + 3/5)^6$. p_1, p_2 and p_2 are the last three terms in the expansion, giving the probabilities of 4, 5 and 6 away wins. $1 - p_1 - p_2 - p_3$ is the sum of the remaining terms, giving the total probability of 0, 1, 2 or 3 away wins.

The simplest form of the binomial distribution occurs when $p = q = 1/2$. Then, in ${}^nC_r p^r q^{n-r}$, the factor $p^r q^{n-r}$ is always $(1/2)^n$. For fairly small values of n the binomial coefficients, i.e. the values of ${}^nC_r$ may be tabulated in the form known as Pascal's triangle:

$$\begin{array}{lccccccccc} n=1 & & & & 1 & & 1 & & & \\ n=2 & & & 1 & & 2 & & 1 & & \\ n=3 & & 1 & & 3 & & 3 & & 1 & \\ n=4 & 1 & & 4 & & 6 & & 4 & & 1 \end{array}$$

and so on. Except for the 1's each number is found as the sum of the two above it. By dividing the numbers in the nth line by 2^n, we obtain the probabilities of 0, 1, 2 . . . n successes in n trials, when $p = q = 1/2$. In this case if we were to draw the histogram or relative frequency polygon of the distribution, it would be symmetrical. For all values of p other than 1/2, the distribution is skew.

One important application of the binomial distribution is to problems of Sampling Inspection, treated in Chapter VIII. A large consignment of goods may be accepted on the basis that a random sample contains fewer than an assigned number of defectives. This gives rise to such questions as the following.

Ex. vi

If the proportion of defective items in a large bulk is 8%, what is the probability that a sample of 30 items will include fewer than 3 defectives?

Here $p = 0{\cdot}08$ = the probability that an item is defective;
$n = 30$ = the number of trials.

The probabilities of 0, 1 and 2 defectives are

$$(0{\cdot}92)^{30},\ 30(0{\cdot}92)^{29}(0{\cdot}08),\ \frac{30\,.\,29}{1\,.\,2}\,(0{\cdot}92)^{28}(0{\cdot}08)^2$$

or 0·082, 0·214 and 0·270.

The required probability is the sum of these, namely 0·566.

3.5.2. Mean and standard deviation of the binomial distribution

We have a variate r which takes the values 0, 1, 2 . . . n with probabilities as given by Bernoulli's Theorem. Treating probabilities as relative frequencies we can find the mean value of r and the s.d. The *mean* is found to be np, i.e. this is the average number of successes in n trials if the probability of success is p in each trial. The s.d. is $\sqrt{npq}$, i.e. this is the r.m.s. deviation of the number of successes from the mean np. The proof of these results will now be given.

The probability f_r of the value r is the coefficient of t^r in the expansion of $(q + pt)^n$, which is called the generating function of the distribution. The usefulness of introducing the additional letter t will become evident in what follows.

$$(q + pt)^n = f_0 + f_1t + f_2t^2 + \ldots + f_nt^n \quad \cdot \quad \cdot \quad (6)$$

where $f_0 = q^n$ and generally $f_r = {}^nC_r\, p^rq^{n-r}$
If we put $t = 1$ in equation (6) we obtain

$$(q + p)^n = \Sigma f$$

i.e. $\Sigma f = 1$, as is otherwise obvious, since the total probability is 1.
Differentiating both sides of equation (6) with respect to t,

$$np(q + pt)^{n-1} = 0 \cdot f_0 + 1 \cdot f_1t^0 + 2f_2t^1 + \ldots + nf_nt^{n-1} \quad . \quad (7)$$

Putting $t = 1$,

$$np = \Sigma rf$$
$$= \text{the mean value of } r.$$

(The mean value is $\Sigma fr/\Sigma f$ in which $\Sigma f = 1$.)
Multiply equation (7) by t and again differentiate with respect to t, obtaining

$$npt(q + pt)^{n-1} = 0 + 1\, f_1t^1 + 2f_2t^2 + \ldots + nf_nt^n$$

$$np(q + pt)^{n-1} + n(n - 1)p^2t(q + pt)^{n-2}$$
$$= 0 + 1^2f_1 + 2^2f_2t^1 + \ldots + n^2f_nt^{n-1} \quad . \quad (8)$$

Putting $t = 1$,

$$np + n(n - 1)p^2 = \Sigma r^2f$$
$$= \mu_2', \text{ the second moment about the origin.}$$
$$\mu_2 = \sigma^2 = \mu_2' - (np)^2$$
$$= np - np^2$$
$$= np(1 - p) = npq.$$

This is the variance of the distribution and the standard deviation is $\sqrt{npq}$.

Ex. vii

If a coin is tossed 100 times in succession, or 100 coins tossed simultaneously, the average number of 'heads' is $100 \times 1/2$, since $p = 1/2 =$ the probability of heads for each toss of a coin. If the experiment were repeated a large number of times we expect an r.m.s. deviation from this average number of 50 heads of amount $\sqrt{npq} = \sqrt{100\,(1/2)\,(1/2)} = 5$.

3.5.3 Expected Frequencies

Suppose we regard the n trials of Bernoulli's Theorem as constituting a single experiment. Let the experiment be repeated N times. Then the terms of the binomial distribution give the probabilities, or theoretical *relative frequencies* in the long run, that the result of an experiment shall be 0, 1, 2 . . . n successes. The total frequency (number of experiments) being N, the expected *frequencies* of 0, 1, 2 . . . n successes are the terms of $N(q + p)^n$.

Note that the change from relative frequency to frequency does not affect the scale of the variate, which is the number of successes, and the mean and s.d. of the (theoretical) frequency distribution are still np and $\sqrt{npq}$ respectively.

Ex. viii

If, as in Ex. vi, the proportion of defectives in a large bulk is 8% and 1,000 samples of 30 are drawn, how many samples are expected to include 0, 1, 2, etc., defectives?

The probabilities of 0, 1, 2, etc., defectives have simply to be multiplied by 1,000. Using the results of Ex. vi and continuing the calculations, we obtain the table:

Number of defectives in sample:	0	1	2	3	4	5	6	7	8
Expected number of samples:	82	214	270	219	129	58	21	6	1

The expected numbers have been given to the nearest whole number, but there is no reason why such mathematically calculated frequencies should not be given fractional values.

3.5.4 Distribution of proportion of successes

If there are r successes in n trials, the proportion of successes is r/n and the probability of this result is ${}^nC_r p^r q^{n-r}$. The variate r takes the values 0, 1 . . . n, so that the range of the variate increases with n. If r/n is made the variate, its range is from 0 to 1 for all values of n. The mean and s.d. of r/n may be obtained from those of r by dividing by n, since we have merely changed the scale of the variate. Hence in n trials the mean value of the proportions of successes is $np/n = p$ with s.d. $\sqrt{npq}/n = \sqrt{pq/n}$. Now as n increases the s.d. decreases, tending to zero as n tends to infinity. This means that *for large values of* n, *it becomes very improbable that the proportion of successes in* n *trials will differ much from the probability,* p, *of success in a single trial.*

To illustrate this numerically, let $p = 1/3$, $q = 2/3$ and give n the values 50, 128 and 288. For all these values of n the mean value of the proportion of successes r/n is 1/3, but the s.d. of the distribution of r/n takes the values 1/15, 1/24, 1/36. If we draw a frequency polygon or histogram for each of the three values of n, the values taken by the discrete variate r/n are in each case sufficiently close together for us to represent the distribution by a smooth curve. In the figure below, the vertical scale is such that the area under the curve remains constant, to represent the total probability 1. As n increases the probability that the value of r/n lies within a chosen range of p also increases: it can, in fact, be made as near 1 as we please by making n large enough.

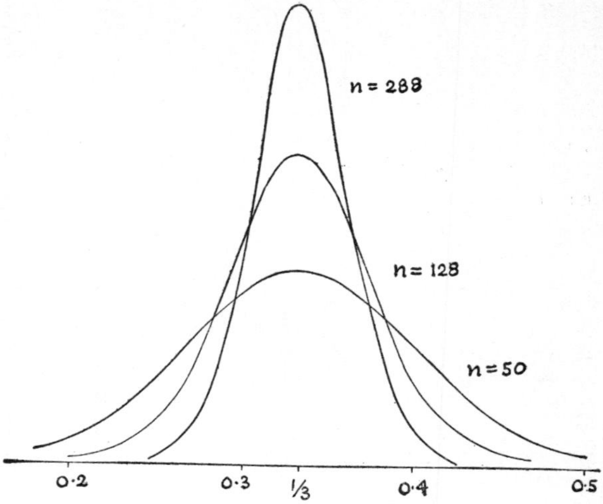

FIG. 8

3.6 Continuous probability distribution

The possible values of a continuous variate, x, may be represented by the points of a finite or infinite straight line. Any one value of x, corresponding to one point of the line, is only one of an infinite number of possible values, so that there is no measurable probability that x takes a particular value. Instead, we may say that there is a small probability δp that x lies in a small interval δx. If there is equal probability for all equal intervals, δx, the probability distribution is uniform. In this case the whole range of x must, in general, be finite, since the total probability over the range is finite and equal to 1. For instance, if a length is measured to the nearest tenth of an inch, the error in the measurement, that is, the deviation of the recorded length from the true length, may with uniform probability lie anywhere in the range $-$ 0·5 to 0·5 tenths. In a large number of such measurements, we should expect to find, if the measurements were repeated to a high degree of precision, say to the nearest thousandth of an inch, that the frequency distribution of errors is rectangular. Similarly if a point is said to be chosen 'at random' in a finite straight line, this is taken to mean that probability is uniformly distributed over the finite range.

When the distribution is not uniform, the range being finite or infinite, the element δp of probability in an interval δx is different at different points of the range and so depends on the value of x. Hence $\delta p = f(x)\delta x$ where $f(x)$ is some function of x. Here δx may be made indefinitely small so that we may write in the notation of differentials

$dp = f(x)dx$. The function $f(x)$, being the limit as $\delta x \to 0$ of $\delta p/\delta x$, which is the probability *per unit length* in the interval δx, is a probability *density*.

The notions involved here will be much used in later developments and extended to two or more variates, corresponding to points in two or more dimensions. A probability distribution of two variates x and y will be expressed by a differential relation $dp = \phi(x, y)dxdy$ where $\phi(x, y)$ is the probability density at the point (x, y) of a plane. Just as a probability distribution of one variate may be represented graphically by area under a curve, so that of two variates may be represented by volume under a surface whose equation is $z = \phi(x, y)$. When the number of variates increases to three or more, however, we can no longer visualize a graphical representation of the distribution, but must be content with the statement that there is a probability density which is a function of all the variates. One may visualize the three-dimensional case by imagining the whole relevant space coloured, the depth of colour being proportional to probability density. The colour will be pale in regions of low probability and deep where the probability density is high.

At the present stage we shall consider only some examples of uniform probability in one or two dimensions.

Ex. ix

A long convoy of military vehicles, each 20 ft. long and with a gap of 60 ft. between the tail of one vehicle and the head of the next, comes to rest on a straight road. What is the probability that the drive, 10 ft. wide, of a house at the side of the road is wholly or partially blocked by one of the vehicles?

In every 80 ft. of road a stretch of 20 ft. is blocked. Let the blocked portion be at the left-hand end. Then the drive is not blocked if its right-hand gate-post is anywhere in a stretch of 50 ft. at the right-hand end. Assuming a uniform probability distribution of the position of the gate-post over the whole 80 ft., there is a probability of 3/8 that the drive is overlapped.

Ex. x

A man cycles 5 miles to work every day along a rough road. He finds that there is a chance of 0·05 that he will have a puncture on any day, and that a puncture is equally likely on any part of his journey. If he has a puncture, he can either repair it, which takes him 9 minutes, and cycle on, or else he can wheel his bike for the rest of the way at 4 miles per hour. Assuming that the chance of two punctures on one journey is negligible, and that he cycles at 10 m.p.h., show that it is worth while to repair the puncture if he is more than 1 mile from his destination, but otherwise it is better to walk.

If he realizes this, find the mean time taken for the journey, and the chance that on any particular day it will last longer than 35 minutes.

(R.S.S.)

If he has a puncture and mends it, the journey takes 39 min.

If he has a puncture after x miles and walks the rest of the way, the journey takes $6x + 15(5-x)$ min. It is worth while to repair the puncture if

$$39 < 75 - 9x,\ 9x < 36,\ x < 4,$$

in which case he is more than 1 mile from his destination. There are now three possibilities:

(i) He has no puncture and the journey takes 30 min. The probability of this is 0·95.
(ii) He has a puncture in the first 4 miles and repairs it, the journey taking 39 min. The probability of this is $0{\cdot}05 \times 4/5$, since the probability of a puncture is uniform over the 5 miles.
(iii) He has a puncture in the last mile and walks the rest of the way. The journey takes $75 - 9x$ min., where x may with uniform probability have any value between 4 and 5. The mean time in this event is $75 - 9(4{\cdot}5) = 34{\cdot}5$ min. The probability of event (iii) is $0{\cdot}05 \times 1/5$.

$$\begin{aligned}\text{The mean time taken} &= (0{\cdot}95)\,(30) + (0{\cdot}04)\,(39) + (0{\cdot}01)\,(34{\cdot}5)\\ &= 30{\cdot}405 \text{ min.}\end{aligned}$$

The journey takes longer than 35 min. if a puncture occurs either in the first 4 miles or at a point in the fifth mile such that $75 - 9x > 35$, $x < 4\frac{4}{9}$. The probability of this is

$$\frac{0{\cdot}05 \times 4\frac{4}{9}}{5} = 2/45.$$

It may be noted that the probability distribution of time of journey is partly discrete and partly continuous. Probabilities 0·95 and 0·04 are concentrated at times 30 and 39, while the remaining probability 0·01 is uniformly distributed over the range 30 to 39. Treating probability as weight, the mean time can be found as the position of the centre of gravity and the probability that the journey takes longer than 35 minutes is the weight beyond $t = 35$.

Ex. xi

Of two lengths of timber, one may with uniform probability vary from 15 to 17 ft., the other from 11 to 12 ft. What is the probability that the total length exceeds 28 ft.?

$$\text{The total length} = 26 + x + y \text{ ft.},$$

where x may vary from 0 to 2 and y from 0 to 1, each with uniform probability. If x and y are made Cartesian co-ordinates of a point in a plane, the probability density is uniform over a rectangle.

Now in the plane all points for which $x + y > 2$ lie above and to the right of the straight line whose equation is $x + y = 2$. This line cuts off a quarter of the area of the rectangle in which our point (x, y) may lie, with uniform probability over the whole rectangle. Hence there is a probability of 1/4 that $x + y$ exceeds 2 and the total length exceeds 28 ft.

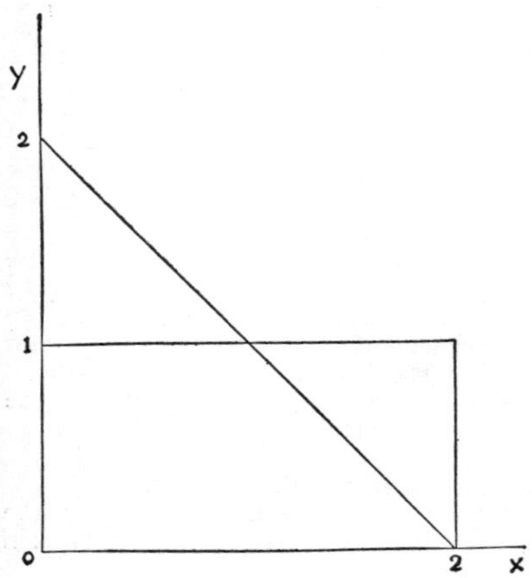

FIG. 9

Ex. xii **Bertrand's problem**

We conclude with a somewhat academic problem because it illustrates the fundamental difficulty of the Theory of Probability, that of classifying certain events as equally probable or, what is equivalent, determining the probability density at any point in the range of the variables.

The problem is: if a chord of a circle is chosen at random, what is the probability that its length is greater than the side of an equilateral triangle inscribed in the circle?

Three solutions are suggested:

(*a*) Suppose by a random chord we mean one such that either end of the chord may with uniform probability be at any point of the circumference. Then for every position of one end, the chord is greater than the side of the equilateral triangle if the other end lies on an arc which is 1/3 of the whole circumference. The required probability is 1/3.

(*b*) Suppose that all directions of the chord are deemed equally probable and that, independently of the direction, all distances of the chord from the centre of the circle, from zero to the radius of the circle, are equally probable. Then for every direction of the chord the required probability is 1/2, corresponding to chords at a distance from the centre of less than half the radius.

(*c*) The length of a chord is determined entirely if we know the radius of the circle and the position of the middle point of the chord. Suppose that the latter point may with uniform probability be anywhere within the circle. Then the chord is greater than the side of the equilateral triangle if the middle point is within a concentric circle, of half the radius. The required probability is the ratio of the area of this smaller circle to that of the given circle, namely 1/4.

We have thus three different answers corresponding to three different interpretations of a 'random' chord. There are theoretical reasons for preferring the second solution, and when some engineering students were invited to draw and measure what they felt to be random chords of a circle of diameter 6 in., out of 280 chords precisely half were greater in length then $3\sqrt{3}$ in. The students were not informed of Bertrand's problem and its solutions.

Suggestions for further reading:

A. C. Aitken, *Statistical Mathematics*, Chapter 1 (Oliver and Boyd).

G. Yule and M. G. Kendall, *Introduction to the Theory of Statistics*, Chapter 10 (Griffin).

M. G. Kendall, *Advanced Theory of Statistics*, Chapters 2, 7 (Griffin).

EXAMPLES ON CHAPTER III

1. (i) What is the chance of an ace being drawn in a random selection of one card from a pack of 52 playing cards?
   (ii) What is the chance of four aces being drawn in a random selection of four cards from the pack?

2. What is the chance that some one of the four hands, of thirteen cards each, in a deal at bridge shall contain (i) four aces, (ii) only court cards (i.e. aces, kings, queens or knaves), (iii) cards all of the same suit, (iv) only spades?

3. Four players draw a card each from a pack. If one player draws a ten, what is the chance that his card is higher than any of the other three cards, aces counting high?

4. If n persons take seats at random at a round table, find the probability that two specified persons are seated next to one another.

5. If the odds against horse A in one race are 5 to 2 and against horse B in another race are 4 to 1, what are the odds against the 'double' event that both horses win? What is the chance that one at least will win?

6. The probability that A can solve a problem is 2/3 and the probability that B can solve it is 4/5. If both try, what is the probability that the problem is solved?

7. In how many ways can 13 matches be classified as 7 home wins, 3 away wins and 3 draws?
If home wins are considered certain in four particular matches and an away win in one particular match, in how many ways can the remaining 3 home wins, 2 away wins and 3 draws be allocated?

8. In a set at tennis, A is leading by 5 games to 3. What is his chance of winning the set if his chance of winning each game is 1/2? The set is won when either player has won at least 6 games and leads by 2 games.

9. A man aged 40 invests in a deferred annuity which will commence 20 years later. His wife is aged 38. Of all men aged 40, 4/5 still survive after 20 years and of all women aged 38, 9/10 survive 20 years. What is the probability that one at least of the man and his wife will be alive when the annuity becomes payable?

10. Dice-players call throwing a total of 8 with two dice, by means of two 4's, the 'hard' way. Why is this? In what other ways can a total of 8 be thrown and what are their probabilities?

11. Of a large number of mass-produced articles, one-tenth are defective. Find the probability that a random sample of 20 will contain (*a*) exactly 2 defective articles, (*b*) at least 2 defective articles. (L.U.)

12. A binomial frequency distribution is given by the coefficients of powers of t in the expansion of $N(1 - p + pt)^n$. Show that the arithmetic mean of the distribution is np.
In sampling a large number of parts made by a machine the mean number of defectives in a sample of 20 is 2. Out of a thousand such samples how many would be expected to contain at least three defective parts? (L.U.)

13. If a hurdler has a probability of 7/8 of clearing each hurdle, what is the probability that he will knock down fewer than three of a flight of ten hurdles?

14. Calculate the mean and s.d. of the frequency distribution obtained in Ex. viii, p. 41. What would the values of the mean and s.d. have been if the frequencies had not been rounded off to the nearest whole number?

15. In a book of values of a certain function there is one error on the average in m entries. Prove that when r values are turned up at random (with the possibility that any value may be selected more than once), the chance of all being accurate is $(m - 1)/r$ times as great as that of having only one error included in the selection.
Find r/m, in terms of m, in order that there may be a nine to one chance that the selection is free from any errors. In this case prove that as a very large set of tabulated values approaches perfection in accuracy, r increases to a limiting value of nearly 10·5 per cent. of the size of m.
(L.U.)

16. If the chance that any one of 10 telephone lines is busy at an instant is 0·2, what is the chance that 5 of the lines are busy? What is the most probable number of busy lines and what is the probability of this number? What is the probability that all the lines are busy?

17. For the binomial distribution, it may be proved that $\mu_3 = npq(q - p)$. What is the value of β_1? Show that, if p is held constant, $\beta_1 \to 0$ as $n \to \infty$. What is the significance of this result?

18. If each of two numbers may with equal probability be any of the integers from 0 to 9, what is the probability (i) that their sum is greater than 15, (ii) that their difference is greater than 6?

19. Answer Question 18 if the two numbers may each, with uniform probability, lie at any point of the continuous range 0 to 10.

20. A bar 13 ft. long is placed at random across and at right angles to a set of parallel rails which are 4 ft. apart. Find the probability that the bar rests on (i) three, (ii) four rails.

21. A chain is made of 10 links. The tensile strength of each link is equally likely to have any value between 50 and 60 lb. What is the probability that the chain will break under a load of 51 lb.? (R.S.S. Cert.)

22. If 500 vehicles an hour traverse a particular stretch of road, taking 1/2 min. to pass, what is the probability that the road is empty at a given instant? Assume that all times of passing are equally likely.

23. Two callers are expected at any time between 11 a.m. and noon. If each may with uniform probability arrive at any time within the hour, what is the probability that they will arrive within 5 minutes of one another?

24. A fairground game consists in rolling pennies on a table ruled in squares. Assuming no skill, what is the probability that a penny comes to rest lying wholly within a square, if the side of each square is $1\frac{1}{2}$ in. and the diameter of a penny $1\frac{1}{4}$ in.?
A player scores 2 successes in 10 trials. What is the probability of this, on the basis of pure chance?

CHAPTER IV

POISSON'S DISTRIBUTION. THE NORMAL DISTRIBUTION

Traffic problems. Poisson's distribution. Sum of Poisson variates. Distribution of time-interval; mean free path. Telephone traffic. The normal distribution and its uses. Sum and difference of normal variates.

4.1 Some traffic and allied problems

Suppose that over a suitable period of the day the traffic on a road is moderate, averaging 2 vehicles per minute. The vehicles may be supposed to arrive at random and independently. Over short intervals of time, such as successive minutes, we shall not expect exactly equal numbers of vehicles and certainly not 1/30 of a vehicle in each successive second. Yet in any second one or more vehicles may arrive. If we take $p = 1/30$ as the prior probability of a vehicle in each second, then a period of one minute may be regarded as affording 60 trials of an event with probability 1/30; it might also be regarded as affording 120 trials in each of which the probability is 1/60, that of a vehicle arriving in any half-second. In either case we could use the binomial distribution to calculate the probabilities of 0, 1, 2, etc., vehicles in a minute. The results of the calculations would agree very closely. For instance, the probability of no vehicle is q^n.

$$\begin{aligned} \text{Now } \log (q^n) &= n \log q \\ &= n \log (1 - p) \\ &= - np, \text{ approximately} \end{aligned}$$

if p is small and the logarithms are Napierian. When p was reduced from 1/30 to 1/60 and at the same time n increased from 60 to 120, the value of np was held constant so that the value of q^n is practically unchanged.

Mathematically, the meaning of this is that as $p \to 0$ and $n \to \infty$ in such a way that np remains constant, the binomial distribution tends to a special form, known as *Poisson's distribution*. Its formula is obtained in the next section.

In practice we may expect Poisson's distribution to reflect conditions, such as those of the traffic problem, in which a given mean number of incidents is the result of a large number of opportunities of occurrence of an incident, the probability of a single occurrence being small.

Another traffic problem of the same type is that of demands on a telephone route. The calls come independently from a large number of subscribers and sufficient alternative lines (e.g. between exchanges) must be provided to make the probability that all lines are occupied very small. From a knowledge of the mean number of calls to be expected in a period, the probabilities of 0, 1, 2 . . . calls can be calculated.

The traffic problem could be put in the form: with a given mean traffic rate what is the probability distribution of the time-interval between successive vehicles or calls? This form of the problem, with distance in place of time, arises in the kinetic theory of gases. The molecules of a gas are in rapid motion and frequent collision. Some physical properties of gases are explained in terms of the 'mean free path' of a molecule, i.e. the mean distance travelled between successive collisions.

Perhaps the best instance of large n and small p is afforded by radioactivity. A small mass of radium contains many millions of atoms. In an assigned interval of time a moderate number of atoms, constituting a very small proportion of the number in the mass, will change as the result of emitting α- or β-particles. We cannot tell why particular atoms change. If we treat the occurrence as one of 'pure chance', the Poisson distribution provides a mathematical model, to which the variation in the number of particles emitted in the chosen time-interval may be expected to conform.

In Chapter III it was seen that the binomial distribution is applicable to sampling of mass-produced articles for defectives. When the proportion, p, of defectives is small enough we can use, as an approximation to the binomial distribution, the Poisson distribution corresponding to the mean number, np, of defectives to be expected in a sample of size n.

Accidents are, in general, events of low probability but with repeated opportunities of occurrence, so that their relative frequencies may be expected to conform to the Poisson distribution.

4.2.1 The Poisson distribution. Its mean and variance

In the binomial distribution the mean is np and the variance npq. If np is held constant, equal to μ, as $p \to 0$ and $n \to \infty$ then $q \to 1$, so that the variance $\to \mu$. It is an important property of the Poisson distribution that

$$\text{the variance } \sigma^2 = \text{the mean } \mu.$$

This is used as a first test that observed frequencies conform to the Poisson law, which is that if the mean value of the variate (number of successes) is μ, then the probabilities of

0, 1, 2, 3 . . . successes are

$$e^{-\mu}, \mu e^{-\mu}, \frac{\mu^2}{2!}e^{-\mu}, \frac{\mu^3}{3!}e^{-\mu} \ldots \quad . \quad . \quad . \quad (1)$$

We could express the binomial distribution in the form: the probability of r successes is the $(r + 1)^{th}$ term in the expansion of $(q + p)^n$. Similarly we may express Poisson's distribution in the form: the probability of r successes is $e^{-\mu}$ times the $(r + 1)^{th}$ term in the expansion of e^{μ}.

*That Poisson's distribution is the limiting form, under the conditions stated, of the binomial distribution, may be shown as follows.

The p.g.f. (probability generating function) of the binomial distribution

$$\begin{aligned} &= (q + pt)^n \\ &= \{1 + p(t-1)\}^n, \text{ since } q = 1 - p, \\ &= \left\{1 + \frac{\mu}{n}(t-1)\right\}^n \end{aligned}$$

A well-known limit is

$$\left(1 + \frac{x}{n}\right)^n \to e^x, \text{ as } n \to \infty$$

Hence if μ is held constant and $n \to \infty$, the p.g.f. of the binomial distribution tends to the form

$$\begin{aligned} e^{\mu(t-1)} &= e^{-\mu}\, e^{\mu t} \\ &= e^{-\mu}\left\{1 + \mu t + \frac{\mu^2 t^2}{2!} + \ldots + \frac{\mu^r t^r}{r!} + \ldots\right\} \end{aligned}$$

The probability of the value r (the number of "successes", or occurrences of the event in question) is the coefficient of t^r, namely $e^{-\mu}\mu^r/r!$, as stated in (1), p. 49.

We can now confirm directly that the mean and variance of Poisson's distribution are both μ. The generating function is

$$e^{-\mu}\left(1 + \mu t + \frac{\mu^2}{2!}t^2 + \ldots + \frac{\mu^r}{r!}t^r + \ldots\right) = e^{-\mu}e^{\mu t}.$$

This may be written

$$f_0 + f_1 t + f_2 t^2 + \ldots = e^{-\mu}\, e^{\mu t}, \quad . \quad . \quad . \quad (2)$$

where f_r is the probability of r successes.

Differentiating with respect to t,

$$f_1 + 2f_2 t + 3f_3 t^2 + \ldots = \mu e^{-\mu} e^{\mu t} \quad . \quad . \quad . \quad (3)$$

Multiplying by t,

$$f_1 t + 2f_2 t^2 + 3f_3 t^3 + \ldots = \mu e^{-\mu} t e^{\mu t}$$

Differentiating again,

$$f_1 + 2^2 f_2 t + 3^2 f_3 t^2 + 4^2 f_4 t^3 + \ldots = \mu e^{-\mu} e^{\mu t} + \mu^2 e^{-\mu} t e^{\mu t} . \quad (4)$$

Putting $t = 1$ in equations (2) and (3), we have $\Sigma f = 1$.

and $\Sigma rf = \mu$

proving that the mean is μ.

Putting $t = 1$ in equation (4), we have

$\Sigma r^2 f = \mu + \mu^2$ as the second moment about the origin. Hence

$$\sigma^2 = (\mu + \mu^2) - \mu^2$$
$$= \mu.$$

*

Ex. i

If the mean traffic rate is 2 vehicles per minute, find the probabilities of 0, 1, 2, 3 . . . vehicles per minute.

We require the terms

$$e^{-2}\left\{1,\ 2,\ \frac{2^2}{2!},\ \frac{2^3}{3!}\ \cdots\right\}$$

$= 0{\cdot}1353,\ 0{\cdot}2707,\ 0{\cdot}2707,\ 0{\cdot}1805,\ 0{\cdot}0902,\ 0{\cdot}0361,\ 0{\cdot}0120$, etc.

These are theoretical relative frequencies. If we observed the traffic for 200 minutes we should expect a frequency table *approximating* to the following:

No. of vehicles in a minute:	0	1	2	3	4	5	6	7	8 or more
Frequency:	27	54	54	36	18	7	2	1	1

Ex. ii

If the proportion of defectives in a bulk is 4%, find the probability of not more than one defective in a sample of 10.

The mean number of defectives in samples of size 10 is 4% of 10, or 0·4. The probability required is the sum of the probabilities of 0 and 1 defective, namely

$$e^{-0\cdot 4}(1 + 0{\cdot}4) = \underline{0{\cdot}9385.}$$

4.2.2. Some numerical considerations

In calculating terms of power series it is usually easiest to get each term from the preceding one, multiplying by the appropriate factor. Thus in the Poisson series the $(r + 1)^{th}$ term is merely μ/r times the rth term. It follows that successive terms are increasing so long as $r < \mu$ but decreasing when $r > \mu$.

If $\mu < 1$, the first term is the greatest.

If $\mu > 1$ and μ is not an integer, the greatest term is the $(r + 1)^{th}$, where r is the largest integer contained in μ. Since the $(r + 1)^{th}$ term is the probability of r successes, the most probable number of successes is the largest integer contained in μ.

If μ is an integer, the probabilities of $\mu - 1$ and μ successes are equal and are the two greatest terms of the series.

Since Poisson's distribution depends only on the single parameter μ, it is easier to tabulate than the binomial distribution. The British

Standards publication *Quality Control Charts* gives a short table of the probabilities of 0, 1, 2 . . . defectives in a sample for 15 different values of the mean μ. There is a warning that the table is not to be used unless p, the proportion of defectives in the bulk, is less than 1/10. The reason is that Poisson's distribution is a satisfactory approximation to the binomial, only for sufficiently small values of p. The following table will give some indication of the way in which a binomial distribution approaches the limiting form.

$r =$	0	1	2	3	4
Binomial: $p = 1/5, \quad n = 20$	0·012	0·058	0·137	0·205	0·218
Binomial: $p = 1/10, n = 40$	0·015	0·065	0·142	0·200	0·205
Poisson: $\mu = 4$	0·018	0·073	0·146	0·195	0·195

$r =$	5	6	7	8	9
Binomial: $p = 1/5, \quad n = 20$	0·174	0·109	0·054	0·022	0·007
Binomial: $p = 1/10, n = 40$	0·164	0·107	0·057	0·026	0·010
Poisson: $\mu = 4$	0·156	0·104	0·060	0·030	0·013

4.3 Sum of Poisson variates

Reverting to our original traffic problem with a mean rate of 2 vehicles per minute, it is evident that if we counted the vehicles in intervals of 2 minutes we should expect a Poisson distribution with mean 4; in intervals of 5 minutes, with a mean 10. But the number of vehicles in 5 minutes is merely the sum of the 5 numbers, each liable to vary according to Poisson's law, of vehicles in each separate minute. We may express this in the form: if r_1 and r_2 are two Poisson variates with means μ_1 and μ_2, then $r_1 + r_2$ is a Poisson variate with mean $\mu_1 + \mu_2$.

It is instructive to demonstrate this result mathematically. With the notation of §4.2.1, the probabilities of the values 0, 1, 2 . . . of the first variate are f_0, f_1, f_2, etc., in the generating function

$$f_0 + f_1t + f_2t^2 + \ldots = e^{-\mu_1}e^{\mu_1}.$$

Similarly $F_0, F_1, F_2 \ldots$ are the probabilities of the values 0, 1, 2 . . . of the second variate, where

$$F_0 + F_1t + F_2t^2 + \ldots = e^{-\mu_2}e^{\mu_2}.$$

Suppose we multiply the two series together. Then, in the product, the coefficient of t^r will be

$$f_0F_r + f_1F_{r-1} + f_2F_{r-2} + \ldots + f_rF_0.$$

Now this is precisely the probability that the sum of the two variates take the value r. For the sum is r if the first variate takes the value 0 and the second r; or if the first is 1 and the second $r - 1$; or 2 and $r - 2$ and so on. But the compound probabilities of these events are the products f_0F_r, f_1F_{r-1}, etc., and the total probability is their sum. Hence the probability of the value r of the sum of the two variates is the coefficient of t^r in the product of the two generating functions and this product is

$$e^{-\mu_1} e^{\mu_1 t} e^{-\mu_2} e^{\mu_2 t} = e^{-(\mu_1+\mu_2)} e^{(\mu_1+\mu_2)t}$$

which is simply the generating function of a Poisson variate with mean $\mu_1 + \mu_2$.

It is a property of generating functions (g.f.) that to find the g.f. of the sum of two independent variates we have only to form the product of their g.f.'s.

4.4 Distribution of time-interval between incidents

Given that, on the average, μ vehicles arrive per unit time, what is the probability that a time t elapses between consecutive vehicles? Divide the time t into a large number n of small intervals δt. Then the probability of a vehicle in each interval δt is $\mu\delta t$ and in each interval there is a probability $1 - \mu\delta t$ that no vehicle arrives. We require the probability that there is *no* vehicle in n successive intervals, but there is a vehicle in the $(n + 1)^{th}$ interval. The compound probability is

$$(1 - \mu\delta t)^n \mu\delta t = \delta p, \text{ say.}$$

Since $t = n\delta t$, this may be written

$$\frac{\delta p}{\delta t} = \mu\left(1 - \frac{\mu t}{n}\right)^n.$$

Now as $\delta t \to 0$, $n \to \infty$ and

$$\frac{dp}{dt} = \mu \times \lim\left(1 - \frac{\mu t}{n}\right)^n$$

$$= \mu e^{-\mu t}$$

The time t between consecutive vehicles therefore has a probability distribution given by

$$dp = \mu e^{-\mu t} dt \quad . \quad . \quad . \quad . \quad (5)$$

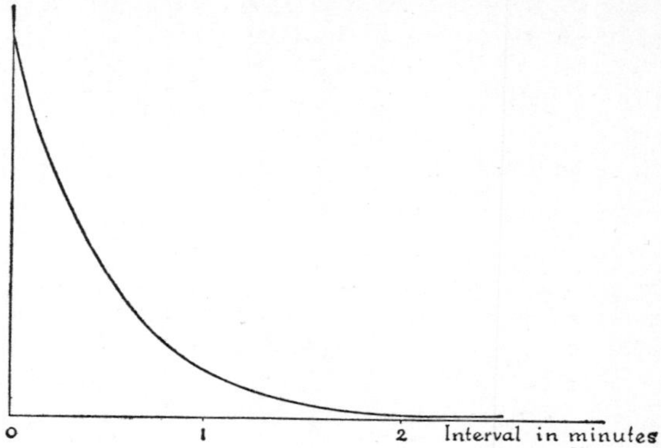

Probability distribution of time interval between vehicles, with mean traffic rate of 2 vehicles per minute.

FIG. 10

This is a continuous distribution in which t may vary from 0 to infinity, with probability density $\mu e^{-\mu t}$.

The reader may easily verify that the total probability,

$$\int_0^\infty \mu e^{-\mu t} dt = 1.$$

The mean time between vehicles

$$= \int_0^\infty t\mu e^{-\mu t} dt$$

$$= \Big[-te^{-\mu t}\Big]_0^\infty + \int_0^\infty e^{-\mu t} dt$$

$$= 0 + \frac{1}{\mu} = \frac{1}{\mu}.$$

This is to be expected; if, on the average, μ vehicles arrive per unit time then, on the average, there is a time interval $1/\mu$ between vehicles. If this mean time interval is denoted by a, equation (5) may be written

$$dp = \frac{1}{a} e^{-t/a} dt, \quad 0 < t < \infty, \quad . \quad . \quad . \quad (6)$$

which may be called the Poissonian law of traffic incidence. We shall reserve the name 'Poisson's distribution' for that given by formula (1).

Replacing time t by distance x, the above argument can be used to show that the probability distribution of the distance travelled between successive collisions of a molecule of a gas is

$$dp = \frac{1}{l}e^{-x/l}dx, \quad 0<x<\infty, \quad . \quad . \quad . \quad (7)$$

where l is the mean free path.

Ex. iii

If breakdowns in a power supply occur according to the Poissonian law, with an average of one breakdown per month, what is the probability that an interval of (i) less than a week, (ii) between 10 and 11 weeks, elapses between consecutive breakdowns? If records were kept for 5 years, how often may consecutive breakdowns be expected to be less than a week apart?

We shall take a month to mean 4 weeks. Taking 1 week as unit of time, $\mu = 1/4$ or $a = 4$. The probability that $t<1$ is the area under the distribution curve between $t = 0$ and $t = 1$, namely

$$\int_0^1 \frac{1}{4}e^{-t/4}\,dt = \left[-e^{-t/4}\right]_0^1$$

$$= -e^{-1/4} + 1 = 0{\cdot}22 \text{ approximately.}$$

The probability that t lies between 10 and 11

$$= \int_{10}^{11} \frac{1}{4}e^{-t/4}dt = e^{-10/4} - e^{-11/4} = 0{\cdot}018 \text{ approximately.}$$

In 5 years, or 65 months of 4 weeks, the expected number of breakdowns is 65, with 64 intervals between them. The number of intervals which may be expected to be less than a week = 0·22 of 64 = 14.

Note that since t is a continuous variate we cannot speak of the probability that the interval between breakdowns is precisely one week, but only that the interval lies in a chosen range.

4.5 Telephone traffic

Probability problems arising in telephone traffic are treated at length in *Probability and its Engineering Uses* by T. C. Fry. The formulae used vary according to the technical details of the telephone system concerned. One formula, known as Erlang's, gives what may be called a truncated Poisson distribution. By it, if the expected number of busy channels is μ, and there are n channels in all, then the probability that r channels are busy is

$$\frac{\mu^r}{r!}\Bigg/\left\{1 + \mu + \frac{\mu^2}{2!} + \frac{\mu^3}{3!} + \ldots \frac{\mu^n}{n!}\right\} \quad . \quad . \quad (8)$$

The series in the denominator is that for e^μ terminated at the $(n + 1)^{th}$ term. If μ were small enough the denominator could be replaced by e^μ, so that Erlang's law becomes Poisson's. The Erlang formula reflects the fact that there cannot be more channels busy than are available, although with a mean traffic μ there are by Poisson's law small probabilities of indefinitely large numbers of calls.

* Denote expression (8) by $E_{r,n}(\mu)$, the probability of r busy channels out of n with mean traffic μ. In a system with two selector stages and n channels in each, a call, having found one of the $n - r$ free channels in the first stage, must find a *corresponding* free channel in the second stage. If the traffic rate is the same in the two stages it is μ/n per channel. The call is blocked if all the particular $n - r$ channels in the second stage are busy, the probability of which is $(\mu/n)^{n-r}$. Now the total probability of congestion of the system is

$$E_L = \sum_{r=0}^{n} E_{r,n}(\mu) \cdot \left(\frac{\mu}{n}\right)^{n-r}$$

$$= \sum_{r=0}^{n} \mu^n/r!n^{n-r}\left(1 + \mu + \frac{\mu^2}{2!} + \ldots + \frac{\mu^n}{n!}\right)$$

$$= E_{n,n}(\mu) \sum_{r=0}^{n} \frac{n^r n!}{n^n r!} = \frac{E_{n,n}(\mu)}{E_{n,n}(n)} \quad . \quad . \quad . \quad . \quad (9) \; *$$

Ex. iv

Calculate the probability of congestion with 10 channels in each of 2 selector stages and a mean traffic of 3 in each stage.

$$E_L = \frac{E_{10,10}(3)}{E_{10,10}(10)} = \frac{3^{10}/10!}{1 + 3 + \ldots 3^{10}/10!} \div \frac{10^{10}/10!}{1 + 10 + \ldots + 10^{10}/10!}$$

$$= 0{\cdot}0038 \text{ approximately.}$$

The reader should now work Questions 1 to 9 at the end of the chapter.

4.6.1 The normal distribution

The next probability distribution to consider is the most important in statistical theory. It is a bell-shaped symmetrical distribution of a continuous variate x which may range from $-\infty$ to $+\infty$. If the mean of the distribution is chosen as origin of x, the distribution is defined by

$$dp = \frac{1}{\sigma\sqrt{2\pi}} e^{-x^2/2\sigma^2} dx \quad . \quad . \quad . \quad . \quad (10)$$

If the variate is measured from any origin, such that μ is then the mean value of x, the equation will be

$$dp = \frac{1}{\sigma\sqrt{2\pi}} e^{-(x-\mu)^2/2\sigma^2} dx \quad . \quad . \quad . \quad (11)$$

This is the general equation of a normal distribution, and is completely specified by the mean μ and the standard deviation σ. Such a distribution is referred to as normal (μ, σ). The distribution (10) is normal $(0, \sigma)$.

It will be found that some frequency distributions of Chapter II, namely those of Exs. ii, v, vi and Examples on Chap. II, 2, 4, 10, are practically normal. It is clear that such an equation as (10) would not easily be discovered by trial, but must have a theoretical basis which should be an indication of the type of variate whose frequency distribution is likely to be approximately normal. This will be discussed later. We first show that the normal distribution is another limiting form of the binomial distribution, as $n \to \infty$. In Poisson's the mean was held constant and the range of the variate, which remained discrete, was extended to infinity over non-negative values. In the distribution of r/n (§3.5.4) the range is kept constant and as n increases without limit the variate becomes effectively continuous, the standard deviation tending to zero, as $n \to \infty$. It was notable that, as indicated in Fig. 8, the skewness which, for any value of p other than 1/2, must be a characteristic of the binomial distribution for moderate values of n, tends to zero as n increases. See also Examples on Chapter III, 17.

We now determine a third limiting form, as $n \to \infty$, by controlling the horizontal scale in such a way that the s.d. remains finite and non-zero; it may, in fact, be held constant. This process yields the normal distribution.

* We must suppose the ordinates of the binomial distribution erected at equal horizontal intervals c. Then $\sigma^2 = npqc^2$ and as $n \to \infty$, c is to tend to zero in such a way that σ remains constant; in other words, we make c vary inversely as $\sqrt{n}$. Then the range nc will vary as $\sqrt{n}$ and tend to infinity. The mean npc will also tend to infinity, but since we are interested in the tendency to symmetry about the mean, as n increases, we shall maintain the origin at the mean. This new origin is denoted by O′ in the figure.

We may suppose np an integer, since even if p is irrational we can

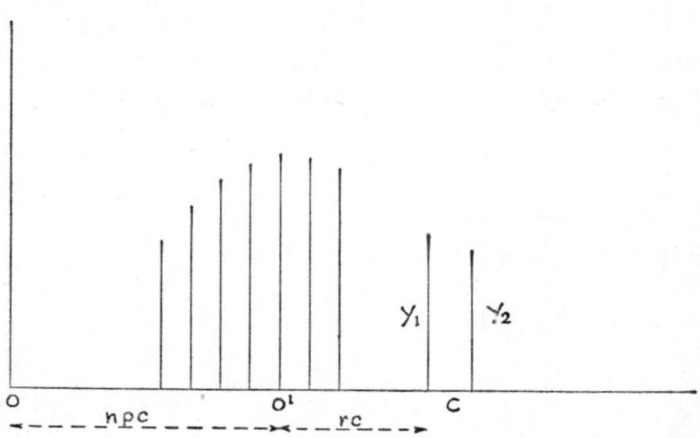

FIG. 11

approximate to it as closely as we please with a rational value. Let y_1 and y_2 be the ordinates of the binomial distributions at r and $r + 1$ intervals beyond the mean.

$$y_1 = {}^nC_{np+r}p^{np+r}q^{nq-r}$$

$$y_2 = {}^nC_{np+r+1}p^{np+r+1}q^{nq-r-1}$$

$$\frac{y_2}{y_1} = \frac{n - np - r - 1 + 1}{np + r + 1}\frac{p}{q} = \frac{(nq - r)p}{(np + r + 1)q}$$

$$\log y_2 - \log y_1 = \log\frac{npq - rp}{npq} - \log\frac{npq + (r + 1)q}{npq}$$

$$= \log\left(1 - \frac{r}{nq}\right) - \log\left(1 + \frac{r+1}{np}\right)$$

$$= -\frac{r}{nq} - \frac{r^2}{2n^2q^2} - \frac{r^3}{3n^3q^3} \cdots - \frac{r+1}{np} + \frac{(r+1)^2}{2n^2p^2} - \frac{(r+1)^3}{3n^3p^3} + \cdots$$

provided r is small enough compared with np and nq.

Now $\log y_2 - \log y_1$ may be written $\delta(\log y)$ or $\delta y/y$ and $c = \delta x$. Dividing both sides of our equation by c, we have

$$\frac{1}{y}\frac{\delta y}{\delta x} = -\left(\frac{r}{nq} + \frac{r+1}{np}\right)\frac{1}{c} + \left\{\frac{(r+1)^2}{n^2p^2} - \frac{r^2}{n^2q^2}\right\}\frac{1}{2c}\cdots$$

$$= -\frac{(r+q)c}{npqc^2} + \frac{(r+1)^2q^2 - r^2p^2}{2n^2p^2q^2c^4}c^3\cdots$$

In the limiting curve, as $n \to \infty$, the limit of the r.h.s. will be the value of $(1/y)(dy/dx)$ at the limiting position of any point within the horizontal interval c between the ordinates y_1 and y_2. One such point is at $x = (r + q)c$, where x is measured from O′. Remembering that, as $n \to \infty$ and $c \to 0$, $npqc^2$ remains constant and equal to σ^2, we have in the limit

$$\frac{1}{y}\frac{dy}{dx} = -\frac{x}{\sigma^2}, \quad . \quad . \quad . \quad . \quad . \quad (12)$$

for the kth term on the r.h.s. is of the form $u_k(x) \cdot c^{k-1}/\sigma^{2k}$ where $u_k(x)$ is a polynomial of degree k in x and the factor $c^{k-1} \to 0$ as $c \to 0$ if $k > 1$. Integrating equation (12)

$$\log y = -\frac{x^2}{2\sigma^2} + A$$

$$y = Be^{-x^2/2\sigma^2}$$

where B is a constant to be determined by the fact that the total area under this limiting curve is to be unity, the total probability. The range of x is now infinite on both sides of the origin O′ and the curve is symmetrical about the y-axis.

$$\therefore 2B \int_0^\infty e^{-x^2/2\sigma^2} dx = 1.$$

The substitution $x = \sqrt{2}\sigma u$ reduces this equation to

$$2\sqrt{2}\sigma B \int_0^\infty e^{-u^2} du = 1.$$

The integral is a well-known one (though not elementary) and its value is $\sqrt{\pi}/2$. Hence $B = 1/\sigma\sqrt{2\pi}$. Our y is the ordinate, namely the probability *density*, for any x, so that the required probability distribution is given by equation (10), the mean being taken as origin. *

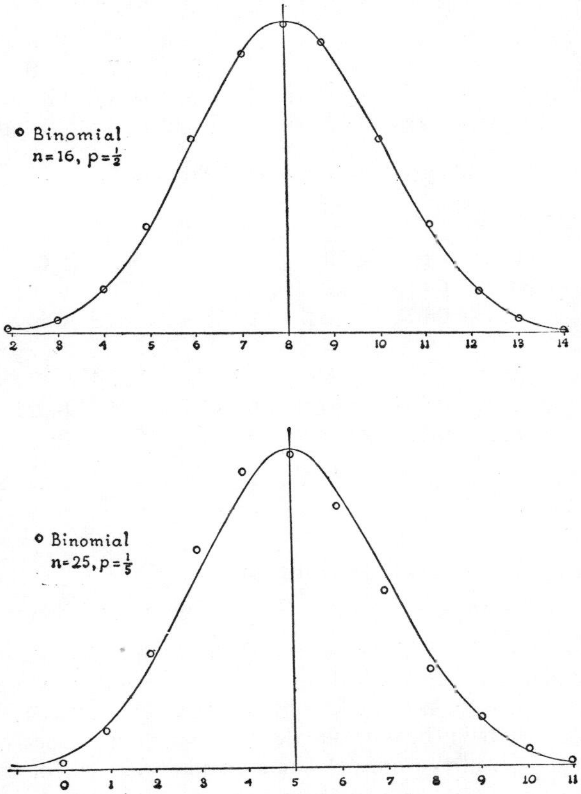

FIG. 12

Since the normal distribution is a limiting form of the binomial distribution as $n \to \infty$, it may be used as an approximation to the binomial distribution if n is sufficiently large. The rapidity of approach to the normal, as n increases, depends on the value of p. The approach is most rapid when $p = q = 1/2$ and there is symmetry throughout. The approach is slow if p (or q) is nearly zero; in fact if p (or q) is of the order $1/n$ the normal distribution is in no sense an approximation to the binomial. This was implied by the conditions $c \to 0$, $npqc^2 = \sigma^2$ (constant) of the proof. Fig. 12 illustrates the approximations when $p = 1/2$, $n = 16$; $p = 1/5$, $n = 25$.

It will be seen that if $p = 1/2$, the binomial probabilities are so close to those given by the normal curve as to be almost indistinguishable graphically, even when n is as small as 16. With $p = 1/5$ the discrepancies are visible, when $n = 25$. Below is a table showing the binomial probabilities and normal ordinates, for $n = 16$, $p = 1/2$ and $n = 36$, $p = 1/5$.

$n = 16$, $p = 1/2$.

r:	2	3	4	5	6	7	8	9 etc.
Binomial:	0·002	0·009	0·028	0·067	0·122	0·175	0·196	0·175
Normal:	0·002	0·009	0·027	0·065	0·121	0·176	0·200	0·176

Both distributions are symmetrical about $r = 8$.

$n = 36$, $p = 1/5$

r:	1	2	3	4	5	6	7
Binomial:	0·003	0·013	0·036	0·075	0·119	0·154	0·165
Normal:	0·006	0·016	0·037	0·068	0·109	0·147	0·166

r:	8	9	10	11	12	13	14
Binomial:	0·150	0·117	0·079	0·046	0·024	0·011	0·005
Normal:	0·157	0·125	0·086	0·043	0·023	0·012	0·003

The mean is now at $r = 7{\cdot}2$. The binomial distribution exhibits slight skewness.

4.6.2 Uses of the normal distribution

The principal uses of the normal distribution are as follows.

(i) For suitable values of n and p it is a good approximation to the binomial distribution. Taking np as μ and $\sqrt{npq}$ as σ, we can use tables of the normal distribution to find the probability that r, the variate in the binomial distribution, lies in a chosen range.

(ii) Gauss derived the normal law as a plausible law of errors of direct measurement by a method given later, which is based not on the binomial distribution but on the 'principle of maximum likelihood'. This principle has wider applications in statistical theory.

(iii) Certain frequency distributions which occur in practice are very nearly normal. The most important for the engineer is the distribution of a chosen dimension of a mass-produced article.

(iv) Statistical theory involves the investigation of 'sampling' distributions which give the probabilities that different values will occur as the result of drawing samples at random from a population. Many of these sampling distributions tend to the normal as the size of sample increases.

It is natural to enquire whether we can relate uses (ii) and (iii) to (i), which has been demonstrated mathematically. The variations in the reading of a single quantity by an instrument such as a theodolite or screw-gauge and the variations in size of the product of an automatic machine may be explained as the sum of many very small variations due to a slight lack of uniformity in pressure, temperature, humidity, density and hardness of material, pressure on the thumb-screw of an instrument and so on. We could easily show that if n such causes each produced with equal probability a small negative or positive deviation from the mean of the same magnitude, the total deviations from the mean would occur with probabilities given by the binomial expansion of $(1/2+1/2)^n$. The assumptions of equal probability and equal magnitude involved cannot be justified. There is however a general theorem, known as the Central Limit theorem, to the effect that a variate, of variance σ^2, which is the sum of n variates whose variances are of the *order* σ^2/n, has a distribution which tends to the normal as n increases. As regards (iv), it enables us to say that, though the normal law might be inexact for single articles in mass-production, it will be very accurate if the variate, instead of being a measured dimension of one article at a time, is the *average* of that measurement for a sample of fixed size which is usually quite moderate; say from 5 to 10.

4.6.3 Numerical values

It is an aid to drawing the normal curve to know that it has inflexions at $x = \pm\sigma$, the mean being taken as origin, and that the tangents at these inflexions meet the x-axis at $(\pm 2\sigma, 0)$. The reader could easily construct an approximate table of probabilities as follows. Take any convenient horizontal length to represent σ and any convenient height for the ordinate at the mean. Then at $\pm\sigma, \pm 2\sigma, \pm 3\sigma$ the heights of the curve are $e^{-1/2}$, e^{-2} and $e^{-4\cdot 5}$ times the height at the mean. The curve can hardly be plotted over a wider range than $\pm 3\sigma$ from the mean. It is, of course, obvious that errors of measurement, etc., do not range from

$-\infty$ to $+\infty$; we can nevertheless represent their distribution as normal because the probabilities associated with the infinite 'tails' of the normal distribution are negligibly small. To find a probability from the graph, such as the probability that the deviation from the mean is in the range 0 to $+\sigma$, divide the area under the graph in this range by the area under the whole graph. Quite a good estimate of the area under a curve can often be obtained by drawing a straight line which appears to approximate to the arc, cutting off as much area from below the curve as it introduces above the curve.

A table of probabilities for the normal distribution is given in Table 1, but some particular values will be noted here. Let $P(x)$ denote the probability of a deviation numerically less than x, i.e. in the range $-x$ to $+x$ from the mean. Particular values of $P(x)$ are

x	0	σ	2σ	3σ	4σ
$P(x)$	0	0·6827	0·9545	0·9973	0·99994

and values of x corresponding to certain values of $P(x)$:

$P(x)$	0·5	0·95	0·998
x	$0{\cdot}6745\sigma$	$1{\cdot}96\sigma$	$3{\cdot}09\sigma$

Thus only about 6 in 100,000 values of a normal population will differ from the mean by more than 4 times the standard deviation; 95% of values will lie in the range $\pm 1{\cdot}96\sigma$ from the mean.

We need not consider how the tabulated values are calculated, but the first stage of the process is important. By the substitution $x = \sigma t$ the probability differential

$$dp = \frac{1}{\sigma\sqrt{2\pi}}\, e^{-x^2/2\sigma^2} dx$$

becomes

$$dp = \frac{1}{\sqrt{2\pi}}\, e^{-t^2/2} dt \qquad . \quad . \quad . \quad . \quad (13)$$

The distribution is now normal (0, 1) and is said to be *standardized*, by the substitution $t = x/\sigma$.

$$P(x) = \int_{x=-\sigma t}^{\sigma t} dp = \frac{1}{\sqrt{2\pi}} \int_{-t}^{t} e^{-t^2/2} dt = \sqrt{\frac{2}{\pi}} \int_{0}^{t} e^{-t^2/2} dt$$

This integral is called the probability integral. It cannot be evaluated in terms of elementary functions. A further substitution $t = u\sqrt{2}$ reduces it to $\dfrac{2}{\sqrt{\pi}} \displaystyle\int_0^u e^{-u^2}\, du$.

This function of u, called Erf u, is the 'error' function, so called because of its association with the normal distribution, regarded as the law of errors.

Ex. v

Limit gauges are used to reject all components in which a certain dimension is either greater than 1·510 in. or less than 1·490 in. A machine is producing these components with the dimension normally distributed about a mean 1·503 in. with s.d. 0·002 in. What proportion of the product will be rejected?

The gauges reject components for which the dimension is either more than 7 thou., i.e. $3{\cdot}5\sigma$, greater than the mean, or more than 13 thou., i.e. 6·5 , less than the mean.

From Table 1, the probability of a deviation from the mean greater than $+\,3{\cdot}5\sigma$ is 0·0002. The probability of a negative deviation numerically greater than $6{\cdot}5\sigma$ is negligible. Hence only 0·02% of the product will be rejected, and this as being too large.

Ex. vi

Women's shoes are manufactured in sizes 2, 3, 4 . . . 8 the length advancing in steps of $\frac{1}{4}$ in. Size 5 is suitable for a foot of length ranging from $9\frac{1}{4}$ to $9\frac{1}{2}$ in. If the lengths of women's feet are normally distributed with mean 9·4 in. and s.d. 0·25 in. how many pairs of shoes of each size are required out of every 10,000 pairs manufactured?

$9\frac{1}{2}$ in. differs from the mean by 0·1 in. $= 0{\cdot}4\sigma$ and a step of $\frac{1}{4}$ in. is a step of σ. The accumulated probabilities under the normal curve from the negative extreme are as in the third column of the following table. By taking the differences between successive probabilities and multiplying by 10,000, we find the required number of pairs of each size.

Length	*Deviation from Mean*	*Probability*	*Diff.* × 10,000	*Size*
$8\frac{3}{4}$	$-\,2{\cdot}6\sigma$	0·0047	47	2
9	$-\,1{\cdot}6\sigma$	0·0548	501	3
$9\frac{1}{4}$	$-\,0{\cdot}6\sigma$	0·2743	2195	4
$9\frac{1}{2}$	$0{\cdot}4\sigma$	0·6554	3811	5
$9\frac{3}{4}$	$1{\cdot}4\sigma$	0·9192	2638	6
10	$2{\cdot}4\sigma$	0·9918	726	7
$10\frac{1}{4}$	$3{\cdot}4\sigma$	0·9997	79	8

(Except for the mean, fictitious figures have been assumed. We have also assumed that all women with feet less than $8\frac{3}{4}$ in. in length will be content with size 2, while 3 women in every 10,000 requiring a size larger than 8 are not catered for.)

Ex. vii

Records show that in the long run 40% of trainees fail to pass a proficiency test. What is the probability that of a random group of 100 trainees, more than 50 fail to pass?

Here we use the normal distribution as the suitable approximation to the binomial. The 'event' of failing the test has a prior probability of 0·4 and 100 trials are made. The probability that more than 50 trainees fail to pass, i.e. of more than 50 'successes' in the sense of Bernoulli's theorem, is the sum of terms of $(q + p)^n$ with $q = 0{\cdot}6$, $p = 0{\cdot}4$, $n = 100$, r taking the values 51, 52 . . . 100. When we use the normal distribution, in which the variate is continuous, to represent the binomial distribution, it is best to take the values of r as central values of class intervals of width 1. Hence we require the probability of a value greater than 50·5, of a normal variate with mean np and variance npq, i.e. mean 40 and s.d. 4·90. It is of no importance that the maximum value of r is 100, while the range of a normal variate is infinite; the probabilities become negligible beyond about 4σ from the mean. The deviation of 50·5 from the mean is 10·5 which is very nearly $2{\cdot}14\sigma$. The probability of more than 50 trainees failing to pass is 0·016 approximately. This probability would be doubled if the question were put in the form: what is the probability that the number failing differs by more than 10 from the mean, i.e. what is the probability of fewer than 30 or more than 50 failures?

4.7 Sum and difference of normal variates

In a hand-operated machine there will be two sources of variation in the product, namely the machine and the operator. The machine will be set to produce a mean result and there will be some variation about the mean due to the machine itself; superimposed on this will be variation due to inconsistent performance on the part of the operator. Moreover a personal bias, e.g. a tendency to exert excessive pressure, might affect the mean itself.

Other instances occur in which it is required to know when a variate x which is normally distributed (μ_1, σ_1) is added to another variate y, normal (μ_2, σ_2), what is the probability distribution of the sum $x + y$. We can show that, provided the variates are independent, $x + y$ is also normally distributed, with mean $\mu_1 + \mu_2$ and variance $\sigma_1^2 + \sigma_2^2$, i.e. with standard deviation $\sqrt{(\sigma_1^2 + \sigma_2^2)}$.

* The standardized deviation from the mean, of either x or y, is normally distributed (0, 1). If we write $u = (x - \mu_1)/\sigma_1$, $v = (y - \mu_2)/\sigma_2$, then the probability that u lies in a range du is

$$\frac{1}{\sqrt{2\pi}} e^{-u^2/2} du$$

and that v lies in a range dv,

$$\frac{1}{\sqrt{2\pi}} e^{-v^2/2} dv.$$

With u, v as co-ordinates, the probability that the point (u, v) lies in a rectangle du by dv in the u, v plane is

$$\frac{1}{2\pi} e^{-(u^2+v^2)/2} du dv.$$

(By hypothesis the variates x, y and so u, v are independent. This justifies the multiplication of the two probabilities. The procedure really requires further discussion, for which see §5.1.)

The probability *density* at the point (u, v) is the probability per unit area there, namely

$$\frac{1}{2\pi} e^{-(u^2+v^2)/2}.$$

We could represent such a probability distribution by a surface, or by a contour-map of the surface in the u, v plane. Now the contours of equal density, or equal height of the surface, are curves for which

$u^2 + v^2$ is constant, i.e. circles with the origin as centre. Hence the surface is a surface of revolution.

Let
$$z = x + y - \mu_1 - \mu_2$$
$$= \sigma_1 u + \sigma_2 v$$

We require the probability that z lies in a range z to $z + dz$, i.e. we require the volume under the surface standing on a strip of the u, v plane between the straight lines $\sigma_1 u + \sigma_2 v = z$ and $\sigma_1 u + \sigma_2 v = z + dz$.

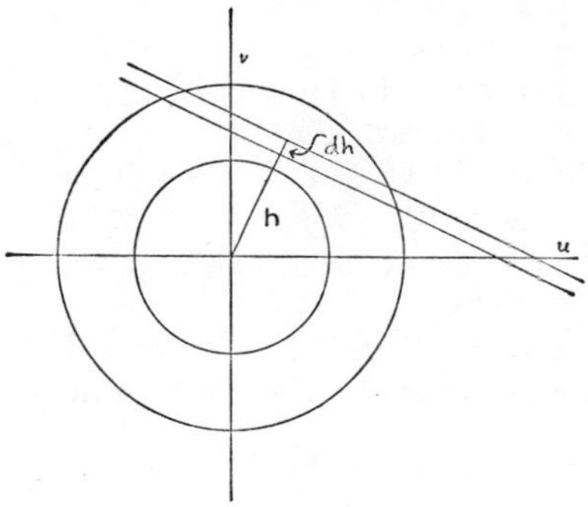

FIG. 13

If h is the perpendicular distance from the origin to the first line

$$h = \frac{z}{\sqrt{(\sigma_1^2 + \sigma_2^2)}}$$

and the distance between the lines is

$$dh = \frac{dz}{\sqrt{(\sigma_1^2 + \sigma_2^2)}}$$

Now by the symmetry of the surface of revolution, the probability in the strip (or volume standing on it) is of the same form as the probability in a strip of width du, or dv, namely

$$dp = \frac{1}{\sqrt{2\pi}} e^{-h^2/2} dh$$

Substituting for h and dh in terms of z,

$$dp = \frac{1}{\sigma\sqrt{2\pi}} e^{-z^2/2\sigma^2} dz$$

where
$$\sigma^2 = \sigma_1^2 + \sigma_2^2$$

Hence the distribution of z is normal $[0, \sqrt{(\sigma_1^2 + \sigma_2^2)}]$. But $x + y = \mu_1 + \mu_2 + z$. Hence $x + y$ is distributed normally $[\mu_1 + \mu_2, \sqrt{(\sigma_1^2 + \sigma_2^2)}]$. *

To obtain the probability distribution of $x - y$ we should set

$$z = (x - \mu_1) - (y - \mu_2)$$
$$= \sigma_1 u - \sigma_2 v \text{ in the above proof.}$$

The perpendicular distance from the origin to this line is still $z/\sqrt{(\sigma_1^2 + \sigma_2^2)}$ and we find that $x - y$ is normal $[\mu_1 - \mu_2, \sqrt{(\sigma_1^2 + \sigma_2^2)}]$. Note that, although we subtract means, we still add variances.

Consider a machine which fills cans. There will be a mechanism by which the machine responds to the total weight of can and contents by shutting the supply tap. We must expect some variation in the total weight and may assume it to be according to the normal law, the variance depending on the precision of the machine. Independently of this the cans will vary in weight, again we may suppose normally. The weight of the contents, in which the customer is most interested, is thus the difference of two normal variates, with a variance which is the sum of their variances.

The method of proof above is easily extended to establish a similar proposition for a linear function of normal variates that if

$$Z = ax + by + cz + \dots$$

where $a, b, c \dots$ are constants; $x, y, z \dots$ are normal variates with means $\mu_1, \mu_2, \mu_3 \dots$ and variances $\sigma_1^2, \sigma_2^2, \sigma_3^2 \dots$ then Z is normally distributed with variance $a^2\sigma_1^2 + b^2\sigma_2^2 + c^2\sigma_3^2 + \dots$ and mean $a\mu_1 + b\mu_2 + \dots$. First establish the result for the variate $Z_1 = ax + by$ then for $Z_1 + cz$ and so on.

Ex. viii

A resistance of 24,000 ohms is obtained by means of 2 elements rated at 10,000 ohms and 4 rated at 1,000 ohms in series. In bulk, elements of the first type will have resistances normally distributed about a mean of 10,000 ohms with s.d. 200 ohms; those of the second type have resistances normally distributed about a mean of 1,000 ohms with s.d. 50 ohms. Show that there is an even chance that the constructed resistance is within $\pm$ 200 of 24,000 ohms.

The constructed resistance is the sum of 6 independent normal variates. Their sum is normal, with mean 24,000 and s.d.σ, where

$$\sigma^2 = 200^2 + 200^2 + 50^2 + 50^2 + 50^2 + 50^2$$
$$= 90{,}000.$$

There is a probability of 1/2 that a normal variate lies within $\pm$ 0·6745σ of the mean and (0·6745) (300) = 200 approximately.

Ex. ix

The total weight delivered by a filling machine varies normally with s.d. 0·15 oz. The cans have a mean weight 6·8 oz. with s.d. 0·20 oz., the distribution being normal. What mean total weight must the machine deliver if it is desired to ensure a probability of 1/20 that the contents of the can weigh less than their nominal or advertised weight of 32 oz.?

If x oz. = total weight,
y oz. = weight of can,

these are independent normal variates. The weight of the contents is $x - y$ oz., normally distributed with

$$\sigma = \sqrt{\{(0\cdot15)^2 + (0\cdot20)^2\}} = 0\cdot25 \text{ oz.}$$

Let μ be the mean value of $x - y$. Then the value of μ is required to be such that the probability of a value of $x - y$ less than 32 is 1/20.

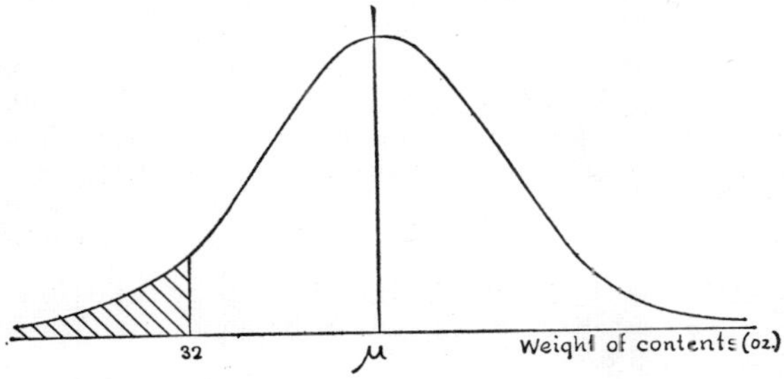

FIG. 14

In the figure the shaded area is to be 1/20.

From tables, $\mu - 32 = 1\cdot65\sigma = 0\cdot41$
$\mu = 32\cdot41$.

The machine will be set to deliver a mean total weight of $32\cdot41 + 6\cdot8$
$= \underline{39\cdot21 \text{ oz.}}$

Suggestions for further reading:

Kendall, *Advanced Theory of Statistics*, Chapter 5 (Griffin).

G. Yule and M. G. Kendall, *Introduction to the Theory of Statistics*, Chapter 10 (Griffin).

E. T. Whittaker and G. Robinson, *Calculus of Observations*, Chapter 8.

EXAMPLES ON CHAPTER IV

1. Calculate the first 6 terms of the Poisson distribution when μ is (i) 1/2, (ii) 1, (iii) 3. Sketch the probability polygon in each case.

2. Certain mass-produced articles, of which 0·5 per cent. are defective, are packed in cartons each containing 100. What proportion of cartons are free from defective articles and what proportion contain two or more defectives? (R.S.S. Cert.)

3. Show that for a Poisson distribution the variance equals the mean, and use this result to test the following distribution for conformity to the Poisson law:

Number of defects per piece of cloth:	0	1	2	3	4	5	
Frequency of pieces:	15	26	21	19	8	3	Total 92

(R.S.S. Cert.)

4. The mean number of calls per hour received at a telephone switchboard is 5. Assuming the Poissonian law, find the probabilities (i) that in a particular hour 5 calls are received, (ii) that more than half an hour elapses between two successive calls.

5. If one in a thousand entries in a book of tables is in error, what is the chance that 50 readings taken from the tables are all correct?

6. Compute the average and standard deviation of the following distribution and verify that approximately $\sigma = \sqrt{m}$.

x:	0	1	2	3	4	5	
f:	20,983	2,615	183	14	2	3	Total 23,800.

Assuming that the frequencies are given by the formula $Nm^re^{-m}/r!$ where $N = 23{,}800$ and r takes successive values 0, 1, 2, 3, 4, 5, compute the expected frequencies from this formula taking m equal to the computed average. (R.S.S. Cert.)

7. A car-hire firm has two cars, which it hires out by the day. The number of demands for a car on each day is distributed as a Poisson distribution, with mean 1·5.
Calculate the proportion of days on which neither of the cars is used, and the proportion of days on which some demand is refused.
If each car is used an equal amount, on what proportion of days is a given one of the cars not in use? What proportion of demands has to be refused? (R.S.S. Cert.)

8. A shopkeeper's sales of a certain article amount to 4 per month on the average, the sales occurring at random and independently. He replenishes his stock once a month. To what number should he make up his stock in order to reduce the chance of running out of stock to less than 1/100?

9. In one hour 1,500 α-particles are radiated by radio-active material, the amount of unchanged material being not appreciably diminished. If the whole period of one hour is divided into intervals of 10 seconds, find the expected numbers of intervals during which 0, 1, 2 . . . α-particles are radiated.

10. Define the binomial distribution and give formulae for the mean and variance of such a distribution.
Answer either (*a*) or (*b*):
(*a*) In sampling from boxes of 200 components, what approximations would you use for the distribution of fractions defective in a box if the average fraction defective over a large number of such boxes was (i) 25%, (ii) 0·5%?
In the second case what is the chance that the box will contain no defective components?
(*b*) In repeated differential white cell counts, each of 100 cells, what approximations would you use for the distribution of the proportion in each count, (i) of neutrophils, if the average proportion over a large number of counts is 55%, and (ii) of basophils, if the average proportion over a large number of counts is 0·5%? In case (ii), what is the chance that there will be no basophils in a particular count? (R.S.S. Cert.)

11. Omitting the constant factor $1/\sqrt{2\pi}$, sketch the normal curve when σ is (i) 1, (ii) 2, (iii) 3, using the same axes and scales in each case.

12. The mean of a normal distribution is 20 and the standard deviation 1·6. Use tables to find the probabilities of a value of the variate (i) greater than 24, (ii) in the range 19 to 21.

13. The data of Examples on Chapter II, 10, may be assumed to constitute a sample from a normal population. Use the computed mean and standard deviation to estimate what proportion of the product may be expected to have a breaking strain greater than 970 lb.wt.

14. The following data give the length of life in hours of a certain type of electric lamp.

Life in hours:	650-	660-	670-	680-	690-	700-	710-	720-	730-	740-
Frequency:	2	5	6	14	26	18	13	10	3	3

Calculate the mean and s.d. of the sample. (Sheppard's correction may be applied. The central value of the group 650- is 654·5, the times having been recorded to the nearest hour.)
Assuming a normal population with the same mean and s.d., what proportion of lamps may be expected to have a burning life of at least 650 hours?

15. Obtain an expression for the moments about the mean in terms of those about an arbitrary point. Give the formulae in full for the first four moments.
Hence or otherwise find the first four moments about the mean of the distribution

$$dF = ae^{-ax}\,dx,\ 0 \leqslant x < \infty;\ a > 0.$$

(R.S.S. Diploma.)

16. Assuming that heights of Englishmen are normally distributed, with mean 67·5 in. and standard deviation 2·5 in., and that all policemen must be of not less than average height (for a man), what would you expect to be the average height of all policemen? (The integral required occurs in §5.3.4 (*b*).)

17. A manufactured product has a weight which is normally distributed with its advertised weight of 1 lb. as mean. It may be regarded as appreciably under the advertised weight if the weight is less than $15\frac{3}{4}$ oz. If the risk of this is to be less than 0·01, what is the maximum allowable value of the standard deviation?

18. A machine makes electrical resistors having a mean resistance of 50 ohms with a standard deviation of 2 ohms. Assuming the distribution of values to be normal, find what tolerance limits should be put on the resistance to secure that not more than 1/1000 of the resistors will fail to meet the tolerances.
Other resistances are made from these by assembling two of them in series. Assuming the pairs to be chosen at random, find what tolerance limits should be placed on the total resistance to secure that not more than 1/1000 of them should fail to meet the tolerances.
(If resistors with resistances R_1 and R_2 ohms respectively are assembled in series, the total resistance is $R_1 + R_2$ ohms.) (R.S.S. Cert.)

19. At one stage in the manufacture of an article, a cylindrical rod with circular cross-section has to fit into a circular socket. Quality control measurements show that the distribution of rod diameter and socket diameter are normal, with constants:

Rod diameter: mean 5·01 cm., standard deviation 0·03 cm.
Socket diameter: mean 5·11 cm., standard deviation 0·04 cm.

If components are selected at random for assembly, what proportion of rods will not fit? Also what is the chance that in a batch of 300 pairs, selected at random, all the rods will fit? (R.S.S. Cert.)

20. A manufacturer sells an article at a fixed price of £1. He guarantees to refund the purchase money to any customer who finds that the weight of his article is less than 8 oz. Actually, the weights of the articles made by the manufacturer are normally distributed with standard deviation 1 oz. His cost of production, £c per article, is related to the mean weight, w oz., of the articles, by the relation

$$c = 0{\cdot}05w + 0{\cdot}30.$$

Graphically, or otherwise, determine approximately the mean weight which the manufacturer should aim at if he wishes to maximize his expected profit per article. (R.S.S. Cert.)

CHAPTER V

SUM OF SQUARES OF NORMAL VARIATES. GOODNESS OF FIT

A target problem. Extension to n *dimensions. Degrees of freedom. The χ distribution and goodness of fit. χ^2 test of Poisson and normal distributions. Other criteria of normality. The null hypothesis and significance. Contingency tables. Maxwell's distribution.*

5.1 Statistical independence

We have seen (§3.3.2) that if A, $\bar{A}$; B, $\bar{B}$ denote success and failure respectively of events A and B respectively, the table of frequencies

	A	$\bar{A}$
B	α	γ
$\bar{B}$	β	δ

has the special feature that $\alpha/\beta = \gamma/\delta$ if the events are independent and that then the table of probabilities

	A	$\bar{A}$	Totals
B	p_1p_2	q_1p_2	p_2
$\bar{B}$	p_1q_2	q_1q_2	q_2
Totals	p_1	q_1	1

has the property that the probability in each cell is the product of the corresponding marginal probabilities.

Suppose now that two variates x, y are each capable of taking a number of different numerical values. We shall say that the variates are independent if the relative frequencies with which x takes its different values are the same, whatever value of y is associated with those values of x. If, for instance, x can take four different values and y three, a table of frequencies of a pair of values (x, y) might, if the variates are independent, take the form

	x_1	x_2	x_3	x_4	Totals
y_1	1	2	5	4	12
y_2	5	10	25	20	60
y_3	3	6	15	12	36
Totals	9	18	45	36	108

Such a table must again have the property that if all the frequencies, including the totals, are divided by the total frequency, then the marginal totals will represent relative frequencies or probabilities of the possible values of the variates, while the cells will contain the probabilities of the possible paired values (x, y) and each cell-probability is the product of the corresponding marginal probabilities.

Next suppose that x and y are continuous variates. We can no longer speak of the probability of either taking a particular value but only of the probability that the value of x lies in an interval δx, or y in an interval δy. The concept of independence requires that the probability that x lies in a chosen interval is the same whatever the interval in which y lies. If this is true, the probability that x lies in an interval δx, being the same for all possible intervals δy, is the relative frequency with which x lies in the chosen δx, for the *whole range* of y. This is a generalization of the theorem of ratios, that

$$\text{if} \qquad \frac{a}{b} = \frac{c}{d} = \frac{e}{f} = \text{etc., then each} = \frac{a + c + e + \ldots}{b + d + f + \ldots}.$$

Hence the element, δp, of probability that x lies in δx does not depend on y; or in differentials

$$dp = f(x)dx$$

gives the distribution of x whatever the value of y. Similarly $dp = g(y)dy$ gives the distribution of y.

If (x, y) is represented by a point in the plane XOY, $f(x)\delta x$ is the proportion of values that fall in the strip AB, of small width δx; $g(y)\delta y$ the proportion that fall in the strip CD. Within CD the same proportion fall in the interval δx as for the whole plane; hence the probability that (x, y) falls in the rectangle common to the two strips is $f(x)\delta x \,.\, g(y)\delta y$. The probability *density*, or probability per unit area of the plane, at (x, y) is $f(x)g(y)$.

Now, in general, if a pair of values (x, y) has a probability distribution, there is a probability density $\phi(x, y)$ and a probability surface (analogous to the probability curve for a single variate) $z = \phi(x, y)$. Our conclusion is that if x and y are statistically independent, $\phi(x, y)$

has the special property that it is the product of a function of x and a function of y, i.e. $\phi(x, y) = f(x)g(y)$.

We may interpret this geometrically by saying that if the probability surface is cut by two pairs of planes parallel to ZOX and ZOY respectively the four z-ordinates will be proportional, i.e. PH/QK = RL/SM.

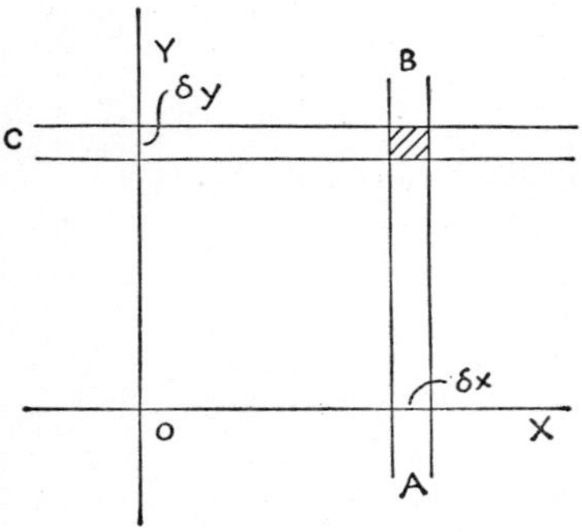

FIG. 15

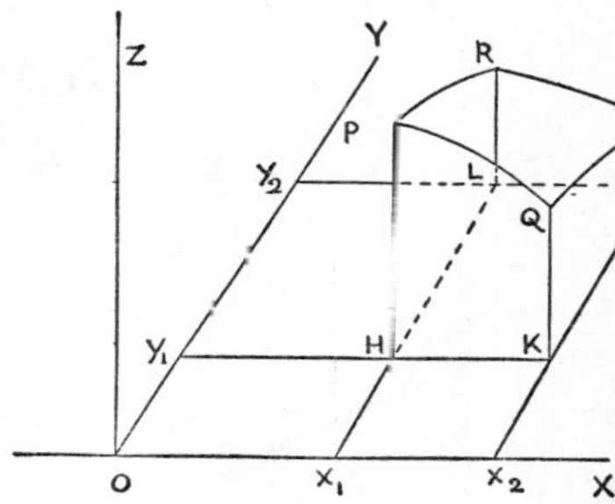

FIG. 16

5.2.1 A target problem

The position of the centre of a hole to be drilled in a rectangular plate is fixed by its distances from two perpendicular edges of the plate. Suppose that in marking out the distance from *either* edge, the liability to error is in accordance with the normal law, with mean zero (so that positive and negative errors are equally likely) and s.d. 1 thou. What is the probability distribution of the displacement of the centre from its correct position?

An alternative form of this problem might be: if a bomber is liable to miss his target by distances which measured (i) east and west, (ii) north and south of the target are independent normal variates with zero mean and equal variances, what is the chance that the bomb falls within any assigned distance of the target? (The variances in the two dimensions are not likely to be equal. They could be made so by standardizing both variates, but circles of equal distances from the target would be transformed into ellipses.)

As in §4.7, where u, v are independent normal variates (0, 1), so, if x, y are such variates, the probability density at (x, y) is

$$\frac{1}{2\pi}e^{-(x^2+y^2)/2} = \frac{1}{2\pi}e^{-r^2/2} \quad . \quad . \quad . \quad (1)$$

where r is the distance of (x, y) from the origin. We require the probability that r lies in any range dr. Now the area between circles with centre 0 and radii r and $r + dr$ is $2\pi r dr$ and in this small ring the probability density is given by (1). Hence the required probability distribution of r is given by

$$dp = 2\pi r dr \frac{1}{2\pi} e^{-r^2/2} = re^{-r^2/2}dr \qquad . \quad . \quad (2)$$

Note that, whereas x and y may take positive and negative values, r is always to be reckoned positive. It may be easily verified that

$$\int_0^\infty re^{-r^2/2}dr = 1, \text{ the total probability.}$$

The graph of the distribution (2) is sketched in Fig. 17.

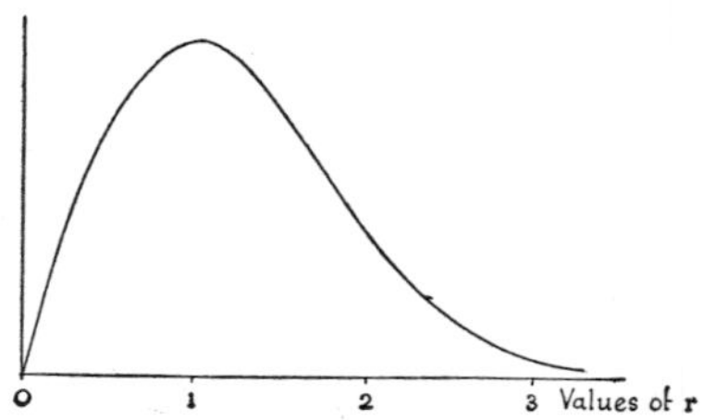

FIG. 17

Ex. i

In the problem of drilling the plate, if we required the probability that the centre of the hole is more than 2 thou. out of position, it is

$$\int_2^\infty re^{-r^2/2}dr = \left[-e^{-r^2/2}\right]_2^\infty = e^{-2} = \underline{0{\cdot}135 \text{ approximately.}}$$

5.2.2 Extension to three and more dimensions

We could envisage the target problem in three dimensions in terms of proximity fuses. Suppose the shell will explode if within a known distance, in some direction or other, of a hostile aircraft and that accuracy of shooting is such that errors, up and down, left and right, short and long, are all normally distributed with zero mean and known equal variances. Then we require the probability distribution of nearest distance between shell and target, of which the three errors are perpendicular components. The derivation is not difficult.

With the common s.d. as unit, the component distributions are

$$dp = \frac{1}{\sqrt{2\pi}}e^{-x^2/2}dx, \; dp = \frac{1}{\sqrt{2\pi}}e^{-y^2/2}dy, \; dp = \frac{1}{\sqrt{2\pi}}e^{-z^2/2}dz.$$

These being independent variates the probability density at (x, y, z) is

$$\frac{1}{(\sqrt{2\pi})^3}e^{-(x^2+y^2+z^2)/2} = \frac{1}{(\sqrt{2\pi})^3}e^{-r^2/2} \qquad . \quad . \quad (3)$$

where r is the distance from the target or origin.

Now the range r to $r + dr$ covers the space between two concentric spheres with these radii, namely $4\pi r^2 dr$. Multiplying (3) by this element of volume we obtain the required probability distribution as

$$dp = \sqrt{\frac{2}{\pi}}\, r^2\, e^{-r^2/2}\, dr. \qquad . \quad . \quad . \quad (4)$$

We consider later a well-known application of this distribution in the Kinetic Theory of gases, namely Maxwell's law for the distribution of resultant velocities of the molecules of a perfect gas. At present the object is to extend the ideas involved to any number of dimensions. The reason is that we thereby arrive at a means of testing the 'goodness of fit' of any observed set of frequencies, provided they are reasonably large, to any corresponding set of hypothetical frequencies. For instance, examples have been given of frequency distributions which appear to be nearly normal. We could calculate the frequencies to be expected on the hypothesis of normality. The question then is, are the discrepancies between the actual and hypothetical frequencies to be regarded as easily explained as chance effects; that is, is the observed set of frequencies to be regarded as a good shot at the multiple target represented by the hypothetical ones?

In the application of this test of goodness of fit an important concept occurs of 'degrees of freedom'. It should help the reader to appreciate the importance of this concept if he is able to follow the arguments below.

* If $x_1, x_2 \ldots x_n$ are n independent normal variates (0, 1) we can imagine an n-dimensional space in which the position of a point is defined by n co-ordinates $(x_1, x_2 \ldots x_n)$. The probability density at such a point is

$$\frac{1}{(\sqrt{2\pi})^n} e^{-(x_1^2+x_2^2+\ldots x_n^2)/2} = \frac{1}{(\sqrt{2\pi})^n} e^{-r^2/2} = \phi(r), \text{ say,}$$

where $r^2 = x_1^2 + x_2^2 + \ldots + x_n^2$.

When n was 2, we had to multiply this density by $2\pi r dr$, the differential of the area of a circle, for $d(\pi r^2) = 2\pi r dr$.

When n was 3, we multiplied by $4\pi r^2 dr$, the differential of the volume of a sphere.

In n dimensions we imagine what is called a hypersphere of radius r. Its volume V is of n dimensions, each proportional to r, i.e. $V \propto r^n$.

Hence $dV \propto nr^{n-1}dr$. Multiplying this by $\phi(r)$ and absorbing n into the proportionality constant we have

$$dp \propto r^{n-1} e^{-r^2/2} dr = K r^{n-1}\, e^{-r^2/2} dr.$$

To fix K we have that the total probability is 1

$$\frac{1}{K} = \int_0^\infty r^{n-1} e^{-r^2/2} dr.$$

To obtain a concise formula for K, it is necessary to use the Gamma function. The substitution $r^2 = 2u$ gives

$$\frac{1}{K} = 2^{(n-2)/2} \int_0^\infty u^{(n-2)/2} e^{-u} du.$$

Now $$\Gamma(n) = \int_0^\infty e^{-x} x^{n-1} dx \text{ so that}$$

$$\frac{1}{K} = 2^{(n-2)/2}\, \Gamma\left(\frac{n}{2}\right)$$

and the distribution of r is

$$dp = \frac{1}{2^{(n-2)/2}\, \Gamma\left(\frac{n}{2}\right)} r^{n-1} e^{-r^2/2} dr, \; (r \geqslant 0) . \quad . \quad (5)$$

*

The reader unfamiliar with the Γ-function need only know, for the purpose of evaluating $\Gamma(n/2)$, the fundamental property $\Gamma(n+1) = n\Gamma(n)$ and the particular values $\Gamma(1) = 1$, $\Gamma(1/2) = \sqrt{\pi}$. But he may disregard the evaluation of K without much loss and in later derivations of probability distributions we shall be content to show the probability function with an undetermined constant factor. For the distribution (5), we could evaluate approximately the probability that r falls in a chosen range by drawing the graph of the function $r^{n-1}e^{-r^2/2}$ and finding, by approximate methods, the ratio of the area under the curve, in that range, to the whole area under the curve.

5.2.3 Degrees of freedom

There is an important corollary to our result. If the variates $x_1, x_2 \ldots x_n$, though otherwise independent, are subject to a restriction of the form

$$a_1x_1 + a_2x_2 + \ldots + a_nx_n = 0 \quad . \quad . \quad . \quad (6)$$

where the a's are constant, then the point $(x_1, x_2 \ldots x_n)$ is constrained to lie in a 'hyperplane', of equation (6), through the origin. Such a hyperplane cuts the hypersphere, $r =$ constant, in a hypersphere of the same radius but of one fewer dimensions; e.g. in three dimensions a plane through the centre of a sphere cuts the sphere in a *circle* of the same radius, so that if a point must lie in the plane, then when the radius varies from r by dr the point can lie only in the ring between two circles instead of in the volume between two spheres.

Hence the effect of a restriction such as (6) is to reduce by 1 the value of n in equation (5). Further, if the n variates are subject to $m(<n)$ restrictions of type (6), each restriction reduces by 1 the number of dimensions of the hypersphere relevant to the volume, dV, used in establishing equation (5). This number of dimensions is called the number of *degrees of freedom*, abbreviated to d.f. The number of d.f. will be denoted by ν. Thus if n variates are subject to m restrictions of type (6), $\nu = n - m$ and the distribution of r becomes

$$dp \propto r^{\nu-1} e^{-r^2/2} dr \quad . \quad . \quad . \quad . \quad (7)$$

5.3.1 Goodness of fit. The χ distribution

Suppose that a total of N observations have been classified in n classes and that the frequencies in the classes are $O_1, O_2 \ldots O_n$, so that $\Sigma O = N$.

Suppose further that on some hypothesis, e.g. that the N observed values are a sample from an infinite normal population, the expected frequencies in the n classes would be $E_1, E_2 \ldots E_n$ where also $\Sigma E = N$. There will be discrepancies between the O's and the E's, and it is desired to know whether, taken as a whole, the discrepancies are to be regarded as large or small. To determine this we require to find the probability that the O's should occur if, on the average and in the long run, the E's are expected.

We have solved the problem already in a simple instance by obtaining the binomial distribution. Here the classes are two in number, successes and failures. On the hypothesis of a prior probability, p, of success in each trial, the expected number of successes in N trials is Np. If the observed number of successes is r we can calculate its probability. In this instance the table of E's and O's is:

	Successes	*Failures*
E's	Np	Nq
O's	r	$N - r$

For the derivation of the required probability distribution, the reader is referred to Kendall: *Advanced Theory of Statistics*, Vol. I, pp. 290, 291. By a method analogous to that used in obtaining the normal distribution as a limiting form of the binomial, it is shown that the *multinomial* distribution involved tends, for sufficiently large values of the O's, to that of r in equation (7), where r^2 is the sum of squares of normal variates (0, 1). Instead of r^2 we have a quantity

$$\chi^2 = \Sigma\frac{(O - E)^2}{E} \quad . \quad . \quad . \quad . \quad (8)$$

and the probability distribution of χ is

$$dp \propto \chi^{\nu-1}e^{-\chi^2/2}\, d\chi, \ (\chi \geqslant 0) \quad . \quad . \quad . \quad (9)$$

Here ν, the number of degrees of freedom, will be $n - 1$, if there are n classes and the only restriction on the variates on the r.h.s. of (8) is that $\Sigma O = \Sigma E$. We shall see that in applications further restrictions occur and ν must be adjusted accordingly.

The method of using χ^2 to test goodness of fit is this. Given the O's, or observed frequencies, calculate the corresponding E's on the hypothesis being tested. Then calculate χ^2 by equation (8). In general, a small value of χ^2 reflects small discrepancies between observed and expected frequencies and indicates a good fit. Tables have been prepared which give, for each ν, the probability P that χ^2 will exceed a particular value. If, for our value of χ^2, P is small, then χ^2 is unduly large and the fit is not good. The choice of the value of P, below which we consider the fit is bad, is necessarily arbitrary; a value often used is 0·05, implying that we consider χ unduly large if it falls in the range corresponding to the right-hand tail of the χ distribution which includes 1/20 of the area under the curve.

A value of χ^2 may be improbably small. The corresponding value of P will be high, i.e. nearly 1, and may be interpreted as due either to a prepared sample of observations or to over-ingenuity in the selection of the hypothetical distribution. For example, a sample of moderate size drawn from a normal population must be expected to show some departure from the smooth rise and fall of the parent distribution. By assuming a more ingenious law for the parent we could make χ^2 very small, but we should consider it more likely that the sample was one of average probability from a normal population than that it was an exceptionally representative sample drawn from the more elaborate hypothetical distribution. Every scientific experimenter prefers to accept a simple law, explaining small deviations therefrom as due to inevitable experimental errors, rather than to invent a complicated formula which his readings would fit exactly.

It was stated above that the derivation of the χ distribution requires that all the observed frequencies shall be 'sufficiently large'. It is also implied in the proof that the standardized deviations of form $(O - E)/\sqrt{E}$ shall be sufficiently small, so that the E's as well as the O's must be sufficiently large. In industrial applications the frequency in a class is often allowed to be as low as 5, though Aitken (*Statistical*

Mathematics) suggests a minimum of 10 and Kendall (*Advanced Theory of Statistics*) suggests 20. When some of the frequencies are below the adopted minimum, it is necessary to pool two or more classes into one. We shall adopt a minimum of 10 as a standard, but where convenience requires it, there is little danger involved in lowering the standard.

5.3.2 χ^2 test of a Poisson distribution

Ex. ii

The following table shows the number of accidents experienced by a number of coal-miners over a period of years.

Number of accidents:	0	1	2	3	4	5	
Number of miners:	35	47	39	20	5	2	Total 148.

Find the frequencies of the Poisson distribution with the same mean as this distribution and test the goodness of fit by the χ^2 test, grouping together the last two frequencies. (A.I.S.)

It will be found that the mean is 1·45 while the variance is 1·36. These are sufficiently nearly equal to afford a first confirmation that the distribution is a likely sample from a Poisson population. It may occur to the reader that 1·36 could be adopted as estimate of the mean of the hypothetical Poisson distribution. This question of choosing between two or more possible estimators of a statistical parameter receives much attention in more advanced statistical theory. Some account of the principles adopted is given in Chapter XIII. Here the mean is to be adopted and the expected frequencies of 0, 1, 2 . . . accidents are the terms

$$148\, e^{-1\cdot45}\left(1,\ 1\cdot45,\ \frac{1\cdot45^2}{2},\ \text{etc.}\right).$$

Calculating these we form the table:

	O	E	$O - E$	$(O - E)^2/E$
0	35	34·7	0·3	0·003
1	47	50·3	− 3·3	0·022
2	39	36·5	2·5	0·171
3	20	17·6	2·4	0·327
4 or more	7	8·9	− 1·9	0·405
	148	148	0	0·928 $= \chi^2$

To determine ν, the number of degrees of freedom, it is sufficient to note that we have 5 values of O or E, but that (i) the totals and (ii) the means of the observed and expected frequencies have been made to agree.

$$\therefore \nu = 5 - 2 = 3$$

If x denotes $(O - E)/\sqrt{E}$,

(i) $\Sigma O = \Sigma E$; $\Sigma(O - E) = 0$

$$\text{i.e. } \Sigma x\sqrt{E} = 0$$

This is one restriction of type (6).

(ii) The means being equal, $\Sigma rO = \Sigma rE$, where r is the number of accidents.

$$\Sigma r(O - E) = 0, \text{ i.e. } \Sigma xr\sqrt{E} = 0.$$

This is the second restriction.

Table 2 now gives the probability of a value of χ^2 exceeding 0·928, when $\nu = 3$, as somewhere between 0·9 and 0·8. The observed frequencies fit those expected on the Poisson hypothesis very well.

5.3.3 χ^2 test of normal distribution

Ex. iii

Test the goodness of fit of the normal distribution to the data of §2.7.3, Ex. vi.

(These data represent corrections to be applied to an earlier catalogue in the light of later and more precise measurements. They therefore afford a fairly large sample by which to test the normal law as the law of errors of measurement.)

A normal distribution is determined by its mean and s.d., and we must accept the sample values of these as the only available estimates of μ and σ of the hypothetical parent distribution.

$$\bar{x} \text{ or } m = \text{estimate of } \mu = -1{\cdot}281$$
$$s = \text{estimate of } \sigma = \sqrt{3{\cdot}109} = 1{\cdot}764.$$

We have now to deal with a continuous variate and the expected relative frequencies in the class-intervals are the corresponding areas under the normal curve. The first thing to do is to standardize the values of the variate at the dividing points between class-intervals so that the tables of the normal probability integral can be used. Now the lowest class-interval is from $x = -7$ to $x = -6{\cdot}1$, since the original values before grouping were given to one decimal place. If X denotes the standardized variate, $X = (x - \mu)/\sigma$. The lowest class-interval includes all observed values below $x = -6{\cdot}05$, i.e. below $X = (-6{\cdot}05 + 1{\cdot}281)/1{\cdot}764 = -2{\cdot}704$. Every advance of 1 in the value of x is an advance of $1/1{\cdot}764$ or $0{\cdot}567$ in the value of X. The reader should now understand the formation of the following table in which P is the probability of a value less than X, i.e. P is the area under the normal curve from $-\infty$ to X. Differences between successive values of P are denoted by ΔP and give the expected relative frequencies in the class-intervals. Since the total of observed frequencies is 450, the values of E are given by $E = 450\Delta P$. The smaller tail-frequencies have been grouped in two alternative ways.

x	X	P	ΔP	E	O	$O - E$		$(O - E)^2/E$	
− 7 on	− 2·704	0·0034	0·0034	1·5	2		5·6		4·35
− 6	− 2·137	0·0165	0·0131	5·9	11	− 1·4		0·07	
− 5	− 1·570	0·0584	0·0419	19·0	12		− 7·0		2·58
− 4	− 1·003	0·1586	0·1002	45·1	35	− 10·1		2·26	2·26
− 3	− 0·436	0·3314	0·1728	77·8	82	4·2		0·23	0·23
− 2	0·131	0·5521	0·2207	99·3	118	18·7		3·52	3·52
− 1	0·698	0·7574	0·2053	92·4	88	− 4·4		0·21	0·21
0	1·265	0·8971	0·1397	62·9	54	− 8·9		1·26	1·26
1	1·832	0·9665	0·0694	31·2	32	0·8		0·02	0·02
2	2·399	0·9918	0·0253	11·4	10		− 1·4		0·17
3	2·966	0·9988	0·0070	3·1	5	1·0		0·07	
4		1·0000	0·0012	0·5	1		2·4		0·16
				450·1	450			7·64	14·76

Total, mean and s.d. of observed and expected distributions have been made to agree. Hence the number of d.f. is three fewer than the number of classes in which the frequencies are compared. When three frequencies are grouped at each tail,

$$\nu = 8 - 3 = 5.\ \chi^2 = 7{\cdot}64.\ P < 0{\cdot}2, \text{ slightly.}$$

When two frequencies are grouped at each tail,

$$\nu = 10 - 3 = 7.\ \chi^2 = 14{\cdot}76.\ P < 0{\cdot}05, \text{ slightly.}$$

It thus appears that the fit is good on the broader grouping but not good on the narrower grouping. Examination of the table shows that this is due to the large discrepancies between E and O in the two class-intervals beginning at $x = -6$ and $x = -5$. A possible explanation is that the earlier star catalogue had a few instances in which the declinations of stars differed from the later, more precisely determined, values by so much that the differences are to be classed as minor 'mistakes' rather than inevitable errors of measurement. The largest corrections are still only of the order of half a second of angle.

5.3.4 Other criteria of normality

(*a*) Graph paper, called Arithmetic Probability paper, can be obtained with a uniform scale for the variate and a non-uniform scale for accumulated relative frequency expressed as a percentage. When the latter is plotted against the former, for a normal distribution the graph is a straight line. If a given frequency distribution is a likely sample from a normal population, and the accumulated relative frequency plotted at suitable intervals of the variate, the resulting points will lie nearly on a straight line. In the example of the last section, if F denotes accumulated frequency up to a value x, we form the following table.

x:	$-6{\cdot}05$	$-5{\cdot}05$	$-4{\cdot}05$	$-3{\cdot}05$ etc.
F:	2	13	25	60 etc.
$\frac{100F}{450}\%$	0·44	2·89	5·56	13·33 etc.

When these were plotted on A.P. paper a fairly straight line crossed the 50% ordinate at

$$x = -6{\cdot}05 + 4{\cdot}8 = -1{\cdot}25,$$

affording a graphical estimate of the mean. The values of x at $\pm\sigma$ from the mean of a normal distribution correspond to accumulated relative frequencies 84·13% and 15·87%, either of which may be used to find a graphical estimate of σ. For the above example it was $\sigma = 4{\cdot}8 - 3{\cdot}0 = 1{\cdot}8$.

(*b*) The *mean deviation* of a normal distribution is very nearly 4/5 of the standard deviation.

The proof is as follows.

$$\text{The m.d.} = \frac{1}{\sigma\sqrt{2\pi}}\int_{-\infty}^{\infty} |x| e^{-x^2/2\sigma^2} dx$$

$$= \frac{2}{\sigma\sqrt{2\pi}}\int_{0}^{\infty} x e^{-x^2/2\sigma^2} dx$$

$$= \sigma\sqrt{\frac{2}{\pi}}\left[-e^{-x^2/2\sigma^2}\right]_0^{\infty}$$

$$= \sigma\sqrt{\frac{2}{\pi}} = \frac{4}{5}\sigma, \text{ nearly} \qquad . \quad . \quad . \quad (10)$$

(*c*) *Skewness.* The normal distribution is symmetrical and the parameter γ_1 (see Chap. II, equation (15)) has the value zero. In a likely sample, the corresponding statistic g_1 should be nearly zero.

In Chap. II, Ex. vi, we obtained β_1, properly denoted by b_1, to be 0·0005, for the same set of data of star-corrections, and the sign of m_3 was positive. Hence $g_1 = +\sqrt{0{\cdot}0005} = 0{\cdot}024$, indicating slight positive skewness.

To decide whether the skewness is significantly different from zero, it is necessary to know the sampling distribution† of g_1 for samples from a normal population. This sampling distribution is nearly normal so that g_1 may be considered significantly large if numerically greater than twice its standard error†, which is $\sqrt{6}/\sqrt{n}$. In the present example $n = 450$ and the s.e. of g_1 is 0·116. The value 0·024 is therefore small.

(*d*) *Kurtosis.* The parameter β_2 (Chap. II, equation (14)) has the value 3 for the normal distribution. Hence a parameter

$$\gamma_2 = \beta_2 - 3 \qquad . \quad . \quad . \quad . \quad (11)$$

is defined, whose departure from zero is a measure of comparison with the normal distribution. The quality which it measures is the sharpness of the peak of the distribution curve and is called Kurtosis. If γ_2 is positive the peak is sharper, if γ_2 is negative the peak is flatter than in the normal curve. The two types are described as leptokurtic and platykurtic respectively.

In the example of the star-corrections the estimate of γ_2 is

$$g_2 = 3{\cdot}40 - 3 = 0{\cdot}40.$$

To test the significance of a value of g_2 requires special tables since the sampling distribution is not normal. The s.e. of g_2 is twice that of g_1. In this example the s.e. is 0·23.

† See next chapter, §6·1.

5.4 Pearson's distributions

We shall restrict our examples of the fitting of a frequency distribution of a continuous variate by means of a theoretical probability distribution to the normal distribution, but other standard probability curves exist, notably those known as Pearson's and referred to in §2.7.1. The χ^2 distribution, with χ^2 as variate, is one of them. Of the family as a whole, it is worth while to indicate the manner in which they arise in statistical theory.

It was shown in §3.3.3, that the compound probability of drawing a particular sample from a finite population will be different, according as the sampling is with or without replacement. The binomial distribution, of which the normal distribution is a limiting form, supposes a constant probability p in each of n drawings. When the wider problem of drawing without replacement is considered, a distribution is obtained which is represented by a hypergeometric series. As the number of drawings increases, the distribution tends to a form in which the probability density function satisfies the differential equation

$$\frac{dy}{dx} = -\frac{(x-a)y}{c_0 + c_1x + c_2x^2} \quad . \quad . \quad . \quad (12)$$

For different values of the four constants the family of curves represented by this equation can be classified in twelve types, and it can be shown that all of them can be specified in terms of the first four moments of the distribution. The normal distribution is a special case, obtained by setting c_1 and c_2 equal to zero.

For a full treatment of Pearson's distributions see Kendall: *Advanced Theory of Statistics*, Vol. 1.

5.5.1 The null hypothesis and significance

The use made of χ^2 so far may be called positive. The experimenter, inspecting his frequency table, considers it likely that the values are a reasonable sample from a population of a particular type and uses the χ^2 test with a view to confirming his hypothesis. The deviations from the expected frequencies should usually have a 'random' quality.

Now sometimes the experimenter wishes to test the reality of a *difference* between two sets of frequencies. Suppose, for instance, that a test is made of the efficiency of inoculation against a certain disease. A comparison is made of the incidence of the disease amongst persons not inoculated and persons inoculated. In both cases it is perhaps reasonable to classify frequencies in three groups: disease not contracted; contracted in a mild form; contracted in a serious form. Inspection of the table of frequencies may suggest that there is a prima facie case for the inoculation as a defence against the disease. But even if the second set of persons had not been inoculated, their proportions in the three groups are likely to differ somewhat from those of the first set, for there are many causes of variation in resistance to infection. (The experimenter may have anticipated and avoided known major causes of variation by choosing comparable sets of persons for the experiment, but minor causes will still operate.) The observed differences between the two sets or samples will be judged *significant*, i.e. as indicating a real effect of the inoculation, if they can be shown to be improbable as the chance effects of sampling. The *null hypothesis* is formed that the effect is *not* real, but that the two samples are samples with the same *expected* relative frequencies in the various classes. The χ^2 test gives the probability P of so marked a difference between the two samples, on this hypothesis. It is necessary to select beforehand a critical low value of P, called a 'level of significance', at which it will be judged that the null hypothesis is to be rejected and the observed effect is real. A standard often adopted is: $P > 0{\cdot}1$, not significant; $0{\cdot}1 > P > 0{\cdot}05$, doubtful; $P < 0{\cdot}05$, significant. A level of significance may be decided on the basis of experience in a particular field of application, or by consideration of the risk involved. If 0·05 is the level adopted, then only 1 in 20 times could the observed effect, or a bigger one, occur on the null hypothesis, so that 19 in 20 times it is correct to reject the null hypothesis. Clearly the economic or other consequences of a wrong decision are greater in some cases than in others and will affect the choice of the critical value of P.

The comparison is not always between two samples, but sometimes between a sample and a large population. For instance, an insurance company may have a very good idea of the number of claims to be expected from car drivers as a whole, but have reason to believe that the proportion of claims is high from a particular class of driver.

When the classes in which frequencies are grouped are not class-intervals in a measured variate, but correspond to some attribute or descriptive quality, the frequency table is called a *contingency table.*

5.5.2 One degree of freedom. 1 × 2 table

In illustrating the use of χ^2 for tests of significance, it is convenient to take first the cases in which $\nu = 1$. The simplest way in which this can occur is that the observations fall into two categories and it is desired to determine whether the ratio of the two frequencies differs significantly from an expected ratio. In genetic theory, for instance, two types of offspring may be expected in a definite Mendelian ratio, but a significant departure from this ratio would indicate the operation of certain causes. Alternatively, the expected ratio might be established by long-term observation, as in the instance of insurance.

When there is such a hypothetical proportion or probability, the test of significance of the proportion found in a sample could be made by means of the binomial distribution. Suppose the hypothetical probability of 'success' is 1/3 and that in a sample of 40 there are 19 successes and 21 failures, instead of the expected $13\frac{1}{3}$ and $26\frac{2}{3}$. (Note that 'expected' values are mean values derived mathematically and need not be whole numbers.) We could calculate the probability of the particular result of 19 successes, but to bring the test of significance into line with the use of χ^2, we must find instead the probability of a result differing from the expected $13\frac{1}{3}$ successes by $5\frac{2}{3}$ or more, whether in excess or defect. That is, the occurrence of 19 successes will be judged significant if the total probability of 19 or more *and* of 7 or less successes is below the chosen level of significance. This requires the summation of the probabilities in two 'tails' of the binomial distribution. To calculate all the terms required is arduous, though tables do exist of the 'incomplete Beta function' from which a partial sum of a binomial series can be found. An alternative method is to use, when justifiable, the normal distribution as a sufficient approximation. As in Ex. vii of Chap. IV, in passing from the discrete binomial variate to the continuous normal variate, it is best to regard the former as a central value. Thus in the present example, 19 is the centre of the interval from $18\frac{1}{2}$ to $19\frac{1}{2}$. We could find the probability that the normal variate with mean $13\frac{1}{3}$ and variance $40\,(\frac{1}{3})(\frac{2}{3})$ exceeds $18\frac{1}{2}$ and double our result to obtain the probability in the two tails.

Now all this can be done in terms of χ^2. With ν d.f. the distribution of χ^2 is that of the sum of squares of ν normal variates (0, 1). When $\nu = 1$, χ^2 is the square of a single normal variate (0, 1), as may be seen from equation (5) which becomes

$$dp = \sqrt{\frac{2}{\pi}}\, e^{-\chi^2/2} d\chi,\ \chi \geqslant 0 \qquad . \quad . \quad . \quad (13)$$

This is the normal distribution with doubled constant to allow for the fact that χ is reckoned positive. The χ^2 table, with the correction of 1/2 unit in the frequency 19, will give the same probability as the normal table and very nearly the same as the sum of the binomial probabilities. The correction of the observed frequencies by 1/2 in the direction of the expected frequencies is known as *Yates' correction*.

Ex. iv

Over a long period a travelling salesman's records show a proportion of successful calls of 1/3. He tries a new technique and achieves 19 successes in 40 calls. Does this signify that the new technique is effective?

On the null hypothesis that the apparent improvement is a chance effect, the table of expected and observed frequencies is

	Successes	*Failures*	*Total*
E . .	$13\frac{1}{3}$	$26\frac{2}{3}$	40
O . .	19	21	40

Using Yates' correction we make the O's read

$$18\tfrac{1}{2} \qquad 21\tfrac{1}{2}$$

$$\chi^2 = \Sigma\frac{(O - E)^2}{E} = \frac{(5\frac{1}{6})^2}{13} + \frac{(5\frac{1}{6})^2}{26} = 3, \text{ very nearly.}$$

The frequencies are in two classes, but totals of E and O have been made to agree so that $\nu = 1$.

The χ^2 table gives P somewhat less than 0·1, which would not usually be reckoned significant. Further experiment should be made. If the improved proportion of successes is kept up it will become significant and the χ^2 test will confirm this. Suppose that in 60 trials the number of successes is 28. The reader should verify that the value of χ^2 becomes 4·84 and P is now distinctly less than 0·05.

With the given observed frequencies, the sum of the binomial probabilities of 19 or more and 7 or less successes is 0·07.

5.5.3 One degree of freedom. 2 × 2 table

Instead of comparing the ratio of frequencies in two categories, occurring in a sample, with a hypothetical expected ratio, we may need to compare the ratios occurring in two samples when no hypothetical ratio is known. In the example which follows, the purpose is to test the effectiveness of inoculation by comparing the proportion attacked by a disease amongst (i) inoculated, (ii) not inoculated persons. A problem identical in principle would be the comparison of two industrial processes in respect of the proportion of defective articles produced.

Ex. v

Are the following figures evidence of the effectiveness of inoculation?

	Attacked	*Not Attacked*	*Totals*
Inoculated .	20	300	320
Not Inoculated .	80	600	680
	100	900	1,000

There is a prima facie case in favour of inoculation, since the proportion of inoculated persons attacked is half that of persons not inoculated. The null hypothesis assumes that this is a chance effect, namely that the chance of being attacked is independent of inoculation. If the marginal totals are divided by the total frequency 1,000, they become relative frequencies and rank as prior probabilities; then on the hypothesis of independence the probability or expected relative frequency in each cell is the product of the marginal probabilities. (See §5.1.) The resulting table of probabilities is

0·032	0·288	0·32
0·068	0·612	0·68
0·1	0·9	1

Multiplying by 1,000 we have the table of E's, or expected frequencies:

32	288
68	612

Comparing the observed frequencies we find $O - E = \pm 12$ in each cell.

$$\chi^2 = \Sigma(O - E)^2/E$$
$$= 12^2\left(\frac{1}{32} + \frac{1}{288} + \frac{1}{68} + \frac{1}{612}\right)$$
$$= 7{\cdot}35.$$

The number of d.f. is the number of O's which can be chosen independently. Now since the marginal totals of E's and O's have been made to agree, as soon as the value of O is specified in one cell, the values in the other three cells are fixed so that there is only one degree of freedom.

With $\chi^2 = 7{\cdot}35$, $\nu = 1$ we find $P < 0{\cdot}01$, so that the figures are very significant of a real effect of inoculation.

Now consider the same problem from a different point of view. On the evidence the proportion of persons attacked in the complete set of 1,000 was 1/10 and the complete set is made up of two samples in one of which the proportion attacked was 1/16; in the other, 2/17. Accepting, on the null hypothesis, a prior probability 1/10 of attack or 'success', we require to know how likely it is that two samples of sizes 320 and 680 should yield proportions of successes differing as widely as 1/16 and 2/17.

In §3.5.4 we found that in n trials with prior probability p in each trial, the proportion, r/n, of successes is binomially distributed with mean p and variance pq/n.

In the present instance samples of size 320 should show a mean proportion of successes 1/10 but are liable to vary about this mean with variance 9/32,000. Samples of size 680 show in the long run the same mean proportion but with variance 9/68,000. Both variates (proportions of successes) are binomial variates which for sufficiently large values of n may be treated as normal variates, with equal means. The *difference* of two normal variates (μ, σ_1) and (μ, σ_2) is a normal variate $\{0, \sqrt{(\sigma_1^2 + \sigma_2^2)}\}$. (See §4.7.) The probability that this difference shall be greater than $2/17 - 1/16$ numerically can now be found from tables of the normal probability integral.

$$\sigma^2 = \sigma_1^2 + \sigma_2^2 = 9/32{,}000 + 9/68{,}000 = 0{\cdot}000414$$
$$x = 2/17 - 1/16 = 0{\cdot}0551$$
$$x/\sigma = 2{\cdot}71.$$

The probability of so large a standardized deviation is 0·0067. The χ^2 test gave $P < 0{\cdot}01$ and closer interpolation would have given $P = 0{\cdot}0067$, so that as in the last section the use of χ^2 with one d.f. is equivalent to the use of the normal distribution as approximation to the binomial. For this reason it is more accurate to apply Yates' correction in a 2×2 table as well as in a 1×2 table. In the above example, the observed frequencies would be adjusted to 20·5, 299·5, 79·5, 600·5, i.e. each is made half a unit nearer expectation. This reduces the value of χ^2 to 6·74 and increases P to 0·0093.

If the frequencies in a 2×2 table are too small to justify the use of χ^2 (or of the normal as approximation to the binomial distribution) the direct calculation of the probability required, namely of so large discrepancies from expectation, is more difficult than in the case of the 1×2 table which required the evaluation of the tail probabilities of a binomial distribution. For the exact method in the case of a 2×2 table the reader is referred to Kendall: *Advanced Theory of Statistics*, Vol. 1, pp. 12, 16.

5.5.4 $h \times k$ table, $1 \times k$ table

A contingency table may consist of h rows and k columns with a total of hk cells. As for the 2×2 table, the marginal totals are used to form a table of probabilities, and hence of expected frequencies, in the cells on the null hypothesis that an apparent bias is a chance effect and that the probabilities in the classes by rows and the classes by columns are independent. The marginal totals of E's and O's are made to agree and consequently the number of d.f. is $(h - 1)(k - 1)$, provided h and k are both greater than 1, for the last column and the last row are determined when the remaining cells have been filled in any manner.

Ex. vi

Candidates for a degree in Physics are required to pass a preliminary examination in Mathematics. The table gives the numbers of candidates, classified according to the grading awarded in the two subjects. Verify that performances in the two subjects are far from independent.

		Class in Physics				
		I	II	*P*	*F*	Totals
Class in Mathematics	I	38	60	50	11	159
	II	24	70	100	27	221
	P	12	36	91	25	164
	Totals	74	166	241	63	544

We form the table of expected frequencies on the hypothesis of independence. In the first cell, for instance, the expected frequency is

$$544 \text{ times product of marginal probabilities}$$

$$= 544 \times \frac{159}{544} \times \frac{74}{544} = \frac{159 \times 74}{544}$$

We obtain

E	21·6	48·5	70·4	18·4
	30·1	67·4	97·9	25·6
	22·3	50·0	72·7	19·0

We might now almost dispense with the χ^2 test, since it is evident that the observed frequencies, as compared with the hypothetical frequencies (E), show a marked bias towards high grading in both subjects or low grading in both. To test whether the apparent bias could be a chance effect, we proceed to evaluate χ^2.

$O - E$	16·4	11·5	− 20·4	− 7·4
	− 6·1	2·6	2·1	1·4
	− 10·3	− 14·0	18·3	6·0
$\dfrac{(O - E)^2}{E}$	12·5	2·7	5·9	3·0
	1·2	0·1	0·0	0·1
	4·8	3·9	4·6	1·9

The total $= \chi^2 = 40{\cdot}7$ and $\nu = (3 - 1)\ (4 - 1) = 6$. We find $\chi^2 = 22{\cdot}46$ corresponds to the level of significance 0·001, so that our larger value of χ^2 is highly significant.

Since the χ^2 test requires that even the smallest frequencies shall be sufficiently large, we need a fairly large number of observations to form such a table. If we had been given actual marks in the two subjects we could have formed a contingency table with more rows and columns, by dividing the ranges of marks into suitable class-intervals. Many cells would now almost certainly contain frequencies too small for the application of the χ^2 test, but we shall see in Chapter XI how to assess the tendency of performances in the two subjects to agree.

Ex. vii

Absentees in three shifts employing equal numbers over a period were as in the following table:

Shift:	I	II	III
Number of absentees:	21	14	31

Is the difference significant?

The null hypothesis is that absenteeism is unrelated to shift so that the expected numbers are equal, namely

22 22 22

$$\chi^2 = \Sigma(O - E)^2/E = \frac{1}{22} + \frac{64}{22} + \frac{81}{22} = 6{\cdot}64.$$

ΣO and ΣE have been made to agree so that $\nu = 3 - 1 = 2$. From tables, $P < 0{\cdot}05$ and the relation of absenteeism to shift is probably real.

If the shifts employed unequal numbers we should divide the total of 66 in the proportions of numbers employed to find the expected frequencies on the null hypothesis.

This is an example of a $1 \times k$ table and we find $\nu = k - 1$. It could have been made a $2 \times k$ table by inserting the numbers observed and expected to be present. This would leave ν unaltered but introduce three new terms in χ^2, but they would be very small terms since the expected numbers present are presumably large.

5.6.1 Limiting form of the χ distribution

Table 2 gives the values of χ^2 corresponding to a number of critical values of the probability P and for values of ν from 1 to 30. As ν increases, the mode of the χ distribution moves to the right and it can be shown that the distribution tends to the normal. The approach to normality is more rapid with $\chi\sqrt{2}$ as variate and it may be assumed with sufficient accuracy that, if $\nu>30$, $\chi\sqrt{2}$ is normally distributed, with $\sqrt{(2\nu - 1)}$ as mean, and unit variance. It is to be remembered in using this normal distribution that the value of $\chi\sqrt{2}$ corresponding to, say, 0·05 as the critical value of P, is such that the tail to the *right* covers an area 0·05. This gives as the criterion of significance

$$\chi\sqrt{2} - \sqrt{2\nu - 1} > 1{\cdot}65, \ (\nu > 30) \quad . \quad . \quad . \quad (14)$$

More generally, for such large values of ν, a value of χ^2 is significant at a level P, if $\chi\sqrt{2} - \sqrt{(2\nu - 1)}$ exceeds the number t such that $2P$ is the probability that a normal variate differs *numerically* from its mean by more than $t\sigma$.

5.6.2 Additive property of χ^2

If χ_1^2, χ_2^2 . . . are values of χ^2 calculated from data relating to a given problem, with ν_1, ν_2 . . . d.f. respectively, and all the data are pooled, the restrictions in filling the cells being retained, then for the pooled data $\chi^2 = \chi_1^2 + \chi_2^2 + \ldots$ with $\nu = \nu_1 + \nu_2 + \ldots$ d.f. The effect is that while the results of single experiments may be of doubtful significance, the χ^2 test for the pooled data gives due weight to the cumulative evidence of several experiments.

Again, the test gives increased weight when experimental values are more numerous without change of degrees of freedom. For if all values of O are, for instance, doubled, so are the values of E; then, since $\chi^2 = \Sigma(O - E)^2/E$, in which the numerator is multiplied by 4 and the denominator by 2, the value of χ^2 is doubled and the value of P becomes smaller.

* 5.7 Maxwell's distribution

The Kinetic Theory of gases and modern Statistical Mechanics are much concerned with probability distributions, one of which, namely Maxwell's law of the distribution of (resultant) velocities of the molecules of a perfect gas, is a particular case of the χ distribution. In equation (4), $r^2 = x^2 + y^2 + z^2$, where x, y and z are independent normal variates. r may be regarded as the resultant of a vector with three orthogonal components x, y, z, and the derivation of equation (4) shows that if the components are independent normal variates of equal variance, then the probability that r lies in any range is a function of r only; in other words, the distribution of r has radial symmetry in the three-dimensional space.

It is conversely true that, given radial symmetry of the distribution of the magnitude of the resultant and independence in the components, the component distributions must be normal. Maxwell's law may be obtained in this way; initial assumptions of symmetry and independence, besides yielding the required distribution of the resultant, establish incidentally the normal law for the components. Assume that the 'chaotic' state of a perfect gas may be expressed by saying that (i) the probability that the resultant velocity of a molecule has a certain magnitude is the same for all directions of the resultant and (ii) the probability of a component velocity of any magnitude is independent of the magnitude of the other two components.

Then if u, v, w are component velocities in three mutually perpendicular directions and r is the resultant velocity, the probability density at a point (u, v, w) of the 'velocity-space' is a function of r only, say $\phi(r)$.

If the components have distributions

$$dp = f(u)du,\ dp = f(v)dv,\ dp = f(w)dw$$

and are independent,

$$f(u)\,f(v)\,f(w) = \phi(r).$$

By logarithmic differentiation,

$$\frac{f'(u)}{f(u)}\,du + \frac{f'(v)}{f(v)}\,dv + \frac{f'(w)}{f(w)}\,dw = \frac{\phi'(r)}{\phi(r)}\,dr$$

But $r^2 = u^2 + v^2 + w^2$. $\therefore\ rdr = udu + vdv + wdw$

Substituting for dr and equating coefficients of du, dv and dw (i.e. equating expressions for the partial derivatives of $\log \phi(r)$ with respect to u, v and w) we obtain

$$\frac{f'(u)}{uf(u)} = \frac{f'(v)}{vf(v)} = \frac{f'(w)}{wf(w)} = \frac{\phi'(r)}{r\phi(r)}$$

We thus have equal to one another functions of u only, of v only, of w only and of r only. This is impossible unless each is a constant.

Hence $\dfrac{f'(u)}{f(u)} = Ku$, where K is constant.

Integrating,
$$\log f(u) = \tfrac{1}{2}Ku^2 + B$$
$$f(u) = Ae^{\frac{1}{2}Ku^2}$$

Now the integral of the probability density throughout the velocity space is 1.

$$A^3\int_{-\infty}^{\infty}\int_{-\infty}^{\infty}\int_{-\infty}^{\infty} e^{\frac{1}{2}Ku^2}\,e^{\frac{1}{2}Kv^2}\,e^{\frac{1}{2}Kw^2}\,du\,dv\,dw = 1.$$

Hence K must be negative; let $K = -\dfrac{1}{\sigma^2}$, then the L.H.S. is the cube of

$$A\int_{-\infty}^{\infty} e^{-u^2/2\sigma^2}\,du$$

which must therefore also be equal to 1.

$$\therefore A = \frac{1}{\sigma\sqrt{2\pi}}$$

Thus the probability distribution of each component is normal $(0, \sigma)$. The r.m.s. value of each component of velocity is σ.

$$\text{Now } \phi(r) = f(u)\, f(v)\, f(w) = \frac{1}{(\sigma\sqrt{2\pi})^3} e^{-r^2/2\sigma^2}$$

The element of volume dV between the spheres with origin as centre and radii r and $r + dr$ is $4\pi r^2 dr$. Hence the probability of a value of r in a range dr is

$$dp = \frac{4\pi r^2}{(\sigma\sqrt{2\pi})^3} e^{-r^2/2\sigma^2} dr = F(r)dr, \text{ say} \qquad . \quad . \quad (15)$$

With $\sigma = 1$ this is the form of equation (4) or of equation (5) when $n = 3$.

This is Maxwell's distribution and may be put in the form

$$\frac{dN}{N} = F(r)dr$$

where N is the total number of molecules in a given mass of gas and dN is the number having a resultant velocity of magnitude between r and $r + dr$.

If C is the r.m.s. value of r

$$C^2 = \frac{1}{N}\int r^2 dN \text{ throughout the mass}$$

$$= \int_0^\infty r^2\, F(r)dr$$

$$= 4\pi A^3 \int_0^\infty r^4 e^{-r^2/2\sigma^2} dr$$

$$= \frac{4\pi}{(\sigma\sqrt{2\pi})^3} \cdot \frac{3}{2} \sigma^5 \sqrt{2\pi}$$

$$= 3\sigma^2,$$

so that the r.m.s. resultant velocity is the resultant of the r.m.s. component velocities. One-third of the K.E. of the gas is associated with each of the three components, an example of the principle of the Equipartition of Energy.

In the Kinetic Theory of gases it is proved that the K.E. of a mass of gas is $3\,RT/2$.

$$\therefore \tfrac{1}{2} NmC^2 = 3\,RT/2$$

$$\therefore \frac{C^2}{3} = \frac{RT}{Nm} = \frac{kT}{m} \qquad . \quad . \quad . \quad . \quad (16)$$

where k is the gas constant per molecule and m the mass of a molecule. Thus the value of σ is known for a given gas at a given temperature.

It is easy to show that the velocity of maximum probability is $\sqrt{\frac{2}{3}}\,.\,C.$

Maxwell's theory was criticized, particularly in respect of assumption (ii), that the component velocities are statistically independent. Boltzmann introduced the broader principle that the equilibrium state of a gas is the state of maximum probability. Assuming that the prior probability of components (u, v, w) is equal in equal portions of the velocity-space, it can be shown that the state of maximum probability, if the ranges of u, v and w are infinite, is that in which the distribution of velocities is Maxwell's. The same principle leads to uniform diffusion as the most probable state of a gas contained in a finite volume. In the derivation of Maxwell's distribution, we had $f'(u)/f(u) = Ku$. If K were zero, it would follow that $f(u) = A$, i.e. that the probability distribution of u is uniform. This is impossible if the range of u is infinite. In the problem of the probability distribution of position of molecules within a finite volume, uniform distribution is the most probable. Moreover, if it is supposed that, for a gas not in equilibrium, the tendency *with time* is towards the state of maximum probability, the second law of Thermodynamics is interpreted in terms of probability and entropy identified with the logarithm of a compound probability. More modern developments, called Einstein-Bose and Fermi-Dirac statistics, meet difficulties in the classical theory by modifications of the assumed prior probabilities. *

Suggestions for further reading:

M. G. Kendall, *Advanced Theory of Statistics*, Chapter 12 (Griffin).

F. Yates, *Contingency tables involving small numbers and the χ^2 test* (Supp., Journal of the Royal Statistical Society, 1, 217).

R. A. Fisher, *Conditions under which χ^2 distribution measures discrepancy between observation and hypothesis* (J.R.S.S., 87, 442).

EXAMPLES ON CHAPTER V

1. Verify the answer to worked Ex. i by interpolation from the χ^2 table. (The probability that $r > 2$ is the probability that $\chi^2 > 4$, and $\nu = 2$.)

2. The position of a point in a room is fixed by measurements from two walls at right angles and from the floor. If all measurements are liable to error normally distributed (0, 1), show that there is a 1 in 10 chance that the point is more than 2·5 units out of position.

3. Apply the χ^2 test to the data of Examples on Chapter IV, **3**, comparing the observed frequencies with the expected frequencies of a Poisson distribution having the same mean and the same total frequency. Pool the last two frequencies.

4. Apply the χ^2 test of goodness of fit to the data of Examples on Chapter II, 1, the hypothesis being that the parent distribution is Poisson's. Pool the last three frequencies.

5. For the data of Examples on Chapter II, 10, calculate the expected frequencies in each class-interval for the normal distribution having the same mean, variance and total frequency. Test the goodness of fit of the observed to the expected frequencies.

6. Test the goodness of fit of the data of Examples on Chapter IV, 14, to the normal distribution.

7. The following data give N, the number of days on which the rainfall exceeded R inches at a certain station over a period of one year:

R:	0·00	0·04	0·10	0·20	0·50	1·00
N:	296	246	187	119	30	3

Test by means of χ^2 whether the data are consistent with the law $\log_{10}N = 2{\cdot}47 - 1{\cdot}98R$. (R.S.S. Cert.)

8. A series of single digits, supposed to be random, is grouped into sets of four. In each set of four the number of digits greater than 5 (i.e. 6, 7, 8 or 9) is recorded. From 200 sets so inspected the following results were obtained:

Number of digits greater than 5:	0	1	2	3	4
Number of sets:	29	71	75	23	2

Would you consider the hypothesis that each of the ten digits has an equal chance of occurrence at each place in the original series to be borne out by these data? (R.S.S. Cert.)

9. Define the binomial distribution and state how tables of the normal probability integral can be used to calculate approximate significance levels for tests based on the binomial distribution.
Indicate how allowance can be made for the discrete nature of the distribution.
An event occurs 12 times out of 16 trials. Does this contradict the hypothesis that the event is just as likely to happen as not to happen? (R.S.S. Cert.)

10. $\begin{array}{c|c} a & b \\ \hline c & d \end{array}$ For the 2 × 2 table, the marginal totals being used to determine the probabilities, prove that

$$\chi^2 = \frac{(ad - bc)^2 (a + b + c + d)}{(a + b)(c + d)(a + c)(b + d)}.$$

11. Of A candidates taking a certain paper, a are successful; of B taking another paper, b are successful. Show how the significance of the difference between the ratios a/A, b/B may be tested (i) by a χ^2 test on a

contingency table and (ii) by comparing the difference with its standard error assessed by means of a binomial distribution. (You may assume that all frequencies are sufficiently large for Yates' correction to be ignored.) Prove algebraically that the value of χ^2 is the square of the ratio of $(a/A - b/B)$ to its standard error. (R.S.S. Cert.)

12. Genetic theory states that children having one parent of blood-type M and the other of blood-type N will always be of one of the three types M, MN, N and that the proportions of these types will on the average be as 1 : 2 : 1. A report states 'of 162 children having one M parent and one N parent, 28·4% were found to be of type M, 42% of type MN and the remainder of type N. The low value of χ^2 demonstrates the truth of the genetic theory.' Calculate the value of χ^2, make the appropriate test of significance and comment on the conclusions quoted. (R.S.S. Cert.)

13. A certain type of surgical operation can be performed either with a local anaesthetic or with a general anaesthetic. Results are as given below:

	Alive	*Dead*
Local . . .	511	24
General . . .	173	21

Test for any difference in the mortality rates associated with the different types of anaesthetic. What qualifications would have to be imposed on any conclusion from such data as to the desirability of one or the other method of anaesthesia? (R.S.S. Cert.)

14. In a number of sections in a factory the following figures of the average numbers employed and of accidents were recorded during a year.

Section:	1	2	3	4	5	6	7	8	9	10
Average Number employed:	20	22	18	105	109	23	148	69	162	93
Number of accidents:	11	10	6	28	29	6	32	15	31	9

Find in which sections the accident rates are significantly greater than in the factory as a whole. (A.I.S.)

15. An advertiser receives 14, 23 and 17 replies to the same advertisement inserted in three newspapers. Use the χ^2 test, taking as hypothesis that the expected numbers of replies is the same for all three papers, to show that there is not sufficient evidence that the second paper is a better advertising medium than the others. If the numbers of replies had been 42, 69 and 51, would the difference be significant at the 5% level?

16. Certain pellets are mass-produced by the automatic compression of a quantity of powder to a predetermined height. The specific gravity of each pellet is required to lie between two given tolerance limits, and the weight of the pellet is also required to lie between two given tolerance limits.
A sample of 500 pellets was taken and each pellet was weighed against two weights equal to the lower and upper weight tolerance respectively. This determines for each pellet whether it is below within or above the

tolerance limits. By immersing the pellets in liquids of specific gravities equal to the lower and upper tolerance specific gravities respectively, it is decided whether the specific gravity of each pellet is below, within or above the tolerance limits.

The following table summarizes the results of this investigation.

		Weight	
	Below tolerance	*Within tolerance*	*Above tolerance*
Specific gravity: Below tolerance	6	63	2
Specific gravity: Within tolerance	18	294	22
Specific gravity: Above tolerance	2	84	9

Do these results indicate that any association exists between the specific gravity and the weight of pellets?

What advantages would be gained by the actual determination of the weight and specific gravity of each pellet? (R.S.S.)

17. Draw the graph of the function $r^2e^{-r^2/2}$ as r varies. The graph will serve as that of Maxwell's distribution with $\sigma = 1$ and $C = \sqrt{3}$. By counting squares or other approximate method of finding areas under the graph, show that the probability of a velocity greater than C is about 0·39 and of a velocity greater than $2C$ is less than 0·01.

18. Using standard physical constants and equation (16), show that the r.m.s. velocity of hydrogen molecules at N.T.P. is $(1{\cdot}84)10^5$ cm. per sec. approximately.

CHAPTER VI

SMALL SAMPLES. STUDENT'S *t*. VARIANCE RATIO

Standard error. S.e. of the mean. Best estimate of variance. Expectation. Distribution of mean and s.d. of sample from a normal population. Student's t*: significance of a single mean: the method of controls: significance of two means. Variance ratio and Fisher's* z*. Bartlett's test. The Fisher-Behrens test.*

6.1. Sampling distributions and standard error

It has been assumed so far that observed samples are fairly large; for instance that the mean and variance of a frequency distribution can be accepted as good estimates of those of the wider population of which the observed values are a sample, and that frequencies in different classes are sufficiently large for the application of the χ^2 test of goodness of fit. The two tests of significance which are the main subject of this chapter relate to small samples from supposedly infinite normal populations. Estimates of the mean and variance of the parent population have to be based on small samples, and it is important to examine the degree of precision with which this can be done.

A single value of a variate is an estimate, of sorts, of the mean value. How good an estimate depends on the variation in the whole population. If you measure a single component produced by a machine you have an idea of the average size, but until you have measured at least two components you can have no idea of the variation. The successive averages of a dimension taken over 1, 2, 3, 4 . . . components give a progressive average. Its reliability as an estimate of the mean for all components increases rapidly in the early stages, but when a fairly large sample has been taken, the inclusion of one more value of the variate has little effect on the average. To give precision to this idea that the size of sample affects the reliability of an estimate, we need to investigate the probabilities involved in the process of random sampling.

A complete population, real or hypothetical, will be characterized by certain parameters such as mean, variance, mode, etc. Corresponding quantities can be calculated from a random sample of values and are, in general, estimates of the parameters. A quantity calculated from a sample is called a *statistic*. It may be used as an estimate of a corresponding population parameter, but also sometimes to estimate a different parameter. For instance, the *median* value of a sample from a normal population can be used as an estimate of the population *mean*, and the *range* of a small sample affords an estimate of the *standard deviation* of the normal population.

For samples of given size each statistic has a *sampling distribution.* Given the parent population, we can find the probability that the statistic will take different values for different samples of the same size. Thus if the mean of a sample of size n is regarded as an estimate of the mean of the population, to find the precision of the estimate we consider the probability distribution of the means of all possible samples of size n. This distribution has itself a mean and a variance. We find that the first of these, namely the mean of the means of all possible samples, is the mean of the parent population. When a statistic, called an estimator, has this property that the mean value of the estimator is the true value of the parameter required, it is called an *unbiased* estimator. Next we consider the variance of the sampling distribution of means, that is we find the variance of the means of all possible samples of size n. It is σ^2/n, where σ^2 is the variance of the parent population, provided the population is large. This confirms what common sense suggests, that variance decreases, and so precision increases, as the size of sample increases.

These two properties of the sampling distribution of the mean are, as we shall see, independent of the nature of the parent distribution, but in general the whole character of a sampling distribution must depend on the parent distribution from which the samples are drawn. Fortunately a large number of sampling distributions tend to the form of the normal distribution as $n \to \infty$. Hence we are able, if n is sufficiently large, to interpret the mean and variance of the sampling distribution in terms of probabilities: for instance we may say that there is a probability of 0·05 that the deviation from the mean (which for an unbiased estimator is the true value of the parameter) exceeds (numerically) 1·96 times the standard deviation of the sampling distribution, for this is a property of the normal distribution.

Owing to its importance as a measure of precision of estimate, the standard deviation, i.e. the square root of the variance, of a sampling distribution, is determined whenever possible and is given the name of *standard error*. The word 'error' is here used in a technical sense and standard error is not to be confused with error of scientific measurement or of computation.

6.2.1 Mean and variance of means of samples

Suppose we have a population of N values, $x_1, x_2 \ldots x_N$. Without loss of generality we may suppose that the mean of the population has been chosen as origin, so that

$$\sum^{N} x = 0.$$

Let a sample of n values, say $x_1, x_2 \ldots x_n$, be chosen at random from the population. There are ${}^N C_n$ such samples.

If m is the mean of a sample and $\overline{m}$ the mean value of m for all possible samples,

$$nm = \sum_1^n x \text{ and } {}^N C_n \overline{m} = \Sigma m$$

Now each x occurs in ${}^{N-1}C_{n-1}$ samples. (There are ${}^{N-1}C_{n-1}$ ways of choosing the $n - 1$ other values after choosing one value.)

$$\begin{aligned} \therefore n\Sigma m &= {}^{N-1}C_{n-1} \sum_1^N x \\ &= 0 \\ \Sigma m &= 0 \\ \overline{m} &= 0 \end{aligned}$$

Thus the mean of means of samples is the mean of the population.

If σ^2 is the variance of the population,

$$N\sigma^2 = \sum_1^N x^2 \qquad . \quad . \quad . \quad . \quad (1)$$

since the origin was taken at the mean.

Let σ_m^2 denote the variance of the m's. Since their mean is also zero,

$$\begin{aligned} {}^N C_n \sigma_m^2 &= \Sigma m^2, \text{ over all samples} \qquad . \quad . \quad . \quad . \quad (2) \\ &= \frac{1}{n^2}\Sigma(x_1 + x_2 + \ldots + x_n)^2, \text{ over all samples} \\ &= \frac{1}{n^2}{}^{N-1}C_{n-1}\sum_1^N x^2 + \frac{2}{n^2}{}^{N-2}C_{n-2}\,\Sigma x_j x_k \qquad . \quad . \quad . \quad (3) \end{aligned}$$

since the square of each x occurs in ${}^{N-1}C_{n-1}$ samples, while the product $2x_jx_k$ $(j \neq k)$ occurs in ${}^{N-2}C_{n-2}$ samples.

$$\text{Now } \left(\sum_1^N x\right)^2 = \sum_1^N x^2 + 2\Sigma x_j x_k$$

Substituting for $2\Sigma x_j x_k$ in equation (3),

$$\begin{aligned} {}^N C_n\, \sigma_m^2 &= \frac{1}{n^2}\left\{{}^{N-1}C_{n-1} - {}^{N-2}C_{n-2}\right\}\sum_1^N x^2 \\ &= \frac{1}{n^2}{}^{N-1}C_{n-1}\left\{1 - \frac{n-1}{N-1}\right\}N\sigma^2 \\ \therefore \sigma_m^2 &= \frac{1}{n}\left\{1 - \frac{n-1}{N-1}\right\}\sigma^2 \qquad . \quad . \quad . \quad . \quad (4) \\ &= \frac{N-n}{n(N-1)}\sigma^2 \qquad . \quad . \quad . \quad . \quad . \quad (5) \end{aligned}$$

If N is large compared with n, we may say

$$\sigma_m^2 = \sigma^2/n \quad . \quad . \quad . \quad . \quad . \quad . \quad (6)$$

i.e. the s.d. of means of samples is $1/\sqrt{n}$ of the s.d. of the population. *This is the standard error of the mean.*

Ex. i

A normal population has mean 67·5 and s.d. 2·5. What is the probability that (i) a single value, (ii) the mean of 10 values drawn from the population will be within 1 unit of the mean of the population?

(i) The probability of a value within $0{\cdot}4\sigma$ of the mean is 0·31.
(ii) The sum, and hence the mean, of two or more normal variates is normally distributed. The s.d. of means of samples of size 10 is $2{\cdot}5/\sqrt{10} = 0{\cdot}79 = \sigma_m$. 1 unit $= 1{\cdot}27\sigma_m$ and the probability that the mean of the sample is within this range of the mean of the population is 0·80.

Ex. ii

The s.d. of a machine process is well-established as 2 thousandths of an inch. The first five measured dimensions after a new setting are 2·1045, 2·1050, 2·1060, 2·1035, 2·1055 in. Between what limits may it be expected with 95% probability that the mean dimension will lie?

The mean of the sample is 2·1049. The standard error of this estimate $= \sigma_m = 2/\sqrt{5}$ thou. The distribution of the mean may be assumed normal (see §4.6.2) and there is a probability of 0·95 that 2·1049 lies within $1{\cdot}96\sigma_m$ of the true mean. We say, with 95% probability of being right, that the mean of the population is in the range $2{\cdot}1049 \pm 0{\cdot}00175$ in.

In these examples it is assumed that σ for the parent population is known. We next consider how to estimate σ on the evidence of the variance of a sample.

6.2.2 Mean of variances of samples

Let s^2 denote the variance of a sample, i.e. the mean square deviation of the values of a sample from their own mean m.

Then
$$s^2 = \frac{1}{n}\sum_1^n x^2 - m^2$$

Summing for all samples,

$$\Sigma s^2 = \frac{1}{n}\,{}^{N-1}C_{n-1}\sum_1^N x^2 - \Sigma m^2$$

where, as before, N is the total number of values in the population. The mean value of s^2

$$= \frac{\Sigma s^2}{{}^{N}C_n}$$

$$= \frac{1}{N}\sum_{1}^{N} x^2 - \sigma_m^2, \text{ using equation (2)}$$

$$= \sigma^2 - \sigma_m^2$$

$$= \left(1 - \frac{1}{n}\right)\sigma^2, \text{ if } N \text{ is large} \quad . \quad . \quad . \quad . \quad (7)$$

$$= \frac{N(n-1)}{n(N-1)}\sigma^2, \text{ exactly} \quad . \quad . \quad . \quad . \quad (8)$$

Equation (7) shows that s^2, the variance of a sample, is not an unbiased estimator of σ^2.

s^2 is an unbiased estimator of its mean value, $(n - 1)\sigma^2/n$. Hence $ns^2/(n - 1)$ is an unbiased estimator of σ^2.

Now with any origin, if m is the mean of the n values in a sample,

$$s^2 = \frac{1}{n}\sum_{1}^{n} (x - m)^2$$

If we denoted by σ_e^2 the estimate afforded by the sample of the parent variance σ^2, then

$$\sigma_e^2 = \frac{ns^2}{n-1}$$

$$= \frac{1}{n-1}\sum_{1}^{n} (x - m)^2 \quad . \quad . \quad . \quad (9)$$

Thus to estimate σ^2 we divide the sum of the squared deviations from the mean of the sample by $n - 1$, instead of by n, as in calculating s^2, the variance of the sample. This result may appear at first surprising, but there are two aspects which make it less so. We could have been sure that s^2 affords on the average too low an estimate of σ^2; for mean square deviation of a group is a minimum when measured from the mean of the group and the value of s^2 for every sample is less than the mean square deviation from any other value than m, the mean of the sample. Since m is not in general the mean of the population, it is clear that the mean value of s^2 will be less than σ^2, the mean square deviation of the population from the mean of the population.

A second aspect of the substitution of $n - 1$ for n is that the mean of a sample is determined by the n values of the sample. Consequently the deviations from the mean are not independent; they are restricted by the relation that their sum is zero. This amounts to a loss of one degree of freedom. In a sample of one there can be no variance; in a sample of two there is only one quantity to determine the variance, namely the difference between the two values, and so on.

To know the standard error of $ns^2/(n - 1)$ as an estimate of σ^2, we should find the sampling distribution. The method of combinations for a finite population of N values becomes elaborate. The resulting formula involves the fourth moment as well as the second moment of the population. If the parent distribution is normal the standard error is $2\sigma^4/(n-1)$. Since the standard error of sampling must exist and the effect of multiplying by $n/(n-1)$ is small if n is large, we may use s^2 as a satisfactory estimate of σ^2, from large samples.

Ex. iii

If the sample values in Ex. ii afford the only available evidence of the variance of the machine process, what is the estimated s.d. in the bulk product?

The deviations from the mean of the sample are

$-4, 1, 11, -14, 6$ ten-thousandths of an inch

$$\sigma_e^2 = (16 + 1 + 121 + 196 + 36)/4 = 92{\cdot}5$$
$$\sigma_e = 9{\cdot}6 \times 0{\cdot}0001 \text{ in.} = \underline{1 \text{ thou., nearly.}}$$

If the s.d. in the bulk product is really 2 thou., as supposed in Ex. ii, the small sample has yielded too small an estimate of the variance of the process. In establishing Quality Control limits (see Chapter VII) a firm estimate of the variance of the process would be based on a sufficient number of samples.

Ex. iv

To illustrate the effects of sampling on a small scale, we take in the following table a small 'population' of 200 values, the different numerical values of the variate being the integers 1 to 9. The mean of the population is 5 and its variance 4. In fact the 'triangular' distribution:

x:	1	2	3	4	5	6	7	8	9
f:	8	16	24	32	40	32	24	16	8

has been shuffled and tabulated and then 10 samples, each of 10 values, have been selected at random.

7	8	3	3	6	7	5	3	2	4	5	9	6	1	2	2	4	2	2	4
6	4	7	4	7	5	6	2	7	6	5	5	2	9	9	6	4	6	7	9
6	7	9	3	4	5	3	5	8	5	5	5	2	4	7	6	7	3	3	5
6	6	6	4	1	6	5	4	8	5	5	4	8	4	7	6	9	3	3	8
6	6	6	8	6	4	4	3	4	5	7	3	3	7	8	4	7	3	5	2
3	5	2	3	5	7	6	8	5	8	3	7	9	1	7	5	4	8	7	7
5	4	5	5	8	7	6	4	4	3	2	5	2	5	6	1	1	4	2	3
3	6	5	5	6	5	4	8	4	4	5	5	4	3	5	2	3	2	6	4
6	5	4	6	5	4	5	5	1	6	5	7	1	6	8	6	6	1	6	8
6	5	4	7	3	2	5	4	9	4	8	7	5	5	7	4	3	8	7	3

												m	s^2	σ_e^2
Sample	1.	8	4	7	6	8	5	7	6	6	4	6·1	1·89	2·10
	2.	6	4	9	6	6	5	5	5	5	2	5·3	2·81	3·12
	3.	3	6	5	1	8	2	4	3	5	4	4·1	3·69	4·10
	4.	4	6	6	7	6	3	8	5	5	9	5·9	2·89	3·20
	5.	2	5	4	8	5	3	4	3	6	5	4·5	2·65	2·94
	6.	5	8	2	1	8	5	8	3	6	4	5·0	5·80	6·44
	7.	6	2	2	8	3	9	2	4	1	5	4·2	6·76	7·51
	8.	2	4	6	7	8	7	6	5	8	7	6·0	3·20	3·56
	9.	5	5	4	3	3	7	5	8	2	5	4·7	3·01	3·34
	10.	3	2	5	8	5	3	5	4	6	7	4·8	3·16	3·51

Mean value of m $= 5{\cdot}06$

Estimated variance of $m = \frac{1}{9}\Sigma(5{\cdot}06 - m)^2 = 0{\cdot}545$

Mean value of σ_e^2 $= 3{\cdot}98.$

It will be seen at once that the values of m show much less deviation from the mean of the population than the values of the population. The mean value of m over 10 samples is very near the true mean and the mean value of σ_e^2 very near σ^2. Regarding the 10 values of m as a sample from all possible values of m, we can form a single estimate of the standard error of m. We know its true value. The variance of m (for all samples) is

$$\frac{N-n}{n(N-1)}\sigma^2 = \frac{200-10}{10.199} \times 4 = 0{\cdot}382.$$

Our estimate is 0·545, which is considerably in excess, but it should be remembered that it is based on one sample of 10 values of m; in our seventh sample of 10 values of the variate we had 7·51 as the estimate of the population variance, but nevertheless obtained a very good estimate of σ^2 from 10 samples. The important thing is that even a single sample is likely to give an estimate of the right order of magnitude.

6.2.3 Expectation

It is convenient to consider at this point the significance of equation (6) as an approximation to equation (5), when N is large. The approximation amounts to the omission of the term $(n-1)/(N-1)$ in equation (4) and this term arises from $\Sigma x_j x_k$. In this sum each x is multiplied by the sum of all the *others*; if each were multiplied by the sum of *all*, the total would be zero (since the mean of the population was chosen as origin). This condition would be satisfied if the sampling were sampling with replacement, or if the n values were drawn, one from each of n identical populations; the n values in a sample would then be independent variates. It transpires, therefore, that they may be regarded as independent if drawn from a very large population.

This principle justifies a simpler method, which is often used, of deriving equations (6) and (7). Let us denote by $E(z)$ the mean value of a quantity z taken over all samples of a given size. It is called the *expectation* of z. If m is the mean of a sample of size n,

$$nm = x_1 + x_2 + \ldots + x_n$$

$$n^2m^2 = \sum_1^n x^2 + \text{product terms.}$$

Averaging over all samples we may assume the mean value of the product terms to vanish. Moreover, in the first term on R.H.S. each sample contributes n values of x^2 and over all samples, all x's will occur with equal frequency. Hence

$$n^2\,E(m^2) = nE(x^2).$$

Since the mean value of m, as of x, is zero, this equation is

$$n^2\sigma_m^2 = n\sigma^2$$

$$\text{i.e. } \sigma_m^2 = \sigma^2/n \text{ as in equation (6).}$$

Note also that if $z = x_1 + x_2 + \ldots + x_n$, then

$$E(z^2) = nE(x^2) = n\sigma^2 \quad . \quad . \quad . \quad . \quad (10)$$

that is, the variance of the sum of n values drawn from a large population is n times the parent variance. This is otherwise obvious, since the variate z is n times the variate m; hence the distribution of z is identical with that of m, except for the scale factor n, and so the standard deviation of z, namely $\sigma\sqrt{n}$, is n times that of m, which is $\sigma/\sqrt{n}$.

A more general form of equation (10) is often useful. Let

$$z = a_1x_1 + a_2x_2 + \ldots + a_nx_n,$$

where the a's are constants and the x's are independent variates drawn from n different populations whose variances are $\sigma_1^2, \sigma_2^2 \ldots \sigma_n^2$.

$$\text{Firstly } E(z) = a_1E(x_1) + a_2E(x_2) + \ldots + a_nE(x_n),$$

i.e. the mean value of z is the same linear function of the mean values of the x's as z is of the x's. Now suppose the mean of each population taken as origin. The mean value of z is then also zero, and

$$z^2 = \sum_1^n a_r^2 x_r^2 + \text{product terms.}$$

Averaging over all samples

$$E(z^2) = a_1^2 E(x_1^2) + a_2^2 E(x_2^2) + \ldots + a_n^2 E(x_n^2)$$

$$\text{i.e. } \sigma_z^2 = a_1^2\sigma_1^2 + a_2^2\sigma_2^2 + \ldots + a_n^2\sigma_n^2 \quad . \quad . \quad (11)$$

This result was obtained in §4.7 for the particular case of a linear function of *normal* variates. It is now seen to be a property of variance, whatever the probability distributions of the component variates x_1, $x_2 \ldots x_n$, provided these are drawn from large populations. Taken in conjunction with the Central Limit Theorem (§4.6.2), it means that we can often assume that z is normally distributed and interpret σ_z accordingly, in terms of probabilities, although the component variates are not normal. If z is the sum of n values drawn at random from any large population of variance σ^2, then $\sigma_z^2 = n\sigma^2$. If m is the mean of n values, so that $m = z/n$, then $\sigma_m^2 = \sigma^2/n$ and for sufficiently large values of n, the distribution of m may be assumed normal with this variance and with mean equal to the mean μ of the parent distribution.

Now consider the expectation of s^2, the variance of a sample of n values drawn from a single parent population. With the mean of the population as origin,

$$ns^2 = \sum_1^n x^2 - nm^2, \text{ for each sample.}$$

Averaging over all samples

$$nE(s^2) = nE(x^2) - n\,E(m^2)$$
$$= n\sigma^2 - n\sigma^2/n = (n-1)\sigma^2$$
$$E(s^2) = (n-1)\sigma^2/n,$$

as in equation (7), from which we deduced that s^2 is an unbiased estimator of $(n-1)\sigma^2/n$.

6.3 Sample from normal population

In a number of tests of significance the null hypothesis takes the form that samples are from a common parent population of the normal form. The rejection of the null hypothesis will be on the basis that sample means and variances show differences which are of low probability. The calculation of the probabilities required depends in the first instance on the determination of the sampling distributions of mean and variance, when the parent population is normal. These sampling distributions lead to further probability distributions which are the basis of Tables 3 and 4 of probabilities associated with 'Students' t and the variance ratio.

The reader who is impatient of the mathematical theory will find in §§6.4.2, 6.4.3, 6.4.4, and 6.5.2 the rules and methods of application of these significance tests. The theory is given in this section and §§6.4.1 and 6.5.1.

* Consider first a sample of three values, x_1, x_2, x_3, from the normal population

$$dp = \frac{1}{\sigma\sqrt{2\pi}} e^{-x^2/2\sigma^2} dx,$$

in which the mean of the population has been chosen as origin of x.

The probability density at a point (x_1, x_2, x_3) in space of three dimensions is proportional to

$$e^{-(x_1^2+x_2^2+x_3^2)/2\sigma^2} = e^{-r^2/2\sigma^2} \quad . \quad . \quad . \quad . \quad (12)$$

All samples having a mean m correspond to points in the plane

$$x_1 + x_2 + x_3 = 3m \quad . \quad . \quad . \quad . \quad (13)$$

The variance s^2 of the sample x_1, x_2, x_3 is given by

$$s^2 = (x_1^2 + x_2^2 + x_3^2)/3 - m^2$$

so that all samples with a variance s^2 correspond to points of the sphere

$$x_1^2 + x_2^2 + x_3^2 = 3(m^2 + s^2) = r^2 \quad . \quad . \quad . \quad (14)$$

To find the probability of a sample with mean in a range dm and standard deviation in a range ds, we must multiply the probability density (12) by the element of volume enclosed by two near parallel planes of type (13) and between two near concentric spheres of type (14).

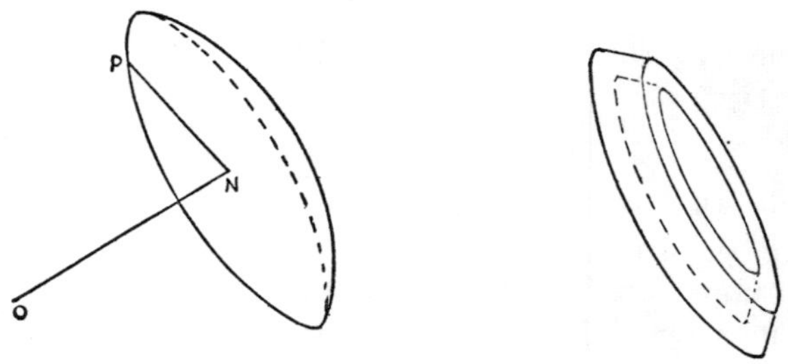

FIG. 18

$$\text{ON} = \text{the perpendicular distance of plane (13) from origin}$$
$$= 3m/\sqrt{3} = m\sqrt{3}$$

$$d(\text{ON}) = d(m\sqrt{3}) = \text{distance between near planes.}$$

$$\text{NP}^2 = \text{OP}^2 - \text{ON}^2 = 3(m^2 + s^2) - 3m^2 = 3s^2.$$

Sphere (14) cuts plane (13) in a circle of radius NP and of area $3\pi s^2$. The required element of volume is $d(m\sqrt{3}) \times d(3\pi s^2)$ and the probability in this element

$$= dp \propto e^{-3(m^2+s^2)/2\sigma^2}\, d(s^2)d(m).$$

This argument may be generalized. If the sample consists of n values $x_1, x_2 \ldots x_n$, represented by a point in n-dimensional space, the factor 3 is replaced by n and the circle of radius NP by a hypersphere of radius $s\sqrt{n}$ and of $(n - 1)$ dimensions. The resultant probability distribution of m and s is

$$dp \propto e^{-n(m^2+s^2)/2\sigma^2} \quad d(s^{n-1})dm$$
$$\propto (e^{-nm^2/2\sigma^2})\,(e^{-ns^2/2\sigma^2}s^{n-2})ds\, dm$$

Since this is of the form $f(m)\, g(s)\, ds\, dm$, it follows that m and s are *statistically independent*, their probability distributions being

$$dp \propto e^{-nm^2/2\sigma^2}\, dm \quad . \quad . \quad . \quad . \quad . \quad (15)$$

and

$$dp \propto s^{n-2}\, e^{-ns^2/2\sigma^2}\, ds \quad . \quad . \quad . \quad . \quad . \quad (16)\ *$$

Equation (15) shows that means of samples of size n from a normal population are themselves normally distributed with variance σ^2/n.

Equation (16) shows that $s\sqrt{n}/\sigma$ is distributed like χ, with $(n-1)$ d.f.

The statistical independence of m and s implies that, of samples with a given m, the proportion having a given s is the same as the proportion of *all* samples (of size n) having that s. There is a converse proposition that if the distributions of mean and variance of sample are independent, then the parent population is normal.

6.4.1 Distribution of Student's *t*.

The distribution (15) is standardized by the substitution $z = m\sqrt{n}/\sigma$, that is, by making the new variate z equal to the ratio of m to its standard error $\sigma/\sqrt{n}$. The distribution is then normal (0, 1) and from the normal probability integral may be found the probability that m exceeds any value. The mean of the population was taken as origin; if, with any other origin, the mean of the population is μ and m is the mean of a sample of size n, then the distribution of $(m-\mu)\sqrt{n}/\sigma$ is normal (0, 1).

To use this result we must suppose known the mean and variance of the parent population. If the only evidence of the parent population is a single sample, its mean is the only evidence of the parent mean and the best estimate of σ^2 is given by equation (9).

Writing under the name of 'Student', W. S. Gosset investigated the distribution of m, standardized by the estimated variance, σ_e^2, which varies from sample to sample. Student's distribution used s^2 as estimate of σ^2 and was later modified by R. A. Fisher, using $ns^2/(n-1)$ in accordance with equation (9).

* The joint distribution of m and s (equations (15), (16)) is

$$dp \propto s^{n-2}\, e^{-ns^2/2\sigma^2}\, e^{-nm^2/2\sigma^2}\, ds\, dm.$$

Take (s, m) as rectangular co-ordinates of a point in a plane. Converting to polar co-ordinates by the substitutions $s = r\cos\theta$, $m = r\sin\theta$, the probability density becomes proportional to

$$r^{n-2}\cos^{n-2}\theta\, e^{-nr^2/2\sigma^2} \quad . \quad . \quad . \quad . \quad (17)$$

and the element of area is $r d\theta dr$.

We define a new variable t, to standardize m by means of the estimated variance σ_e^2.

Thus
$$t = \frac{m\sqrt{n}}{\sigma_e} = \frac{m\sqrt{n-1}}{s} \quad . \quad . \quad . \quad . \quad (18)$$

since
$$\sigma_e^2 = ns^2/(n-1),$$

$$t = \frac{r\sin\theta\sqrt{n-1}}{r\cos\theta} = \sqrt{n-1}\tan\theta \quad . \quad . \quad (19)$$

and is a function of θ only.

To find the probability that t lies in a range t to $(t + dt)$ we must integrate the probability density (17) over the area in which θ varies from θ to $(\theta + d\theta)$, while r may take all values.

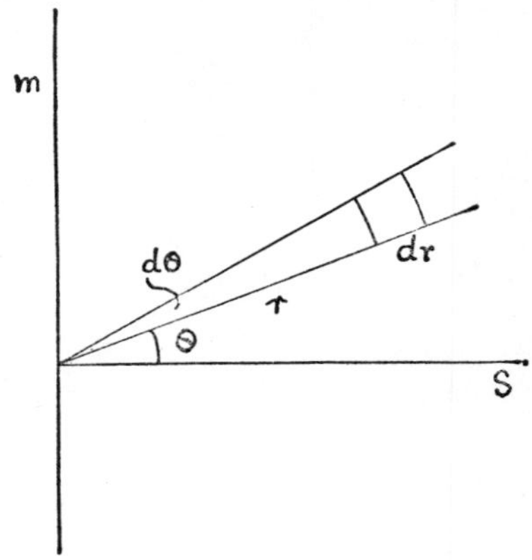

FIG. 19

The required element of probability is

$$dp = K \cos^{n-2}\theta \, d\theta \int_0^\infty r^{n-2} \, e^{-nr^2/2\sigma^2} \, dr$$

$$= K' \cos^{n-2}\theta \, d\theta$$

But $$dt = \sqrt{n-1} \sec^2\theta \, d\theta$$

$$\therefore dp \propto \cos^n\theta \, dt$$

$$= \frac{dt}{(1 + \tan^2\theta)^{n/2}}$$

Writing ν for $(n - 1)$, where ν is the number of degrees of freedom

$$dp \propto \frac{dt}{\left(1 + \frac{t^2}{\nu}\right)^{(\nu+1)/2}} \quad . \quad . \quad . \quad . \quad (20)$$

Note that in the plane of (s, m) s is always positive, m may be positive or negative, θ varies from $-\pi/2$ to $\pi/2$ and t from $-\infty$ to ∞. *

The important point about the variable t is that it is defined in terms of the sample values and its distribution is independent of the unknown variance of the normal parent distribution. If the latter variance were known and used to standardize m, the resulting distribution would be exactly normal; when m is standardized by means of σ_e, however, we obtain distribution (20) which is far from normal for small samples. In the extreme case when $\nu = 1$ the distribution is $dp = dt/\pi(1 + t^2)$ which is known as Cauchy's distribution and has infinite variance. As $\nu \to \infty$ the R.H.S. of (20) tends to $e^{-t^2/2}\,dt$, so that for *large* samples the distribution of t is normal (0, 1).

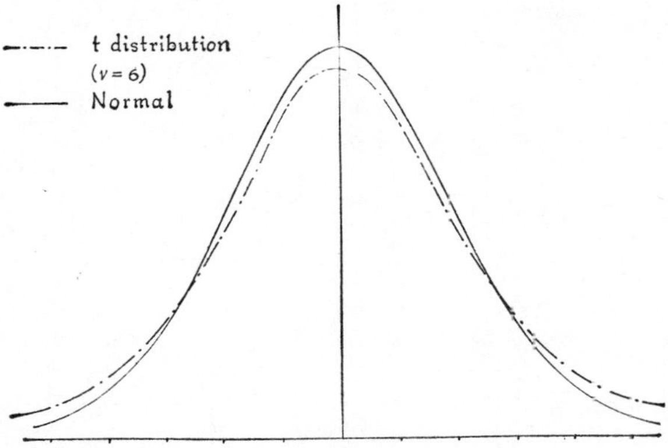

FIG. 20

6.4.2 Significance of a single mean

The first use of t is to test whether the mean m of a sample of n values differs significantly from a hypothetical mean, μ, of the supposed parent population, which is assumed to be normal. Calculate t from the formula

$$t = \frac{(m - \mu)\sqrt{(n-1)}}{s} \quad . \quad . \quad . \quad (21)$$

where s^2 is the variance of the sample. Some writers use s to denote, not the s.d. of the sample, but the modified form when n is replaced by $(n - 1)$ to give the best estimate of σ.

We may write
$$t = \frac{(m - \mu)\sqrt{n}}{\sigma} \quad . \quad . \quad . \quad . \quad (22)$$

provided we understand by σ this best estimate, so that

$$\sigma^2 = \frac{1}{n-1}\sum_1^n (x - m)^2 \quad . \quad . \quad . \quad (23)$$

Ex. v

In a certain factory a machine producing components to a nominal dimension of 1·500 in. is re-set every morning. The following data represent the dimensions of a random sample taken from the first hour's production after re-setting. Is there any evidence that the machine has been set wrongly?

1·502, 1·501, 1·504, 1·498, 1·501, 1·502, 1·500, 1·501
1·501, 1·502, 1·503, 1·501, 1·499, 1·500, 1·501, 1·505.

If x denotes deviation from a trial mean of 1·500 in. in thousandths of an inch,

$$\Sigma x = 21 \quad \text{and } \Sigma x^2 = 73$$
$$\bar{x} = 1\cdot3$$

The trial mean is also the hypothetical mean of the supposed parent population, so that

$$m - \mu = 0\cdot0013 \text{ in.}$$

The estimated $\sigma^2 = \frac{1}{15}\left(73 - \frac{21^2}{16}\right)$ in units (thou.)^2

$$= 3, \text{ approximately}$$
$$\sigma = 1\cdot73 \text{ thou.} = 0\cdot00173 \text{ in.}$$
$$\frac{\sigma}{\sqrt{n}} = \frac{0\cdot00173}{4} = 0\cdot00043$$
$$= \frac{0\cdot0013}{0\cdot00043} = 3, \text{ and } \nu = n - 1 = 15.$$

From Table 3, the probability of so large a value of t is just less than 0·01. There is evidence that the machine has been set high.

It may happen that in applying the t-test it is appropriate to use only one tail of the distribution. For instance, in testing the effectiveness of a new component as a means of reducing fuel consumption, London Transport statisticians observed the consumption, in gallons per 100 miles, for a number of buses. Only a decrease in the average consumption, when compared with normal consumption without the component, is acceptable. Further, an apparent decrease will only be accepted as real if the probability that it has occurred by chance is less than the chosen significance level P. Since the t-distribution is symmetrical it is only necessary to take the level as $2P$ when consulting the table, which is based on two tails.

The example of petrol consumption raises another point. It is measured in gallons per 100 miles, not in miles per gallon. Since either of these variates is proportional to the reciprocal of the other, if one is normally distributed the other has a skew distribution. The t-distribution is based on the assumption of a normal parent population, and it may be important to try to determine which of two or more alternative variates is most likely to be normally distributed, or nearly so. It has been shown (Kendall, *Advanced Theory of Statistics*, Vol. II, §21.13 (Griffin)) that for symmetrical parent distributions not departing too much from the normal in respect of kurtosis (see §5.3.4), the t-test is valid.

6.4.3 The method of 'controls'

If the t-test is used to justify the inference that a particular factor is the cause of a significant change in the mean, we need to be reasonably sure that there is no other factor likely to be the cause of so large a difference. (The assumption of the normal parent population allows that there are many factors likely to cause small random deviations from the mean.) Now in some cases there are likely to be large major variations between the subjects of an experiment; for instance, a group of children made the subject of an experiment in nutrition or medical treatment might give widely different results on account of differences in age, weight, sex, medical history, etc. In such cases the statistician may be able to isolate the effect of which he requires to test the significance by choosing the individuals in pairs, so that each pair agree with one another in respect of the factors (such as age, sex, etc.) which it is desired to eliminate. Two groups are now formed, each consisting of one individual from each pair; one of the two groups is subjected to the treatment which is being tested, and the other group, called the controls, is not treated; alternatively the two groups receive two treatments A and B which it is desired to compare. It may also be possible for one individual to be the subject of each of two treatments and so provide a pair of readings.

Now on account of the pairing and on the null hypothesis that there is no difference as the result of treatment, the two members of a pair should give the same value of the measured variate; or the *difference* between the values of a pair is assumed to be a normal variate with zero mean. We may apply the t-test to the observed values of this normal variate. An illustrative example is given at the end of the next section.

From the examples of significance tests so far considered, it should be clear that the tests are used to give precision to inferences for which there is a prima facie case by mere inspection of the figures. The prima facie case might be so strong as to render the statistical method unnecessary, but often it will occur that figures which might be assumed significant, especially by persons with no experience of statistical methods, represent in fact only such variation as might quite well occur by pure chance. On the other hand, when a value is found to be significant, we do not assert with *certainty* that the suspected effect is real; particularly if the first experiment is with a small sample, a significant value suggests that our suspicions are sufficiently confirmed to warrant further experiments on a larger scale.

6.4.4 Difference of means of two samples

In §6.4.2 we considered the use of Student's t to test the significance of a single mean compared with a given hypothetical mean. In §6.4.3 we are still testing a single mean, namely the *mean of differences* between pairs against a hypothetical mean of zero. Sometimes no hypothetical mean is known, but we have two samples, not necessarily of equal size, and it is desired to know whether the difference of their means is significant. For instance the effectiveness of a treatment could be tested by comparing two samples, one with and the other without treatment. The untreated sample is sometimes called the 'control' sample.

Let the two samples be of sizes n_1 and n_2 and let their means be m_1 and m_2. On the null hypothesis both samples are drawn at random from the same normal population of unknown mean and variance μ and σ^2.

Then the distribution of m_1 is normal $\left(\mu, \dfrac{\sigma}{\sqrt{n_1}}\right)$ and that of m_2 also normal $\left(\mu, \dfrac{\sigma}{\sqrt{n_2}}\right)$.

Hence $m_1 - m_2$ is a normal variate with mean zero and variance

$$\frac{\sigma^2}{n_1} + \frac{\sigma^2}{n_2} = \frac{\sigma^2(n_1 + n_2)}{n_1 n_2} \quad . \quad . \quad . \quad (24)$$

To form t we standardize the variate $m_1 - m_2$, using the estimate of the variance (24) afforded by the two samples. For theoretical reasons it is wrong to pool the two samples as one and use the combined sample variance as estimating $(n_1 + n_2 - 1)\sigma^2/(n_1 + n_2)$. The combined sample variance is not statistically independent of the variate $m_1 - m_2$. Instead, we estimate σ^2 as follows. If s_1^2, s_2^2 are the variances of the two samples

$$n_1 s_1^2 \text{ is an estimate of } (n_1 - 1)\sigma^2$$

$$n_2 s_2^2 \text{ ,, ,, ,, ,, } (n_2 - 1)\sigma^2$$

Adding, we obtain $n_1 s_1^2 + n_2 s_2^2$ as an estimate of $(n_1 + n_2 - 2)\sigma^2$.

This is equivalent to using a weighted mean of the separate best estimates of σ^2 afforded by the two samples, namely $n_1 s_1^2/(n_1 - 1)$ with weight $n_1 - 1$ and $n_2 s_2^2/(n_2 - 1)$ with weight $n_2 - 1$.

Put
$$\sigma_e^2 = \frac{n_1 s_1^2 + n_2 s_2^2}{n_1 + n_2 - 2} \quad . \quad . \quad . \quad . \quad (25)$$

and
$$t = \frac{m_1 - m_2}{\sigma_e} \sqrt{\frac{n_1 n_2}{n_1 + n_2}} \quad . \quad . \quad . \quad (26)$$

standardizing $m_1 - m_2$ by the best estimate of the variance (24). In §6.7 will be given the theoretical justification of (25) as this best estimate. The number of d.f. is $n_1 + n_2 - 2$.

Ex. vi

The following figures give the percentage extension under a given load of two random samples of yarn, the first sample being taken before washing, the second after six washings:

Before washing:	12·3	13·7	10·4	11·4	14·9	12·6	
After six washings:	15·7	10·3	12·6	14·5	12·6	13·8	11·9

Is there any evidence that extensibility is affected by washing?

In another experiment on the same type of yarn, six lengths of yarn were selected at random and each length was cut into two halves. One of the halves was tested for extension without washing, the other after six washings, and the following percentage extensions were obtained.

Length	1	2	3	4	5	6
Before washing:	13·9	12·5	11·0	11·8	10·8	14·6
After six washings:	14·7	12·1	13·2	13·6	11·5	15·4

Can it be said, on the basis of this evidence, that washing has affected extensibility? Compare this experiment with the previous one.

In the second experiment, why is it not correct to deduce that washing affects extensibility from the fact that in five cases out of six the percentage extension after washing was higher? (R.S.S. Cert.)

Expt. 1. We find $m_1 = 12{\cdot}55$

$$n_1 s_1^2 = \text{sum of squared deviations from the mean} = 12{\cdot}855$$

for the first sample, and for the second sample

$$m_2 = 13{\cdot}06,\ n_2 s_2^2 = 18{\cdot}977$$

$$\text{Estimated } \sigma^2 = \frac{12{\cdot}855 + 18{\cdot}977}{6 + 7 - 2} = 2{\cdot}894$$

$$t = 0{\cdot}51\sqrt{\frac{42}{2{\cdot}895 \times 13}} = 0{\cdot}54$$

$$\nu = 11$$

$$P = 0{\cdot}6$$

so that the figures cannot be said to afford evidence that extensibility is affected by washing.

Expt. 2. This is the method of controls and we test whether the mean of differences is significantly different from zero. The observed increases in extensibility are

$$0{\cdot}8,\ -0{\cdot}4,\ 2{\cdot}2,\ 1{\cdot}8,\ 0{\cdot}7,\ 0{\cdot}8.$$

The mean is 0·983 and the unbiased estimate of the population variance 1/5 of 4·208 = 0·842.

The variance of the means of samples of six is

$$0{\cdot}842/6 = 0{\cdot}1403$$

$$t = \frac{0{\cdot}983}{\sqrt{0{\cdot}1403}} = 2{\cdot}94$$

$$\nu = 5$$

$$P < 0{\cdot}05$$

so that the evidence of the second experiment suggests fairly strongly that washing increases extensibility.

The examiner's remaining questions may be said to have been answered in §6.4.3 above.

6.5.1 Ratio of variances of normal samples. Fisher's z

In comparing two samples it may be important to consider not only whether there is a significant difference between their means, but also whether their variances differ significantly. In fact we ought to compare the variances before applying the t-test to the means, for the latter test is based on the assumption of a common parent variance. If, therefore, the sample variances differ so widely that the hypothesis of a common parent variance is rejected, the t-test is invalidated. We shall return to this point in §6.6.3. Apart from the requirements of the t-test, comparison of variances affords a method of testing whether one machine produces a less variable product than another or whether one operator or scientific observer is more precise than another.

If two samples of size n_1 and n_2 have variances s_1^2 and s_2^2, they afford two estimates, σ_1^2 and σ_2^2, of the parent variance, on the null hypothesis that the samples are from normal populations of equal variance. (The parent populations need not have equal means, since the sampling distribution of s does not involve the parent mean μ.) Let σ_1 be the larger of σ_1 and σ_2

then $F = \sigma_1^2/\sigma_2^2$ . . . . (27)

and is called the variance ratio.

Since $\sigma_1^2 = n_1 s_1^2/(n_1 - 1)$ and $\sigma_2^2 = n_2 s_2^2/(n_2 - 1)$ we could write

$$F = \frac{n_1(n_2 - 1)s_1^2}{n_2(n_1 - 1)s_2^2}$$

but in practice it is best to calculate the two estimates of variance in the usual way. It may be noted, however, that if the samples are of equal size so that $n_1 = n_2 = n$, then $F = s_1^2/s_2^2 = ns_1^2/ns_2^2$.

Table 4 shows values of F corresponding to chosen levels of significance. In consulting the table, make $\nu_1 = n_1 - 1 =$ the number of d.f. of the sample giving the larger estimate of variance; $\nu_2 = n_2 - 1 =$ the number of d.f. of the other sample. Two samples are considered to differ significantly in respect of variance if they yield a value of F greater than that given by the table, at the chosen level of significance. An alternative variable to F is Fisher's z defined by

$$z = 1/2 \log_e F = \log_e \sigma_1 - \log_e \sigma_2$$

where again σ_1 is the larger of the two estimates of variance. Tables, whether of F or z, give the probability that either of two samples from normal populations of equal variance, shall give so large an estimate of the common variance, compared with the other sample.

* The derivation of the distribution of F or z may be outlined as follows.

The distribution of s, for samples of size n, is

$$dp \propto s^{n-2}\, e^{-ns^2/2\sigma^2} ds \quad . \quad . \quad . \quad . \quad (28)$$

Let $$x = \frac{\sigma_1^2}{\sigma^2} = \frac{n_1 s_1^2}{(n_1 - 1)\sigma^2} \text{ and } y = \frac{\sigma_2^2}{\sigma^2} = \frac{n_2 s_2^2}{(n_2 - 1)\sigma^2}$$

Substituting in (28), the distributions of x and y are

$$dp \propto x^{(\nu_1-2)/2}\, e^{-\nu_1 x/2} dx$$

$$dp \propto y^{(\nu_2-2)/2}\, e^{-\nu_2 y/2} dy$$

where $\nu_1 = n_1 - 1$ and $\nu_2 = n_2 - 1$.

Transform to polar co-ordinates, using the y-axis as initial line. Then $x = r \sin\theta$, $y = r\cos\theta$ and $F = x/y = \tan\theta$.

Now as in the derivation of the t-distribution, we determine the probability dp associated with the sector of infinite radius and angle $d\theta$. The integration is, however, more difficult, requiring a further substitution $u = \frac{1}{2}(\nu_1 \sin\theta + \nu_2 \cos\theta)r$.

The distribution of F is

$$dp \propto \frac{F^{(\nu_1-2)/2}}{(\nu_1 F + \nu_2)^{(\nu_1+\nu_2)/2}} dF$$

The substitution $z = 1/2 \log_e F$ gives

$$dp \propto \frac{e^{\nu_1 z}}{(\nu_1 e^{2z} + \nu_2)^{(\nu_1+\nu_2)/2}} dz$$

*

6.5.2 Significance of variance ratio

It is desired to test whether two samples differ significantly in variance, for instance whether one operator is more consistent than another, or whether one process yields a more uniform product than another. Given two samples consisting of values obtained by the two methods which are being compared, we take as null hypothesis that the two samples are drawn at random from normal populations of equal variance. Each sample yields an estimate of this parent variance. Call the larger estimate σ_1^2, the smaller σ_2^2, and calculate the ratio $F = \sigma_1^2/\sigma_2^2$. Enter Table 4 with $\nu_1 = n_1 - 1$, where n_1 is the number of values in the sample yielding the larger estimate of variance, and $\nu_2 = n_2 - 1$ where n_2 is the number of values in the other sample. The table gives values of F at certain levels of significance. If the computed value of F is significant at the chosen level, the null hypothesis is rejected, that is, it is inferred that the value of F is too large to be likely as the chance effect of sampling and is significant of a real difference in precision.

Alternatively, if tables of Fisher's z are available, its value is calculated from the equation

$$z = \tfrac{1}{2}(\log \sigma_1^2 - \log_e \sigma_2^2)$$

and its significance similarly tested.

Ex. vii

Two surveyors make repeated careful measurements of the same angle. Their results differ only in the figure for seconds of angle, the values of which are:

Surveyor A:	40	29	37	41	38	35		
Surveyor B:	38	37	40	33	32	39	40	34

Should these results be accepted as evidence that B is the more precise observer?

With fictitious mean 35, the deviations and their squares are

								Totals
A:	x:	5	− 6	2	6	3	0	10
	x^2:	25	36	4	36	9	0	110

										Totals
B:	x:	3	2	5	− 2	− 3	4	5	− 1	13
	x^2:	9	4	25	4	9	16	25	1	93

The estimates of variance are

$$\sigma_1^2 = \frac{1}{5}\left(110 - \frac{10^2}{6}\right) = 18{\cdot}7$$

$$\sigma_2^2 = \frac{1}{7}\left(93 - \frac{13^2}{8}\right) = 10{\cdot}3$$

$$F = \frac{\sigma_1^2}{\sigma_2^2} = 1{\cdot}8.$$

With $\nu_1 = 5$, $\nu_2 = 7$, this value of F is not significant, even at the level 0·20; the smaller variance of B's readings could easily happen as the result of random sampling.

6.6.1 More than one comparison

The tests of significance so far considered have depended on the probability, on some hypothesis, that a single value (e.g. of t or F) obtained by comparison of two sets of figures, should exceed some amount. If we have more than two sets to compare, it is evident that there is an increased probability that one or other of the values obtained by comparing the sets in pairs shall exceed a chosen amount, merely as the result of random sampling. It follows that a value which would be judged significant (i.e. not likely as the result of random sampling) if it represents a single comparison, should not necessarily be so judged if it represents one of several comparisons. Suppose P is the (small) probability that a single value exceeds some amount. If there are r *independent* values, $(1 - P)^r$ is the probability that all of them are *less* than that amount; hence the probability that *one or other* is greater than that amount is

$$1 - (1 - P)^r$$
$$= 1 - 1 + rP, \text{ approximately, since } P \text{ is small,}$$
$$= rP.$$

It is this probability which must still be small if we are to assert that one or other of our set of r values is significantly high. Hence if P is our adopted level of significance for a single value, we shall say that our largest values are significantly high if, as single values, they have a probability less than P/r. In the following example we consider 10 different values of t obtained by comparing each of 10 means with a given hypothetical mean.

Ex. viii

Ten schools submit candidates for an examination in which the overall average mark is 58. The following are the marks obtained by the candidates from the several schools. Is there evidence that the candidates from any one school are above or below average?

School	
A	51, 53, 61, 63, 56, 58.
B	66, 55, 57, 53, 63, 61, 59.
C	58, 52, 67, 54, 62, 60, 55, 65, 66.
D	58, 51, 64, 55, 61, 57, 59.
E	51, 59, 56, 46, 48, 62, 64, 50, 52, 60, 58, 54.
F	56, 48, 46, 40, 42, 44, 64, 62, 60, 58, 49, 53, 50, 54, 51, 53.
G	51, 65, 49, 47, 68, 63, 53, 61, 54, 60, 55, 59, 56, 58.
H	52, 70, 61, 50, 64, 61, 58, 63, 61, 57, 61, 65.
J	66, 55, 57, 59, 73, 61, 77, 75, 71, 63, 64, 66, 69, 68, 66, 70, 62.
K	69, 64, 74, 59, 61, 63, 79, 77, 75, 65, 73, 67, 71, 68, 70.

We find

	A	B	C	D	E	F	G	H	J	K
n	6	7	9	7	12	16	14	12	17	14
$\bar{x}$	57	59	60	58	55	52	57	60	66	69
$\Sigma(x - \bar{x})^2$	106	129	303	105	362	740	481	331	610	492
$\Sigma(x - \bar{x})^2/(n - 1)$	21·2	21·5	37·9	17·5	32·9	49·3	37·0	30·1	38·2	38·0
σ	4·60	4·64	6·16	4·18	5·74	7·02	6·08	5·49	6·18	6·16
$\sigma/\sqrt{n}$	1·88	1·75	2·05	1·58	1·66	1·75	1·63	1·58	1·74	1·65
$\lvert 58 - \bar{x} \rvert$	1	1	2	2	3	6	1	2	8	11
t	0·53	0·57	0·98	1·27	1·81	3·43	0·61	1·27	4·60	6·67
$P <$	0·7	0·6	0·4	0·3	0·1	0·01	0·6	0·3	0·001	0·001

If we adopt the level of significance 0·01 for single values, the level at which we reject the hypothesis of constant school-to-school average is 0·01 ÷ 10 = 0·001. The results of schools J, K are then significantly high. A more precise evaluation of P for school F is 0·004, so that there is a fairly strong suggestion that its average is significantly low.

6.6.2 Bartlett's test

In the above example we were able to compare each sample mean with the given population mean, obtaining 10 independent values of t. If, however, we had, say, four sample variances, themselves affording the only evidence of the parent variances, they could be paired in 6 ways to give 6 values of Fisher's z or of the variance ratio F. These 6 values are not independent; we could obtain three independent values by comparing three of the sample variances with the fourth. If the four samples, each of size 10, give sample variances $s_1^2 = 53{\cdot}5$,

$s_2^2 = 16\cdot4$, $s_3^2 = 27\cdot1$, $s_4^2 = 20\cdot0$, it is clearly the first which is suspect. Here it should be sufficient to note that the largest ratio is s_1^2/s_2^2, giving $F = 3\cdot233$ which would be just significant, as a single value, at the level 0·05; as the worst of three it would not be held to contradict the hypothesis that the samples are from populations of equal variance.

It is clear that there are many ways of choosing $r - 1$ independent ratios of pairs chosen from r values. The situation is met by *Bartlett's test,* which we shall merely state without proof.

Suppose we have k samples, each of size n. Let the sample variances be $s_1^2, s_2^2 \ldots s_k^2$ and denote the arithmetic mean of these by $\overline{s^2}$. Then if

$$A = (n - 1)\left\{k \log_e \overline{s^2} - \sum_{r=1}^{k} \log_e s_r^2\right\}$$

and

$$B = 1 + \frac{k + 1}{3k(n - 1)}$$

the ratio A/B is distributed as χ^2 with $k - 1$ d.f. We are here testing the values of s^2, on the hypothesis of a common parent variance estimated by their average. Adopt a level of significance appropriate to a single value of χ^2 (say 0·05). A large value of χ^2 suggests that the one or more values of s^2 which differ most from the average are significantly different.

6.6.3 The Fisher-Behrens test

The distributions of Student's t and of the variance ratio F are both derived from the hypothesis of a normal parent population.

In using t to test for significance of difference of means, we assume that in any case the two samples are drawn from populations of equal variance. Hence we ought always to test first for significance of the variance ratio. (Whereas the t-test is invalidated if the variances are very different, the z- or F-test is not invalidated if the means are different.) If the two samples are of equal or nearly equal size, there is not much danger of error in applying the t-test even though the ratio of variances is fairly large. There is, however, another test of difference of means, known as the Fisher-Behrens test, which does not assume a common parent variance, as follows:

$$\left.\begin{aligned} \text{Let } \tan\theta &= s_1/s_2 \\ \text{and } d &= (m_1 - m_2)/\sqrt{(s_1^2 + s_2^2)} \end{aligned}\right\} \quad . \quad . \quad (29)$$

where m and s^2 denote mean and variance of a sample.

Tables exist† showing at various levels of significance the values of d, at intervals of 15° in θ for $n_1, n_2 = 6, 8, 12, 24, \infty$.

† Fisher and Yates, *Statistical Tables for Biological, Agricultural and Medical Research* (Oliver and Boyd).

The case when n_2 is 'infinite' may well arise when the mean of a small sample of moderate variance, obtained by precise measurement, is to be compared with a population mean derived from a large amount of information based on rougher measurement; the variance of the population is increased by the variance due to rough measurement.

6.7 Some theoretical considerations

We conclude this chapter by recapitulating and amplifying, for future reference, the principal properties of the variance, s^2, of a sample from a normal population and its relations with χ^2.

* We showed (§5.2.3) that if r^2 is the sum of squares of standardized normal variates with ν d.f., then the distribution of r is

$$dp \propto r^{\nu-1} e^{-r^2/2} dr \qquad (r \geqslant 0)$$

With χ for r this is also the distribution of χ and is usually referred to as such.

Equation (16) shows that $s\sqrt{n}/\sigma$ is distributed like χ with $n - 1$ d.f. Hence s^2 is distributed like $\sigma^2\chi^2/n$ with $n - 1$ d.f. Since we had shown that the mean value of s^2 is $(n - 1)\sigma^2/n$ we may deduce that the mean value of χ^2 is ν. This is easily verified, as follows:

$$dp = k\, \chi^{\nu-1} e^{-\chi^2/2} d\chi$$

Put $\chi^2 = 2z.\ \chi d\chi = dz$, giving

$$dp \propto z^{(\nu/2)-1} e^{-z} dz$$

The mean value of $z = \dfrac{\int_0^\infty z^{\nu/2} e^{-z}dz}{\int_0^\infty z^{(\nu/2)-1} e^{-z}dz} = \dfrac{\Gamma(\frac{\nu}{2}+1)}{\Gamma(\frac{\nu}{2})} = \dfrac{\nu}{2}$

the mean value of $\chi^2 = \nu$ . . . . (30)

Consider next the property that if $\chi^2 = \chi_1^2 + \chi_2^2 + \ldots + \chi_n^2$ where χ_r^2 has ν_r d.f., then χ^2 has $\Sigma\nu_r$ d.f., all the distributions being of the above type. In its context this result was almost self-evident, since in pooling sets of data relating to one problem we merely join their contingency tables into one, retaining whatever restrictions exist on their degrees of freedom. We now give the mathematical derivation.

Let x y be independent χ variates with m, n d.f. respectively. Then their distributions are

$$dp \propto x^{m-1} e^{-x^2/2} dx, \qquad dp \propto y^{n-1} e^{-y^2/2} dy.$$

Put $x^2 = 2u$ $y^2 = 2v$, giving

$$dp \propto u^{(m/2)-1} e^{-u} du, \qquad dp \propto v^{(n/2)-1} e^{-v} dv.$$

Let $2w = z^2 = x^2 + y^2$, so that $w = u + v$.

The probability density at a point (u, v) is

$$\text{proportional to } u^{(m/2)-1}\, v^{(n/2)-1}\, e^{-(u+v)}$$

and we must integrate this over the strip between the lines $u + v = w$ and $u + v = w + dw$, in the positive quadrant. The shaded element is a parallelogram of area $dwdv$.

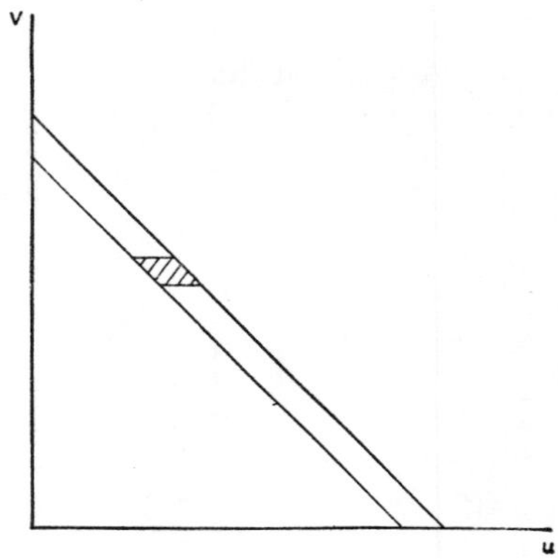

FIG. 21

The resulting element of probability for the strip

$$\propto dw \int_{v=0}^{w} (w - v)^{(m/2)-1}\, v^{(n/2)-1}\, e^{-w}\, dv$$

Put $v = w \sin^2 \theta$, giving

$$dp \propto e^{-w}\, w^{\frac{(m+n)}{2}-1} dw \int_0^{\pi/2} \cos^m\theta \sin^n\theta\, d\theta$$

$$\text{i.e. } dp \propto w^{\frac{(m+n)}{2}-1}\, e^{-w} dw.$$

Hence z is a χ variate with $m + n$ d.f.

By successive application of this result we may extend it to the sum of squares of any number of χ variates. *

In Chapter XII we have occasion to assume a converse form† of this result; knowing that the sum of k terms is distributed as χ^2 with ν d.f. and that $k - 1$ of the terms are distributed as χ^2 with $\nu_1, \nu_2 \ldots \nu_{k-1}$ d.f. we may deduce that the kth term is distributed as χ^2 with $\nu - \nu_1 - \nu_2 - \ldots - \nu_{k-1}$ d.f.

† Referred to as Cochran's Theorem.

A particular application of the direct theorem leads to the estimate of variance, based on two samples, of equation (25). We have

$$n_1 s_1^2/\sigma^2 \text{ is distributed as } \chi^2 \text{ with } n_1-1 \text{ d.f.}$$

$$n_2 s_2^2/\sigma^2 \text{ is distributed as } \chi^2 \text{ with } n_2-1 \text{ d.f.}$$

Hence $(n_1 s_1^2 + n_2 s_2^2)/\sigma^2$ is distributed as χ^2 with $n_1 + n_2 - 2$ d.f. and so with mean $n_1 + n_2 - 2$. It follows that

$$\frac{n_1 s_1^2 + n_2 s_2^2}{n_1 + n_2 - 2}$$

is an unbiased estimate of σ^2. Moreover, it is distributed independently of $m_1 - m_2$, so that the hypotheses assumed in the derivation of the t-distribution are valid for t as given by equation (26).

Suggestion for further reading:
M. G. Kendall, *Advanced Theory of Statistics*, Chapter 21 (Griffin).

EXAMPLES ON CHAPTER VI

1. Ten soldiers visit the rifle range two weeks running. The first week their scores are

   67, 24, 57, 55, 63, 54, 56, 68, 33, 43

   The second week they score, in the same order—

   70, 38, 58, 58, 56, 67, 68, 77, 42, 38.

   Is there any significant improvement? How would the test be affected if the scores were not shown in the same order each time? (A.I.S.)

2. A standard cell, whose voltage is known to be 1·10 volts, was used to test the accuracy of two voltmeters, A and B. Ten independent readings of the voltage of the cell were taken with each voltmeter, and the results were as follows:

A:	1·11	1·15	1·14	1·10	1·09	1·11	1·12	1·15	1·13	1·14
B:	1·12	1·06	1·02	1·08	1·11	1·05	1·06	1·03	1·05	1·08

   From these results is there any evidence of bias in either voltmeter? Also, is there any evidence that one voltmeter is more consistent than the other? (R.S.S. Cert.)

3. A group of 7 seven-week-old chickens, reared on a high-protein diet, weigh 12, 15, 11, 16, 14, 14, 16 ounces; a second group of 5 chickens similarly treated except that they receive a low-protein diet, weigh 8, 10, 14, 10, 13 ounces.
   Calculate the value of t and test whether there is significant evidence that additional protein has increased the weight of the chickens. Criticize the arrangement of the experiment and suggest improvements in design. (R.S.S. Cert.)

4. A group of 8 psychology students were tested for their ability to remember certain material, and their scores (number of items remembered) were as follows:

A	B	C	D	E	F	G	H
19	14	13	16	19	18	16	17

They were then given special training purporting to improve memory and were re-tested after a month. Their scores were then:

A	B	C	D	E	F	G	H
26	20	17	21	23	24	21	18

A control group of 7 students was also tested and re-tested after a month, but was not given the special training. The scores in the two tests were:

	J	K	L	M	N	O	P
(1)	21	19	16	22	18	20	19
(2)	21	23	16	24	17	17	16

Compare the change in each of the two groups by calculating t and test whether there is significant evidence to show the value of the special training.

Is there evidence that the experiment was not properly designed? (R.S.S. Cert.)

5. Two machines are making components to a particular dimension Assuming that the average dimension is a function only of the setting of the machine, do the following observed values indicate that the two machines are set differently?

	Dimension in inches							
Machine A:	2·145	2·142	2·147	2·143	2·148	2·146	2·143	2·144
Machine B:	2·144	2·149	2·147	2·144	2·145	2·143	2·148	2·146

6. If, in Question 5, the two machines should have been working to a nominal dimension of 2·146 in., is there any evidence that either machine is offset?

7. Sickness rates for two factories over a period of six months were as follows:

	Rate per 100	
	Factory A	*Factory B*
January	64	75
February	72	83
March	79	74
April	58	67
May	49	52
June	40	46

By means of the t-test, or otherwise, examine whether there is any significant difference between the two factories. (A.I.S.)

8. Two machines are making components to a particular dimension. Assuming that the average dimension is a function only of the setting of the machine, while the variability is a function only of the operator, investigate whether the following data suggest that one of the operators is a more consistent worker than the other.

Operator A:

0·502, 0·510, 0·506, 0·504, 0·503, 0·509, 0·504, 0·505, 0·503, 0·509,
0·502, 0·507, 0·505, 0·506, 0·511, 0·501, 0·504, 0·506, 0·507, 0·505.

Operator B:

0·507, 0·514, 0·519, 0·505, 0·507, 0·513, 0·516, 0·512, 0·512, 0·514,
0·509, 0·508, 0·518, 0·517, 0·516, 0·509, 0·513, 0·516, 0·511, 0·513.

9. The following are the values, in thousandths of an inch, obtained by two engineers in ten successive measurements with a micrometer of the same dimension.
Is there evidence (*a*) that one engineer is more consistent than the other, (*b*) that one reads consistently higher than the other?

Engineer A: 503, 500, 497, 505, 495, 502, 499, 501, 498, 500,
Engineer B: 502, 497, 492, 498, 496, 499, 495, 497, 496, 498.

10. Examine the following data showing the dimensions of the products from four machines P, Q, R, S, to test whether one machine is significantly more or less consistent than the others.

P:	12·1	12·3	12·6	12·3	12·5	12·2	12·4	12·4	12·3	12·4	12·3	12·2
Q:	12·4	12·3	12·6	12·1	12·5	12·9	12·6	12·5	12·7	12·3	12·5	12·6
R:	12·0	12·4	12·9	13·0	12·0	12·5	12·3	12·6	12·7	12·5	12·6	12·5
S:	12·6	12·9	13·3	12·7	13·4	12·8	13·0	12·7	12·9	12·8	13·2	13·1

11. Apply Bartlett's test to determine whether the four sample variances of §6.6.2 differ significantly.

12. From each of two batches of electric lamps made by the same manufacturer, a random sample of six was chosen and tested by burning out. The results were as follows:

	Lamp No.	1	2	3	4	5	6
Length of	Batch A:	802	959	1,022	1,040	733	897
life in hours	Batch B:	839	961	1,035	896	994	950

Is there any significant difference in average length of life between the two batches?
The average length of life should be at least 1,000 hours to satisfy a certain specification. Is there any reason to suppose, from the given data, that the manufacturer's product is not likely to meet this specification? (R.S.S. Cert.)

CHAPTER VII

QUALITY CONTROL

Control limits for means. Setting up a Quality Control Scheme. Control charts for mean, range, fraction defective.

7.1 Variability of machine product

Even if no wear or other fault occurred in a machine, the dimensions of a machine product would still be liable to variation, and this basic variability may usually be assumed to be normal, or Gaussian. A normal distribution is completely specified by its *mean* and *standard deviation*; their values, when the distribution curve is that of a dimension of a machine product, correspond to the *setting* and the *precision*, respectively, of the machine. If the values of the mean and s.d. were well established the normal probability integral could be used to determine the proportion of values of the dimension which fall within a chosen range.

It is a primary purpose of the application of statistical theory called Quality Control, to obtain warning of the development of faults in the machine by comparing the proportions of observed values falling within chosen ranges of the mean with those expected on the hypothesis of a normal distribution. For this purpose *inner and outer control* limits are chosen which are sometimes called 'warning' and 'action' limits respectively. On the normal hypothesis,

One in forty of observed values would be above $\mu + 1{\cdot}96\sigma$;
One in forty of observed values would be below $\mu - 1{\cdot}96\sigma$.
These are the inner limits.

One in a thousand of observed values would be above $\mu + 3{\cdot}09\sigma$;
One in a thousand of observed values would be below $\mu - 3{\cdot}09\sigma$.
These are the outer limits.

As will be seen presently, it is preferable in practice to work with the means of small samples of chosen size rather than with individual readings, but as a first illustration of the principles involved we may consider a set of individual readings.

Suppose the established mean and s.d. are 2·000 in. and 0·010 in. respectively, so that the inner limits are 2·0196 in., 1·9804 in. and the outer limits 2·0309 in., 1·9691 in. Consider two sets, (*a*) and (*b*), of 20 readings and 14 readings respectively.

(*a*) 2·0170, 2·0085, 2·0165, 2·0080, 2·0190, 2·0185, 2·0075, 2·0215, 2·0220, 2·0180, 2·0190, 2·0175, 2·0185, 2·0060, 2·0170, 2·0225, 2·0160, 2·0025, 2·0170, 2·0180.

The values and their relation to the control limits are more vividly represented graphically in the diagram.

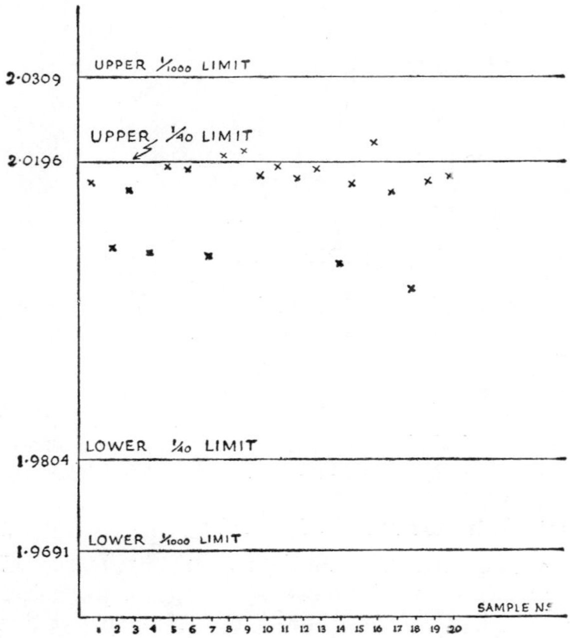

Fig. 22 (*a*)

Three in twenty points are above the upper 1 in 40 limit and all are above the expected mean. There is evidence that the mean dimension has increased, a fault which should be corrected by re-setting. There is no strong indication of a change in standard deviation.

An increase in s.d. without change of mean would result in an increased proposition of readings outside both the upper and lower limits equally. The following set of readings and their graph are strongly indicative of such a change.

(*b*) 1·9950, 2·0005, 1·9725, 1·9850, 2·0225, 2·0100, 1·9905, 1·9785, 2·0325, 1·9910, 1·9810, 1·9685, 2·0210, 2·0275.

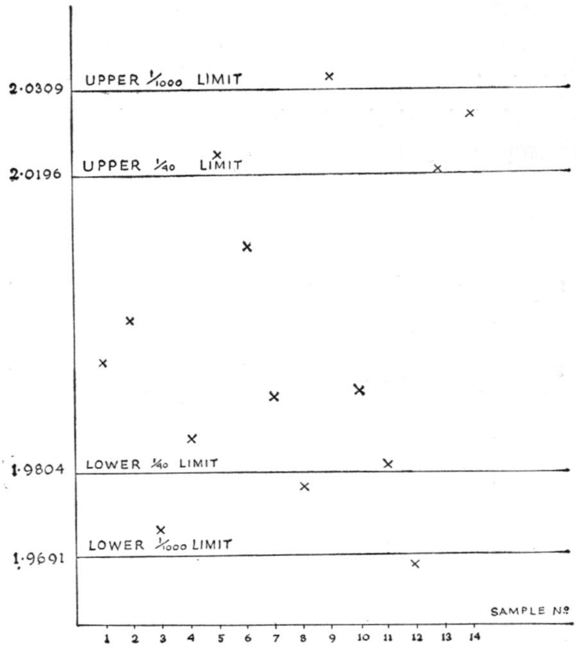

FIG. 22 (*b*)

7.2 Control limits for mean of sample

The use of limits derived from the normal distribution will be more justifiable if the variate is the mean of a sample of given size. By virtue of the Central Limit theorem (§4.6.2), even if the population of individual readings departs appreciably from the normal distribution, the means of samples of moderate size n may be assumed to be distributed normally with the same mean μ as the parent population. The standard deviation of means of samples of size n is $\sigma/\sqrt{n}$ (§6.2.3) so that in the control chart for means,

$$\text{the 1 in 40 limits are } \mu \pm 1{\cdot}96\sigma/\sqrt{n}$$

$$\text{the 1 in 1000 limits are } \mu \pm 3{\cdot}09\sigma/\sqrt{n}$$

Ex. i

A drilling machine gives depths of bore with a mean 0·5230 in. and s.d. 0·0032 in. Calculate control limits for means of samples of 5 and prepare the control chart.

$$\text{The inner limits are } 0{\cdot}523 \pm \frac{(1{\cdot}96)\ (0{\cdot}0032)}{\sqrt{5}} = 0{\cdot}5258,\ 0{\cdot}5202 \text{ (inches).}$$

The outer limits are $0{\cdot}523 \pm \dfrac{(3{\cdot}09)\,(0{\cdot}0032)}{\sqrt{5}} = 0{\cdot}5274,\ 0{\cdot}5186$ (inches).

These are represented in the chart as in Fig. 23.

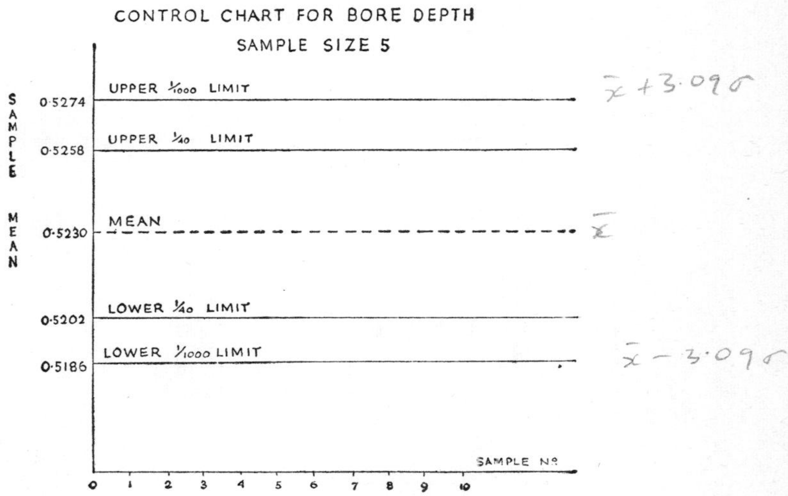

FIG. 23

7.3.1 Setting up a Quality Control Scheme

The first step is to estimate the mean and s.d., under stable conditions of production, of the measurement to be controlled. To do this, as many readings as possible, say of the order of one hundred, should be made, to obtain an accurate estimate. The mean is easily calculated. Instead of calculating the s.d. directly from the readings, it is usual to base the estimate of σ on the relation between σ and the *range* of samples of size n from a normal population. In the subsequent sampling this will have the double advantage of easy calculation together with an estimate of σ which is independent of any change of mean which occurs without change of variance. The sampling distribution of range is one of the more difficult to derive mathematically, but data sufficient for the purposes of Q.C. have been tabulated. For instance, the expected or mean range of samples of size n from a normal population (μ, σ) is $b_n\sigma$ where b_n is given by the following table.

n:	2	3	4	5	6	7	8	9	10
b_n:	1·128	1·693	2·059	2·326	2·534	2·704	2·847	2·970	3·078

Hence σ can be estimated from the mean range, $\bar{w}$, of a number of samples, by the relationship

$$\sigma = \bar{w}/b_n = a_n\bar{w}, \text{ where } a_n = 1/b_n$$

n:	2	3	4	5	6	7	8	9	10
a_n:	0·8862	0·5908	0·4857	0·4299	0·3946	0·3698	0·3512	0·3367	0·3249

Ex. ii

The following fifty readings are random samples from a normal (10, 1) population and are grouped in successive fives. Check the accuracy of the relation $\sigma = a_n\bar{w}$ by evaluating σ by the usual methods and also by determining the mean of the twenty ranges.

Sample	*Readings* (x)					*Range* w
1	9·16	10·68	10·11	11·05	10·09	1·89
2	10·83	11·35	11·01	9·46	10·28	1·89
3	11·37	9·23	9·11	10·61	7·46	3·91
4	11·55	9·08	9·80	11·11	8·48	3·07
5	9·82	8·61	10·99	9·99	9·78	2·38
6	10·36	11·27	10·45	10·59	11·52	1·16
7	10·35	8·09	10·50	9·08	9·70	2·41
8	10·36	11·28	9·31	8·36	10·42	2·92
9	12·82	10·05	9·05	11·17	11·22	3·77
10	10·14	7·87	9·24	10·24	8·93	2·37
						$\bar{w} = 2·577$

$$\Sigma x^2 = 5126·4242, \ \Sigma x = 503·38$$

$$\sigma_e^2 = \frac{1}{49}\left\{\Sigma x^2 - \frac{(\Sigma x)^2}{50}\right\} = 1·196 \qquad \sigma_e = 1·094$$

$$a_n\bar{w} = (0·4299)\,(2·577) = 1·107$$

The value obtained from $a_n\bar{w}$ thus differs by 1·3% from the best estimate of σ afforded by the whole set of readings.

Returning to the problem of starting a Q.C. Scheme, we denote the mean of the initial large set of observations by $\bar{X}$. The control limits for means of samples of size n are to be

$$\bar{X} \pm \frac{1·96\, a_n\bar{w}}{\sqrt{n}} \text{ and } \bar{X} \pm \frac{3·09\, a_n\bar{w}}{\sqrt{n}}$$

or

$$\bar{X} \pm A_{0·025}\bar{w} \text{ and } \bar{X} \pm A_{0·001}\bar{w}.$$

For convenience the quantities $A_{0·025}$ and $A_{0·001}$ have also been tabulated.

n:	2	3	4	5	6	7	8	9	10
$A_{0·025}$:	1·23	0·67	0·48	0·37	0·32	0·27	0·24	0·22	0·20
$A_{0·001}$:	1·94	1·05	0·75	0·59	0·50	0·43	0·38	0·35	0·32

The grand mean, $\bar{X}$, may be derived from the means of samples as in the following example.

Ex. iii

Calculate 1/40 and 1/1000 limits for the mean diameter of cylindrical cavities with samples of size 4 from the following data.

Cavity diameters (inches)

Sample	1	2	3	4	5	6	7	8	9	10	11	12
	0·096	·097	·101	·101	·090	·103	·108	·099	·110	·106	·108	·094
	·099	·101	·101	·112	·118	·107	·103	·113	·096	·106	·096	·105
	·092	·104	·102	·108	·105	·103	·103	·102	·104	·113	·106	·080
	·103	·106	·087	·098	·090	·112	·095	·087	·089	·104	·107	·093
$\bar{x}$:	0·0975	·1020	·0977	·1020	·1007	·1063	·1022	·1003	·0998	·1072	·1043	·0930
w:	0·011	·009	·015	·014	·028	·009	·013	·026	·021	·009	·012	·025

$$\bar{w} = 0{\cdot}0160,\ \bar{X} = \text{mean value of } \bar{x} = 0{\cdot}1013.$$

The inner limits are $0{\cdot}1013 \pm (0{\cdot}48)\ (0{\cdot}016) = 0{\cdot}1090,\ 0{\cdot}0936$ in.
The outer limits are $0{\cdot}1013 \pm (0{\cdot}75)\ (0{\cdot}016) = 0{\cdot}1133,\ 0{\cdot}0893$ in.

7.3.2 Choice of size of sample

The s.d. of the means of samples of size n is $\sigma/\sqrt{n}$ which varies inversely as $\sqrt{n}$. This fact may be used to choose the sample size in such a way that a specified shift in the mean of the population shall be almost certainly detected from the chart. Suppose we say that detection is almost certain if the shift of mean would result in 1 in 5, instead of 1 in 40, values of the sample mean as the expected proportion outside an inner control limit. Express the specified shift of mean as $K\sigma$. If the mean were to increase by this amount the upper inner control limit will be at

$$\frac{1{\cdot}96\sigma}{\sqrt{n}} - K\sigma \text{ from the new mean.}$$

Now 1 in 5 values of a normal variate exceed the mean by 0·84 of the s.d. of the distribution (see Table I). Choose n so that

$$\frac{1{\cdot}96\sigma}{\sqrt{n}} - K\sigma = 0{\cdot}84\frac{\sigma}{\sqrt{n}}$$

$$\sqrt{n} = 1{\cdot}12/K.$$

Ex. iv

If $\sigma = 0{\cdot}012$ and it is desired that a shift of 0·005 in the value of the mean of the population shall cause 1 in 5 values of mean of sample to fall outside an inner control limit, what should be the size of sample?

$$0{\cdot}005 = \frac{5}{12}\sigma.$$

Putting $K = 5/12$ in the equation above

$$n = \left(\frac{1{\cdot}02 \times 12}{5}\right)^2 = 5{\cdot}99.$$

Take samples of 6.

Note that K also varies inversely as $\sqrt{n}$; as the size of sample increases the shift which will be detected on a chosen criterion decreases in proportion to $1/\sqrt{n}$.

7.4. Control chart for range

In many industrial processes, σ, or the associated $\bar{w}$, is a constant characteristic of the process and does not vary unless some serious fault develops. It is nevertheless desirable to keep a check on the variance of the process, if only because the limits on the chart for means depend on σ and must be recalculated if σ changes. This may be done by constructing a control chart for range beneath the chart for means. As each sample is drawn, its mean and range are marked on the joint chart.

The principle of the control chart for range is that if samples of size n are drawn from a normal population it is possible to calculate the proportion of samples in which the range exceeds chosen multiples of the expected mean range. In practice the control limit on sample range is made tighter than on sample mean, since a shift in range is not only important in itself but also in relation to the chart for means. E. S. Pearson has calculated percentage points for the distribution of range of samples from a normal population. For the purpose of the control chart for range, numbers D_n, F_n are tabulated below such that, if $\bar{w}$ is the mean range of samples of size n, then *once in twenty* times the range of a sample may be expected to exceed $D_n\bar{w}$, and once in twenty times to be less than $F_n\bar{w}$.

Sample size:	2	3	4	5	6	7	8	9	10
D_n:	2·45	1·96	1·76	1·66	1·59	1·54	1·51	1·48	1·45
F_n:	0·08	0·25	0·37	0·44	0·49	0·53	0·56	0·59	0·60

Ex. v

Using the data of Ex. iii (diameters of cylindrical cavities, sample size 4), construct the complete control chart for mean and range, plotting the first twelve points.

The control limits for mean were calculated in Ex. iii.

For range, $\bar{w} = 0{\cdot}016$.

The upper 1 in 20 limit is $(1{\cdot}76)\ (0{\cdot}016) = 0{\cdot}0282$ in.

The lower 1 in 20 limit is $(0{\cdot}37)\ (0{\cdot}016) = 0{\cdot}0059$ in.

We note that the twelfth point in the mean chart is just on the lower inner control limit. We expect one in forty points to fall below this limit; the occurrence of one point on the limit in a run of twelve is no cause for alarm.

It has been stated that a shift of mean should usually be remedied by re-setting. If the range chart indicates an increase in variance the cause can only be traced from experience. Factors such as deterioration in workmanship, heterogeneity of raw material, slackness in machine ad-

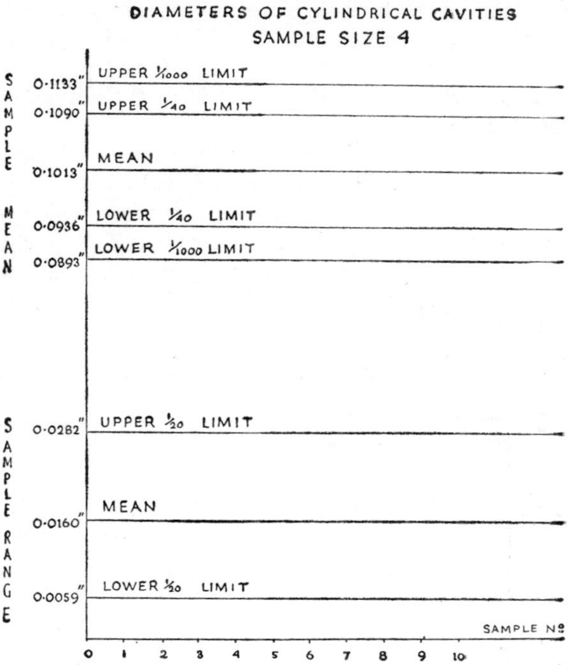

FIG. 24

justment, may all cause increase in variance. In such a case an attempt is made to eliminate the cause. In the happy event of a decrease in variance, indicated by an unexpectedly large number of points below the lower ($F_n\bar{w}$) limit of the range chart, an attempt would be made to ascertain and perpetuate the factors causing the improvement. In whichever direction the variance changes, control limits on the charts for mean and range must be recalculated.

7.5 Modified control limits

No mention has so far been made of the drawing tolerances imposed on the measurement which is the subject of the control chart. Let T_u and T_l be the upper and lower tolerances. Then no single value is to exceed T_u or to be less than T_l. The normal distribution, however, asserts that there is a small probability that the deviation from the mean will exceed any given quantity, however large. We must neglect the tails of the distribution and in fact assume it will be sufficient if the tolerances are wide enough apart to enclose outer control limits for a single value (see §7.1). Thus if

$$T_u - T_l < 6{\cdot}18\sigma$$

the machine is not precise enough for its job. It may be necessary to inspect 100% of the product, but the drawing of samples and maintenance of a control chart may still serve a useful purpose by giving warning of faults which cause departure from the normal distribution of stable production.

If $T_u - T_l > 6{\cdot}18\sigma$, there is room for the mean μ of the population to vary without appreciable danger of values transgressing the tolerances, for μ may be allowed to take any value which is less than $T_u - 3{\cdot}09\sigma$ and, at the same time, greater than $T_l + 3{\cdot}09\sigma$. The control chart for means may be modified to allow for this latitude in μ.

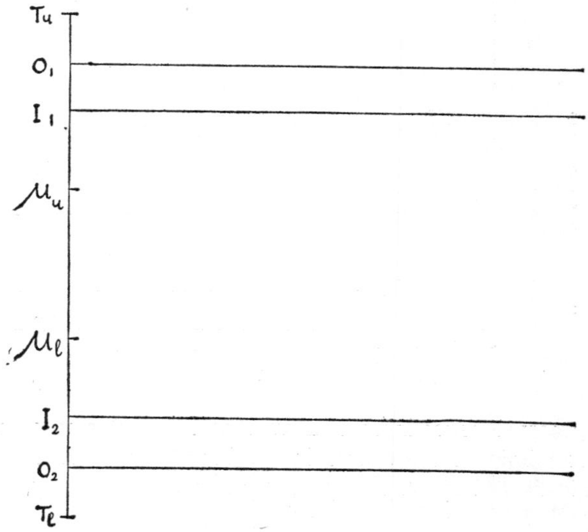

FIG. 25

$\mu_u = T_u - 3{\cdot}09\sigma = T_u - 3{\cdot}09a_n\bar{w}$ is the highest allowable value of μ

I_1 = upper 1 in 40 limits for mean of sample

$= \mu_u + A_{0{\cdot}025}\,\bar{w}$

O_1 = upper 1 in 1000 limit

$= \mu_u + A_{0{\cdot}001}\,\bar{w}$

Similarly $I_2 = \mu_l - A_{0{\cdot}025}\,\bar{w}$ and $O_2 = \mu_l - A_{0{\cdot}001}\,\bar{w}$ are the lower inner and outer limits, determined by the lowest allowable value of μ, namely $\mu_l = T_l + 3{\cdot}09\,a_n\bar{w}$.

In *A First Guide to Quality Control for Engineers* (H.M.S.O.) further numbers are tabulated by means of which O_1 and O_2 may be determined directly from T_u and T_l. It is however further recommended that an additional margin of safety be introduced by using yet another table which has the effect of drawing O_1 and O_2 a little closer together. We shall be content to use our tabulated values of a_n, $A_{0\cdot025}$, $A_{0\cdot001}$ and determine the modified control limits as in Fig. 25.

Ex. vi

With the data of Ex. iii on cylindrical cavities, suppose the tolerances are 0·080 in. and 0·145 in. Determine whether the machine is meeting its requirements, and if so calculate modified control limits.

We had $\overline{w} = 0{\cdot}016$ with samples of 4.

$T_u - T_l = 0{\cdot}065$ and must be greater than $6{\cdot}18\sigma$ for the machine to meet its requirements.

$$6{\cdot}18\sigma = 6{\cdot}18\, a_4\overline{w} = 6{\cdot}18\,(0{\cdot}486)\,(0{\cdot}016) = 0{\cdot}048$$
$$\mu_u = T_u - 3{\cdot}09\sigma = 0{\cdot}145 - 0{\cdot}024 = 0{\cdot}121$$
$$\mu_l = 0{\cdot}080 + 0{\cdot}024 = 0{\cdot}104$$
$$I_1 = 0{\cdot}121 + (0{\cdot}48)\,(0{\cdot}016) = 0{\cdot}129$$
$$I_2 = 0{\cdot}104 - (0{\cdot}48)\,(0{\cdot}016) = 0{\cdot}096$$
$$O_1 = 0{\cdot}121 + (0{\cdot}75)\,(0{\cdot}016) = 0{\cdot}133$$
$$O_2 = 0{\cdot}104 - (0{\cdot}75)\,(0{\cdot}016) = 0{\cdot}092$$

Comparing these results with those of Ex. v and Ex. iii we observe that although the interval between the control limits has been widened, the outer lower limit of 0·092 is higher than it was before. The explanation is that the mean $\overline{X}$ of Ex. v is below μ_l; in fact, if the tolerances are as now given, the machine needs re-setting.

7.6 Control chart for fraction or number defective

In some tests (for example, of ammunition, breaking strains or life lengths) the examination is destructive, and in other tests the measurement of a quality is lengthy or expensive. In such cases it is easier or even essential to express the results as *good* and *bad*, or *go* and *not go*. Quality control takes the form of control of the fraction defective, the object being to obtain warning of an increase in this fraction above the level established, on the basis of stable production, when the control scheme is introduced.

If $p\%$, sufficiently small, is the proportion defective in the bulk and n, sufficiently large, the size of sample, then the probability P of c or more defectives in a sample is given by the Poisson distribution as

$$P = 1 - e^{-m}\left(1 + m + \frac{m^2}{2\,!} + \ldots + \frac{m^{c-1}}{(c-1)\,!}\right)$$

where $m = np/100 =$ expected number of defectives in a sample. Curves are available (see *Quality Control Charts*, B.S.600R) representing this relation between P, m and c.

We may adopt $P = 0{\cdot}1$ as a control limit by drawing a line on the control chart at $c =$ number of defectives. We then expect, in stable conditions, that c or more defectives will occur once in ten samples. The table below shows the values of c corresponding to some values of n and p.

Number of defectives, c, to be taken as warning of possible deterioration.

Percentage Defective p

n \	0·5	1	2	3	4	5	6	7	8	9	10
5	*	*	1	2	2	2	2	2	2	2	2
10	*	1	2	2	2	2	3	3	3	3	3
20	1	2	2	3	3	3	4	4	4	5	5
30	2	2	3	3	4	4	5	5	6	6	6
40	2	2	3	4	4	5	5	6	7	7	8

* No values can be obtained here.

Ex. vii

The following data refer to visual defects found at inspection of the first 10 samples of size 20. Use them to obtain a control limit for number defective in samples of 20. Represent the first 10 samples in a chart.

Sample No.:	1	2	3	4	5	6	7	8	9	10
No. of defectives:	1	0	2	1	1	0	2	0	1	1

The total number of defectives is 9 in 200 items. The estimate of p for the bulk is therefore

$$p\% = 4{\cdot}5\%$$

The table gives $c = 3$ and the control chart is shown in Fig. 26.

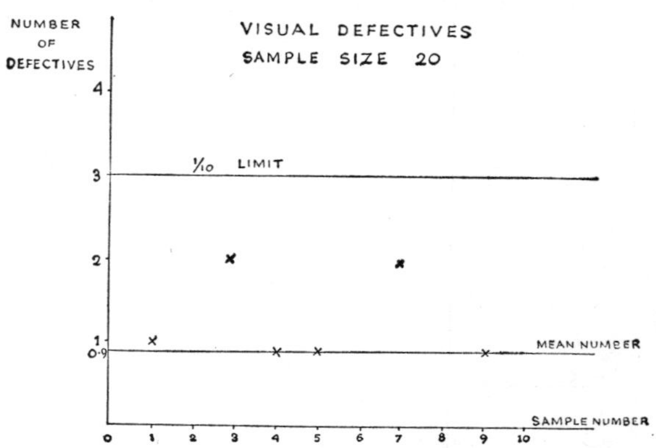

FIG. 26

7.7 Cut limits

It is evident that if the percentage defective is very low and it is desired to control the process at that level, then unless very large samples are taken the percentage defective could increase considerably before a single defective is found in a sample. To avoid increasing the size of sample unduly an alternative device is artificially to increase the number of defectives by narrowing the tolerance width.

Suppose that a measured dimension is in control, the process producing 0·25% high values and 0·25% low, when tested by go and not-go gauges. Assuming a normal distribution of the dimension, this means that the gauges are set at $\pm$ 2·8σ from the mean. With this setting a considerable change in the mean could occur before an appreciable number of defectives would be found, by failure to pass one or other of the gauges, in a sample of moderate size. If the gauges are re-set at $\pm$ 2σ from the mean, the proportion (in stable production) failing one or other gauge is increased to 4·6%; with gauges set at $\bar{x} \pm 1{\cdot}5\sigma$ it is 13·4%.

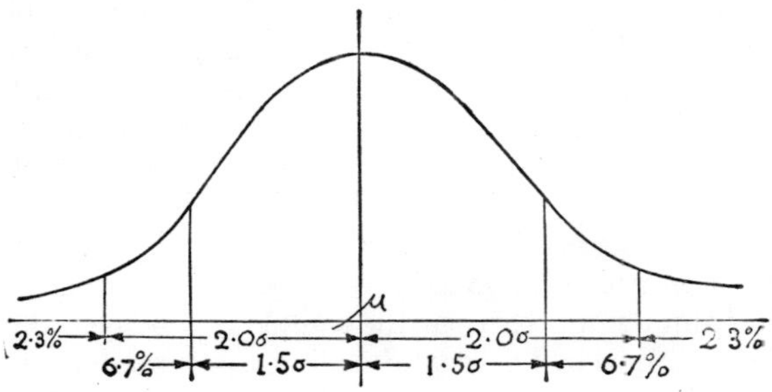

FIG. 27

The setting most sensitive to a change in mean is at approximately $\bar{x} \pm \sigma$ independently of n (the actual setting varies slightly with n). At this setting, in stable production, we expect about 32% defective (16% high and 16% low).

With a sample size of n, the chance of no high defectives is $(0{\cdot}84)^n$; the chance of 0 or 1 high defectives is

$$(0{\cdot}84)^n + n(0{\cdot}16)\,(0{\cdot}84)^{n-1}$$

the chance of 0, 1 or 2 high defectives is

$$(0{\cdot}84)^n + n(0{\cdot}16)\,(0{\cdot}84)^{n-1} + \frac{n(n-1)\,(0{\cdot}16)^2\,(0{\cdot}84)^{n-2}}{2!} \text{ etc.,}$$

or, in tabular form:

Chance of at most K High Defectives in a sample of n

n \ K	0	1	2	3	4	5	6	7
5	0·418	0·817	0·968	0·997	1·000	1·000	1·000	1·000
10	0·175	0·508	0·794	0·937	0·985	0·996	0·998	1·000
15	0·073	0·282	0·561	0·791	0·923	0·978	0·996	1·000

Hence for samples of size 10, the 1/5 high limit would be at 2 high 'defectives' and the (approximately) 1/500 high limit would be at 6 'defectives', and similarly for the low limits, where in this case 'defectives' are individuals that fail the modified gauges. High and low 'defectives' should be plotted on separate charts and a check can then be made on the direction of any shift in mean.

It is, of course, of first importance that the standard deviation should remain constant and a regular check on it should be made.

7.8 The difficulties of random sampling

One of the most important and difficult matters in random sampling is to ensure that sampling is at random. It is interesting to watch an operative sampling (supposedly) at random and, when it is possible, to note the positions or characteristics of the objects sampled. If the product is batched in trays and a random sample has to be made from the tray, many operators go for the corners while others select from the centre. Should the product be assembled in a chance manner on the trays, then such a bias in selection will not matter, but if there is likely to be any variability in systematic manner over the tray (e.g. a tray of pellets filled from a hopper), then the sampling will give a false result.

Even when the product is exhibited in purely random order, some inspectors develop an intuitive sense which enables them to pick out defectives for their (random) sample. This phenomenon has earned the name of 'foreman's fingers'.

One way of ensuring that sampling is truly at random is to use tables of random numbers. These tables consist of digits in an order which has been tested and found free from any systematic effects. A section of the table in Fisher and Yates's *Statistical Tables for Biological, Agricultural and Medical Research* (Oliver and Boyd) is found in Table 6. By assigning a number to each unit liable to be tested, reference to the table will ensure that a random selection is made. It should be stressed that it is not sufficient to glance at the table and just select any number that catches one's eye. Some people's eyes will catch 0's easily and other's go for 9's, etc.; it is best before selection to state where in the table of random numbers one is going to look or to work systematically through them.

Suggestions for further reading:

B.S.600R, *Quality Control Charts* (British Standards Institution).
A First Guide to Quality Control for Engineers (H.M.S.O.)
W. B. Rice, *Control Charts* (Wiley).

EXAMPLES ON CHAPTER VII

1. A machine manufactures plugs to a given diameter. Calculate control limits for mean diameter and range for this machine from the following data (ten samples of five plugs). Draw the charts for mean and range, plotting the first ten points of each.

Plug Diameters (inches)

Sample 1	2	3	4	5	6	7	8	9	10
0·488	·494	·498	·492	·491	·501	·497	·495	·504	·500
·489	·494	·515	·509	·509	·501	·492	·499	·505	·493
·505	·499	·487	·514	·499	·505	·496	·493	·513	·506
·494	·489	·505	·491	·509	·482	·503	·497	·493	·493
·500	·505	·511	·501	·498	·520	·492	·487	·503	·502

2. A specified dimension of a machined component is 0·05 in. Five components are measured every half-hour, and the following table gives the results for 20 consecutive samples, as deviations from the specified values in units of 0·0001 in. Make control charts for the mean and the range, marking on them appropriate limits. Discuss what evidence the chart gives of unstable production and recalculate limits for a part of the series for which the range appears to be in control.

Sample No.

1	2	3	4	5	6	7	8	9	10
+ 10	− 11	− 5	+ 7	− 4	− 1	+ 8	+ 11	− 3	+ 8
+ 17	− 1	− 14	+ 4	+ 1	+ 3	+ 9	+ 4	− 6	0
− 1	+ 10	+ 3	+ 5	+ 6	+ 12	+ 2	+ 8	− 4	+ 2
− 1	+ 10	+ 14	+ 4	+ 2	− 5	+ 3	+ 13	− 3	− 10
− 15	+ 9	− 5	+ 1	+ 7	+ 2	− 4	+ 11	− 8	− 3

Sample No.

11	12	13	14	15	16	17	18	19	20
− 1	− 4	+ 1	− 3	− 5	+ 4	0	− 4	− 2	− 8
− 8	+ 7	+ 4	− 4	+ 4	+ 12	− 3	− 15	− 2	− 2
− 1	− 1	+ 7	+ 11	+ 2	− 6	− 5	− 4	− 14	− 4
0	− 1	+ 5	+ 3	+ 13	+ 5	+ 8	− 6	− 5	− 2
− 4	− 7	+ 1	0	+ 1	+ 5	+ 13	− 4	+ 1	0

(R.S.S. Cert.)

3. If $\sigma = 0{\cdot}22$ in. find n so that an increase of 0·10 in. in the mean of the population will cause 1 in 4 values of the mean of samples of size n to fall above the inner control limit.

4. Calculate ordinary and modified control limits for the hardness of metal blocks, tolerances 900 VDH and 1100 VDH, by taking the following data in samples of 5. Draw the charts and plot the first ten points.

Block Hardness (VDH)

(1)	(2)	(3)	(4)	(5)	(6)	(7)	(8)	(9)	(10)
1006	998	1005	1010	983	987	1004	1008	997	981
1011	1026	1012	988	980	1009	1010	993	987	993
989	1019	1009	992	1003	985	1003	999	1004	995
1000	1002	1019	998	1004	987	1011	999	999	1007
989	998	1002	1000	991	1002	1003	999	993	986

5. In the above process subsequent data were obtained as follows. Plot the points on both charts and at a point at which you think a change in the process has occurred state whether you think it is in mean or variability. Recalculate the limits from the remaining data and plot the points.

Block Hardness (VDH)

(11)	(12)	(13)	(14)	(15)	(16)	(17)	(18)	(19)	(20)
1012	1008	1002	1009	1022	1009	1009	1023	1035	1036
1001	1001	1023	1003	1014	1014	1007	1022	1008	1015
1014	996	1010	1021	1017	1057	1006	1016	1020	1022
993	991	1014	1021	1023	1007	1009	1018	1010	1037
986	996	998	999	1014	1048	1000	1010	1022	1024

6. In the process of Questions 4 and 5 the testing gave the following data for samples 21 to 30. At what point do you think a change in the process occurs? At the point recalculate the limits based on the remaining data and test whether modified control charts can still be used.

Block Hardness (VDH)

(21)	(22)	(23)	(24)	(25)	(26)	(27)	(28)	(29)	(30)
1024	1020	1021	1030	1038	1049	1030	1055	1024	1031
1000	1060	1047	1015	1052	1031	1045	1025	1067	1020
1019	1035	1031	1039	1040	1050	1033	1031	1049	1041
1030	1039	1059	1042	1067	1062	1050	1042	1060	1050
1015	1047	1042	1040	1025	1032	1075	1060	1036	1069

7. Calculate control limits for fraction defective charts based on the following defectives found at visual sample examination of insulating plugs, sample size 10.

Sample No.:	1	2	3	4	5	6	7	8	9	10
No. of defectives:	0	0	1	1	0	2	0	1	0	0

8. In Question 7, the subsequent data were:

Sample No.:	11	12	13	14	15	16	17	18	19	20
No. of defectives:	1	1	2	3	2	2	1	2	0	2

At what stage do you consider the process level has changed? Recalculate the limits, based on the remaining data.

9. Calculate cut limits for a process in which the mean diameter of cylinder blocks is 0·114 in., the standard deviation 0·0012 in. and the tolerances 0·125 in. and 0·103 in.

10. Calculate cut limits for go, no-go gauges based on the following cone lengths (tolerances 9·94 cm. and 10·06 cm.).

Cone Lengths (*cm.*)

10·013	10·012	9·992	9·988	9·984	9·992	10·014	10·006	10·000
10·008	10·014	10·007	9·996	10·001	10·002	9·994	10·011	9·980
9·999	10·016	10·001	10·009	10·019	10·035	9·980	10·017	10·013
9·992	9·991	9·983	10·005	10·002	10·000	10·014	10·002	10·013
9·986	10·009	10·009	9·995	9·993	10·004	9·992	9·997	10·001
9·994	10·010	9·989	9·987	9·978	10·010	9·998	10·010	10·009
10·000	9·998	9·987	10·007	9·980	10·011	9·985	9·982	9·999
10·005	10·007	10·008	10·014	10·011	9·982	9·986	9·997	9·994

CHAPTER VIII

SAMPLING INSPECTION SCHEMES

Single sampling. Choice of scheme. Double sampling. Average sample size. Sequential sampling. Lattice diagram and boundary lines. Extension to continuous variable.

8.1 Purpose of sampling inspection

A manufacturer receiving or despatching lots (batches) of a given component is often not in a position to inspect every item. All he can do is to inspect a sample and so form an estimate of the quality of the lot. The type of inspection we are considering consists of grading an item as 'good' or 'bad', and the purpose is not so much to estimate within precise limits the proportion defective in the lot as to guard against receiving or despatching lots likely to contain more than a certain proportion defective.

There are many types of sampling scheme. Three are considered in this chapter, namely single, double and sequential sampling. It will be assumed throughout that the lot size is sufficiently large compared with the sample size for the chance of drawing a defective from the lot to remain constant throughout the sampling.

8.2.1 Single sampling. Operating characteristic

The simplest sampling scheme consists of taking a random sample of chosen size and accepting the lot if the sample contains fewer than a chosen number of defectives. Such a scheme might be: take a random

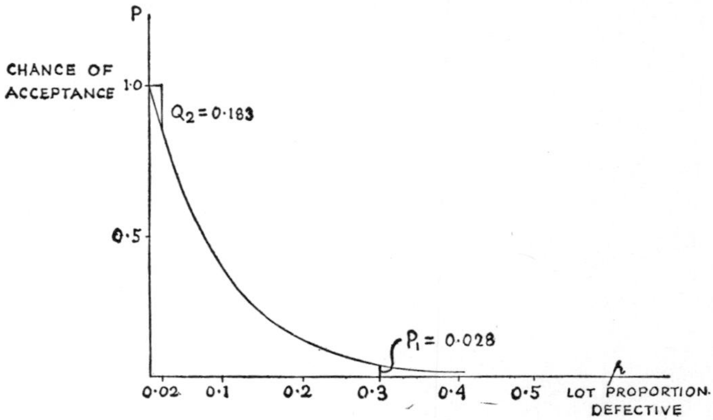

FIG. 28

sample of ten; if all are good, accept the lot, otherwise reject the lot.

The chance of acceptance depends on the (unknown) proportion of defectives in the lot. If p is the proportion defective and P the chance of acceptance, we can determine P as a function of p. The graph of P against p is called the operating characteristic of the scheme.

In our example the chance of acceptance, i.e. of all items in a sample of ten being good, is q^{10} where $q = 1 - p$. Calculating P for various values of p, we find that P becomes negligible, to three decimal places, when p is as high as 0·6, and as low as 0·028 when $p = 0{\cdot}3$.

p:	0	0·02	·04	·06	·08	·10	·15	·20	·25	·30	·40	·50
$P = (1 - p)^{10}$:	1	0·817	·667	·538	·434	·348	·197	·108	·056	·028	·006	·001

Thus, although acceptance implies a sample of ten containing no defectives, there is

a small chance, $P_1 = 0{\cdot}028$, of accepting lots containing 30% defective;
a small chance, $Q_2 = 0{\cdot}183$, of rejecting lots containing only 2% defectives.

These two risks, of accepting lots which should be rejected and of rejecting lots which should be accepted, are inevitable in sampling inspection. Fig. 28 shows the graph of the operating characteristic.

What would be the ideal form of the operating characteristic? We should like to be able to say that *all* lots containing less than a chosen proportion defective would be accepted, while all lots containing more than that proportion would be rejected. The chosen proportion would probably not be the same for consumer as producer, i.e. buyer and seller, but each would like an operating characteristic of the form of Fig. 29.

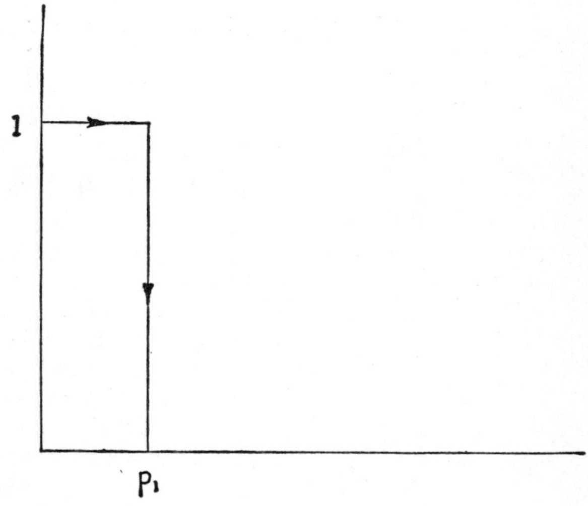

FIG. 29

Here all lots containing less than p_1 defective are accepted and all containing more than p_1 rejected. This ideal form is impossible, but a scheme is efficient if its characteristic curve shows a sharp fall from a high chance of acceptance to a low chance.

Ex. i

A single sampling scheme allows up to and including 2 defectives in a sample of 20. Draw its operating characteristic. What lots have (*a*) more than 95% chance of acceptance, (*b*) less than 10% chance of acceptance?

$$P = \text{chance of 0, 1 or 2 defectives}$$
$$= q^{20} + 20\,q^{19}p + 190\,q^{18}p^2,$$

the sum of the first three terms of a binomial distribution. The graph of P against p is shown in Fig. 30.

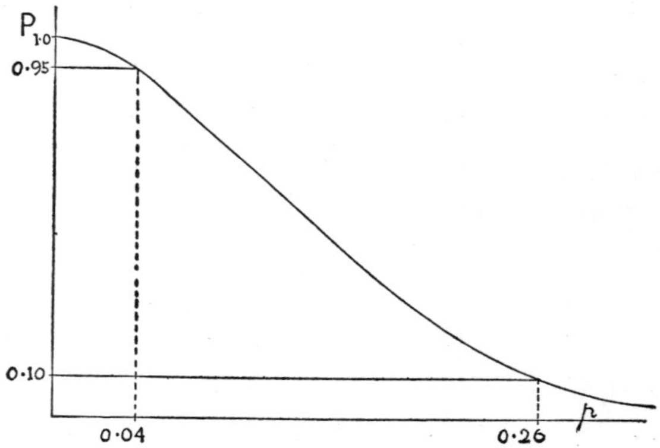

FIG. 30

Lots containing fewer than 4% defective have more than 95% chance of acceptance.

Lots containing more than 26% defective have less than 10% chance of acceptance.

8.2.2 Choice of single sampling scheme

So far we have deduced the operating characteristic of given schemes, but in practice the converse problem is more likely to arise. Given preassigned chances of acceptance for certain fractions defective, what should be the size of sample and the allowable number of defectives?

The equation of the operating characteristic is $P = f(p)$, where $f(p)$ depends on the two parameters: n, the size of sample, and c, the allowable number of defectives. It is therefore to be expected that the curve can be made to pass through *two* chosen points; the conditions that it shall do so give two equations from which to determine n and c.

* Suppose given a proportion p_1 for which the chance of acceptance is to be P_1, a low probability, and a proportion p_2 for which the chance

of acceptance is to be P_2, a high probability. As indicated in Figs. 28 and 30, $p_1 > p_2$, but we shall assume both p_1 and p_2 sufficiently small for Poisson's distribution to be valid as an approximation to the binomial distribution.

$$P_1 = e^{-np_1}\left\{1 + np_1 + \frac{(np_1)^2}{2!} + \ldots + \frac{(np_1)^c}{c!}\right\} \quad . \quad (1)$$

We can obtain a concise expression for the sum on the R.H.S. as an integral. By repeated integration by parts

$$\int_0^x t^c e^{-t}dt = \Big[-t^c e^{-t}\Big]_0^x + \int_0^x ct^{c-1}e^{-t}dt$$

$$= -x^c e^{-x} + c\Big[-t^{c-1}e^{-t}\Big]_0^x + c(c-1)\int_0^x t^{c-2}e^{-t}dt$$

$$= -e^{-x}\left\{x^c + cx^{c-1} + c(c-1)x^{c-2} + \ldots + c!x\right\} + c!\int_0^x e^{-t}dt$$

$$= -e^{-x}\{x^c + cx^{c-1} + \ldots + c!x + c!\} + c!$$

Dividing by $c!$ and transposing,

$$e^{-x}\left\{1 + x + \frac{x^2}{2!} + \ldots + \frac{x^c}{c!}\right\} = 1 - \frac{1}{c!}\int_0^x t^c e^{-t}dt$$

Put $x = np_1$ and substitute in equation (1)

$$P_1 = 1 - \frac{1}{c!}\int_0^{np_1} t^c e^{-t}dt$$

*

$$Q_1 = 1 - P_1 = \frac{1}{c!}\int_0^{np_1} t^c e^{-t}dt \quad . \quad . \quad . \quad (2)$$

Similarly

$$Q_2 = 1 - P_2 = \frac{1}{c!}\int_0^{np_2} t^c e^{-t}dt \quad . \quad . \quad . \quad (3)$$

Given p_1, p_2, P_1, P_2, we require to find n, c. The integrals could be evaluated, given p_1, p_2, n, c by means of the Incomplete Gamma function which has been tabulated, and it is possible to use the tables to find values of n, c which satisfy equations (2) and (3) approximately. (We must, of course, have integers for n, c, so that the solution cannot be exact.)

Graphs have been prepared from which the solutions can be read when $P_1 = Q_2$, that is, when the two risks, of accepting bad lots and rejecting good lots, are made equal.

We can, however, solve the equations by means of the χ^2 table (Table 2) for the integrals involved are associated with the χ^2 distribution. The rule is as follows:

Set $\chi_1^2 = 2np_1$ and $P_1 =$ the level of significance of χ_1^2
$\chi_2^2 = 2np_2$ and $P_2 =$ the level of significance of χ_2^2
with $\nu = 2c + 2$

This means that, taking the two chances of *acceptance* as levels of significance in the χ^2 table, we seek two values of χ^2 at these two different levels of significance which are in the ratio of p_1 to p_2, for the same *even* number of degrees of freedom, ν. Either value of χ^2 then gives n and ν gives c.

Ex. ii

Design a single sampling scheme such that lots containing 2% defective have 95% chance of acceptance, while lots containing 8% defective have 5% chance of acceptance.

Here $p_1 = 0{\cdot}08$, chance of acceptance 0·05
$p_2 = 0{\cdot}02$, chance of acceptance 0·95

In the χ^2 table we require two values of χ^2 in the same line, under significance levels 95% and 5% which are in the ratio 1 : 4. The nearest we can find are

95% 5%

$\nu = 12 \quad \chi_2^2 = 5{\cdot}226 \quad \chi_1^2 = 21{\cdot}026$

Put $2c + 2 = 12, c = 5$

$2np_2 = \chi_2^2$ i.e. $0{\cdot}04n = 5{\cdot}23$, $n = 131$

($2np_1 = \chi_1^2$ gives $0{\cdot}16n = 21{\cdot}03$, $n = 131$)

The required scheme is: allow up to 5 defectives in samples of 131.

As a rule, if the two chances of acceptance are now calculated for the scheme, they will differ slightly from the stipulated probabilities.

Ex. iii

Find a single sampling scheme in which the risk of accepting lots containing 10% defectives is 5% and the risk of rejecting lots containing 1% defectives is 1%.

The two values of χ^2 are to be in the ratio 1 : 10. We find

99% 5%

$\nu = 8 \quad \chi^2 = 1{\cdot}646 \quad \chi^2 = 15{\cdot}51$

The ratio obtained is not so accurate as in Ex. ii. If we use the first value of χ^2 to determine n we shall be strengthening our guarantee against accepting bad lots, for $\chi^2 = 16{\cdot}46$ would correspond to a level of significance lower than 5%.

Suppose we adopt the scheme given by

$$2c + 2 = 8 \quad 0{\cdot}02n = 1{\cdot}646$$

Then $c = 3$, $n = 82$.

If we now calculate the chances of acceptance of lots with 1% and 10% defectives we find them to be 0·990 and 0·030.

The reader may now with advantage work numbers 1 to 6 of the examples at the end of the chapter.

8.3.1 Double sampling schemes

In double sampling, a first sample is not necessarily decisive. Besides acceptance or rejection we allow the possibility of taking a second sample which, taken in conjunction with the first, is then to be decisive. Such a scheme might be: take a random sample of 10; if no defectives are found, accept the lot; if 4 or more defectives are found, reject the lot; but if 1, 2 or 3 defectives are found, take a second sample of 10; if the total of defectives in the two samples is 4 or more, reject the lot, but otherwise accept the lot. The scheme may be shown diagrammatically:

Take a sample of 10

If no defectives, *accept.*

If 4 or more defectives, *reject.*

If 1, 2 or 3 defectives, *take a second sample of* 10

If 3 or less defectives in combined sample of 20, *accept.*

If 4 or more defectives in combined sample of 20, *reject.*

To obtain the operating characteristic of this scheme we proceed to calculate the probability P, that a lot containing a proportion p of defectives will be accepted.

The total chance of acceptance $= P_1 + P_2$, where $P_1 = (1 - p)^{10} = q^{10} =$ chance of acceptance on first sample, and $P_2 =$ chance of acceptance on second sample.

Now $P_2 =$ (chance of 1 in first 10) (chance of 2 or less in second 10)
+ (chance of 2 in first 10) (chance of 1 or less in second 10)
+ (chance of 3 in first 10) (chance of 0 in second 10)

$$= 10q^9p(q^{10} + 10q^9p + 45q^8p^2) + 45q^8p^2(q^{10} + 10q^9p) + 120q^7p^3q^{10}$$

$$= 5q^{17}p(2q^2 + 29qp + 204p^2)$$

$$= 5q^{17}p(2 + 25p + 177p^2)$$

p:	0	0·02	0·04	0·06	0·08	0·10	0·15	0·20	0·25	0·30	0·40
$P = P_1 + P_2$:	1	0·999	0·993	0·971	0·934	0·868	0·657	0·417	0·249	0·104	0·020

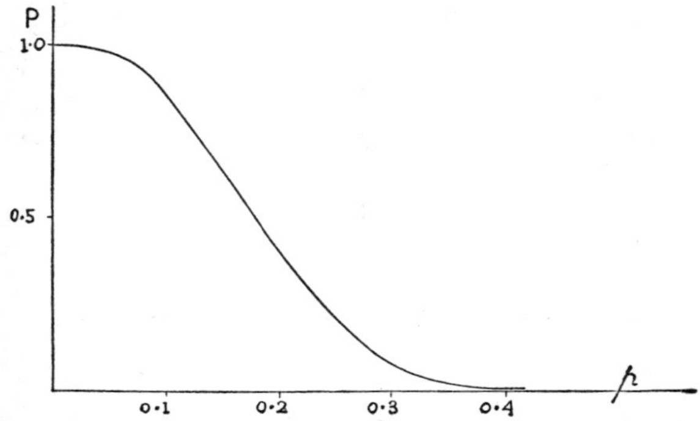

FIG. 31

In this scheme there is a risk of about 10% of accepting lots containing more than 30% defectives and a risk of about 10% of rejecting lots containing fewer than 9% defectives.

8.3.2 Average sample size

With reference to the above scheme, the reader might be inclined to suggest that it would be simpler to take a single sample of 20 and accept the lot if the sample contains 3 or less defectives. The answer is that it is as well to take advantage of the quick decisions afforded by a first sample of 10 containing no defective, which makes it highly probable that the lot is good. The testing which constitutes the inspection may be expensive of time or material. Now a feature of repeated sampling as opposed to single sample is that we do not know beforehand whether the second sample will be necessary, which makes it more difficult to estimate the costs of inspection. In order to make such an estimate, we can calculate how the average sample size depends on the quality of the lot. In double sampling with 10 in each sample, for each value of p there will be an average sample size $10 + 10R$, where R is the probability of taking a second sample, for it is certain that the whole sample taken includes the first 10 and there is a probability R that it will include a further 10.

In the example of §8.3.1,

R = probability of 1, 2 or 3 defectives in first sample

$= 10q^9p + 45q^8p^2 + 120q^7p^3 = 5q^7p(2 + 5p + 17p^2)$

and $\bar{n}$ = average sample size

$= 10(1 + R)$

A feature of the graph of $\bar{n}$ against p is that $\bar{n}$ has a maximum value for a certain value of p, so that we can at least know the maximum amount of sampling to be expected under the scheme.

p:	0	0·05	0·10	0·15	0·20	0·25	0·30	0·35
$\bar{n}$:	10	14·0	16·4	17·5	17·7	17·2	16·2	15·0

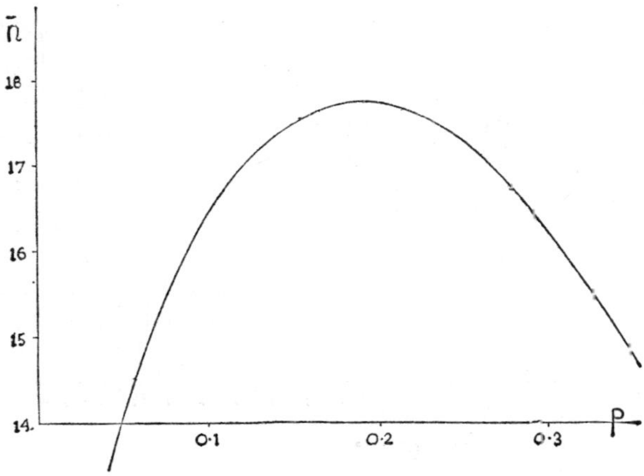

FIG. 32

Ex. iv

For each lot of a certain product, n articles are taken at random by Inspector A. He passes the lot unless there is more than one defective article amongst the n he has chosen. Inspector B follows the same procedure.

If the two inspectors agree, the batch is finally accepted or rejected as the case may be. If they disagree, a further random sample of $2n$ articles is taken. The batch is accepted unless there are more than 2 defectives among the $2n$ chosen. Denoting the proportion of defectives in a lot by p, obtain expressions for

(i) the probability that the lot is accepted;
(ii) the average number of articles examined per batch.

(R.S.S. Cert.)

For the two samples of n, the chances are:

	Pass	*Refer*
Inspector A	p_1	$1 - p_1$
Inspector B	p_1	$1 - p_1$

where p_1 = chance of 0 or 1 defective

$$= q^n + nq^{n-1}p.$$

The inspectors agree and accept with probability p_1^2
The inspectors agree and reject with probability $(1 - p_1)^2$
The inspectors disagree with probability $2p_1(1 - p_1)$.

In the last event the further sample of $2n$ is taken and the probability that it leads to acceptance is

$$p_2 = q^{2n} + 2nq^{2n-1}p + n(2n - 1)q^{2n-2}p^2$$

(i) The total probability that the lot is accepted is

$$p_1^2 + 2p_1(1 - p_1)p_2.$$

(ii) The average number of articles examined is

$$2n(1 + 2p_1(1 - p_1))$$

Both answers may be evaluated in terms of p.

8.4.1 Sequential sampling

A logical extension of double sampling is to a sequence of three or more possible samples. Each sample may consist of a small group or of a single item, and after each sample a decision is made (*a*) to accept the lot, or (*b*) to reject the lot, or (*c*) to continue sampling. The process continues until the decision is either (*a*) or (*b*).

A method of recording the sequence of results obtained in drawing single items, each of which will be classed as either good or bad, consists of a graph known as a lattice diagram. Perpendicular axes are labelled G (good) and B (bad). Starting at the origin, move one unit horizontally whenever a good item is drawn, and one unit vertically whenever a bad item is drawn. Thus the sequence

$$G, G, G, B, G, B, G, G, G, B, B, G, G, G, G, B, G$$

produces the diagram of Fig. 33.

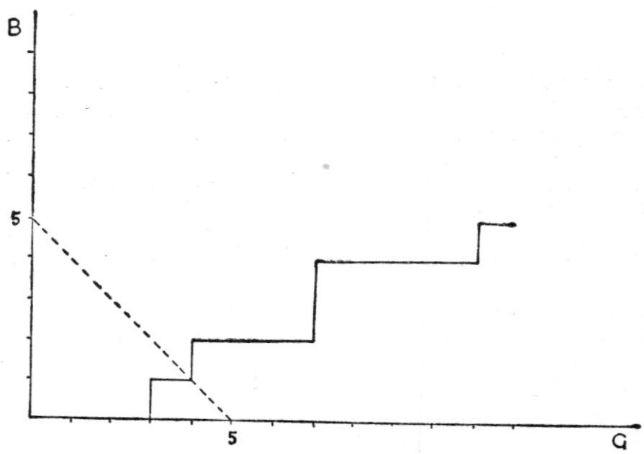

FIG. 33

When n items have been drawn, the point reached is on the line $B + G = n$, whatever the sequence of good and bad items.

The decision at each stage as to acceptance, rejection or continued sampling is made dependent on reaching or crossing one or other of two parallel straight lines, dividing the whole quadrant into three zones.

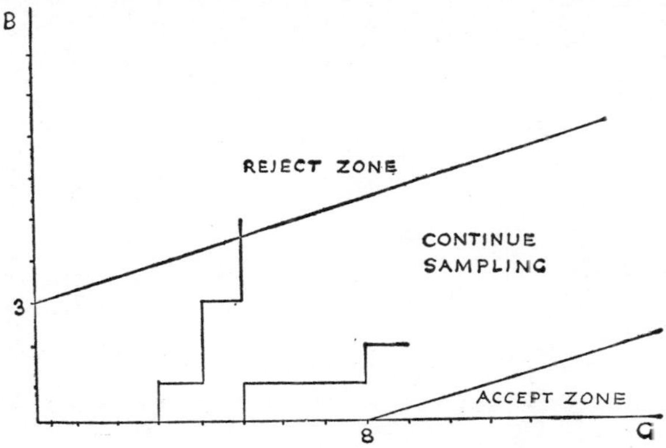

FIG. 34

In the example in Fig. 34,
an initial run of 8 good items results in acceptance;
an initial run of 3 bad items results in rejection;
the run $G, G, G, B, G, B, B, G, B, B$ results in rejection;
the run $G, G, G, G, G, B, G, G, G, B, G$ is indecisive.

8.4.2 Risks of error. Operating characteristic

Before giving formulae by which the boundary lines may be determined, we may note that the principle of their determination is that used (§8.2.2) in the choice of a single sampling scheme. In all cases in which the lot is not completely inspected there are risks of two types of error, namely of accepting a lot which should be rejected and of rejecting a lot which should be accepted. (These might be called the consumer's and producer's risks respectively.) The boundary lines are to be fixed so that there is an assigned small risk P_1 of accepting a lot containing a given proportion p_1 of defectives and another assigned small risk Q_2 of rejecting a lot containing another given proportion p_2 $(< p_1)$ of defectives. Lots with this lower proportion of defectives are to have a high chance, P_2 $(= 1 - Q_2)$, of acceptance.

It can then be shown that the boundaries of the zones are given by the equations

$$\left.\begin{aligned} B &= a_1 + sn \\ B &= a_2 + sn \end{aligned}\right\} \quad . \quad . \quad . \quad , \qquad (4)$$

where $n = B + G$, the total number of items drawn,

and

$$\left.\begin{aligned} a_1 &= \log\frac{1-Q_2}{P_1}\Big/\log\frac{p_1(1-p_2)}{p_2(1-p_1)} \\ a_2 &= \log\frac{1-P_1}{Q_2}\Big/\log\frac{p_1(1-p_2)}{p_2(1-p_1)} \\ s &= \log\frac{1-p_2}{1-p_1}\Big/\log\frac{p_1(1-p_2)}{p_2(1-p_1)} \end{aligned}\right\} \quad . \quad . \quad (5)$$

In the particular case when $P_1 = Q_2$, i.e. the chosen risks of errors of the two kinds are equal, then

$a_1 = a_2 = a$(say) and the equation

$$p = \left\{\left(\frac{P}{1-p}\right)^{s/a} - 1\right\}\Big/\left\{\left(\frac{P}{1-p}\right)^{1/a} - 1\right\} \quad . \quad (6)$$

is that of the operating characteristic, but with P as independent and p as dependent variable.

Ex. v

A sequential sampling scheme is to give a chance of 0·01 of accepting lots with 20% defective and a chance of 0·99 of accepting lots with 10% defective. Sketch the lattice diagram boundaries and the operating characteristic.

Here $P_1 = Q_2 = 0{\cdot}01$, $p_1 = 0{\cdot}20$, $p_2 = 0{\cdot}10$
$a_1 = a_2 = \log 99/\log 9/4 = 5{\cdot}667$
$s = \log 9/8/\log 9/4 = 0{\cdot}1451$

The boundaries are

$$B = -5{\cdot}67 + 0{\cdot}145\,(B + G)$$

i.e. $0{\cdot}855B = 0{\cdot}145G - 5{\cdot}67$

$B = 0{\cdot}170G - 6{\cdot}63$ (accept),

and $B = 0{\cdot}170G + 6{\cdot}63$ (reject).

Hence the lattice diagram with its boundaries is as in Fig. 35.

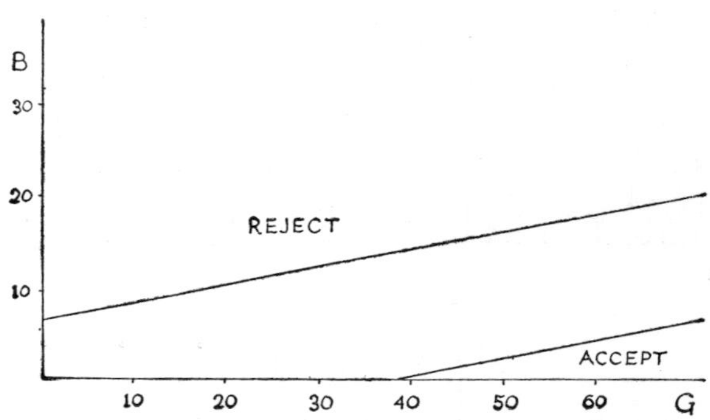

FIG. 35

For the operating characteristic, since $s/a = 0{\cdot}0256$ and $1/a = 0{\cdot}1765$,

$$p = \left\{ \left(\frac{P}{1-P}\right)^{0\cdot0256} - 1 \right\} \Big/ \left\{ \left(\frac{P}{1-P}\right)^{0\cdot1765} - 1 \right\}$$

from which we calculate the values in the following table

P:	0·9999	0·999	0·99	0·9	0·75	0·5†	0·25	0·1	0·01	0·001	0·001
p:	0·065	0·081	0·1	0·122	0·133	0·145	0·157	0·170	0·2	0·23	0·261

† When $P = 0{\cdot}5$, the expression for p is indeterminate, but it is easily shown that as $P \to 0{\cdot}5$, $p \to s = 0{\cdot}145$, in this case.

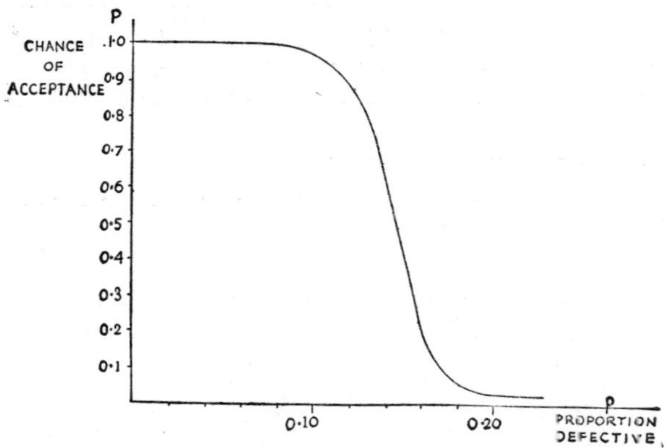

FIG. 36

8.4.3 Average sample size

As in the case of double sampling, it is useful to estimate the amount of sampling which is likely to be necessary. If a batch contains a high proportion, much greater than p_1, of defectives we shall expect the sequence of points in the lattice diagram to move quickly into the 'reject' zone; if the proportion defective is small compared with p_2, the 'accept' zone is likely to be quickly reached. It is for proportions between p_1 and p_2 that sampling may have to be continued for some time before one of the boundaries is crossed.

Denoting by $\bar{n}_p$ the average sample size when the proportion defective is p, it can be shown that

$$\bar{n}_p = \frac{P(a_1 + a_2) - a_2}{s - p} \quad . \quad . \quad . \quad (7)$$

When

$$p = s,\ P = \frac{a_2}{a_1 + a_2} \quad . \quad . \quad . \quad . \quad . \quad (8)$$

and the R.H.S. of (7) becomes indeterminate, its true value being

$$\bar{n} = \frac{a_1 a_2}{s(1 - s)} \quad . \quad . \quad . \quad . \quad . \quad . \quad (9)$$

Ex. vi

Tabulate values of $\bar{n}_p$ against p for the scheme of Ex. v. What are the largest expected size of sample and the value of p for which it occurs?

$$a_1 = a_2 = 5{\cdot}67,\ s = 0{\cdot}145$$

$$\bar{n}_p = \frac{11{\cdot}34P - 5{\cdot}67}{0{\cdot}145 - p}\ (p \neq s).$$

$$\bar{n}_s = \frac{5{\cdot}67^2}{(0{\cdot}145)\,(0{\cdot}855)} = 259{\cdot}3.$$

Using the values of P, p obtained in Ex. v, we now obtain:

p:	0·065	0·081	0·100	0·122	0·133	0·145	0·157	0·170	0·200	0·230	0·261
$\bar{n}_p$:	71	89	124	197	236	259	236	181	101	67	49

In this table, the largest value of $\bar{n}_p$ is 259, when $p = s$.
It is easily verified that $d\bar{n}_p/dp = 0$ when $p = s$, using equation (8).

8.4.4 Barnard's handicap method

The theory and methods of sequential analysis were developed simultaneously by Wald in U.S.A. and Barnard in U.K. In place of the lattice diagram Professor Barnard devised a method using two parameters, a 'handicap' and a 'penalty', determined by the chosen values of p_1, p_2, P_1, Q_2. Inspection commences with an initial score, equal to the handicap. For each good item drawn, 1 is added to the score; for each bad item the score is reduced by subtracting the penalty. If the score reaches twice the handicap, the lot is accepted; if the score reaches zero, the lot is rejected. The score may be kept on a cribbage board. Starting at the centre, the peg is moved one hole to the right for a good item and b holes to the left, where b is the penalty, for a bad item. A decision is reached when the peg has reached H holes, where H is the handicap, either to the left or to the right of the centre. (Charts from which values of H and b can be read were published by the Ministry of Supply under the title *Target-Handicap Charts in Sampling Inspection*, but may not now be available.)

8.5 Extension of sequential sampling to continuous variable

The classification of individual articles as good or bad may be the only practical one, that is to say the quality cannot be measured on a scale, but the fact of a visible (or audible, or other) defect renders an article 'bad'. When an article is graded as good or bad by means of go and not-go gauges, the dimension concerned could be measured accurately. The methods of sequential sampling have been extended to deal with quality as estimated by such measurable quantities. We consider as illustration a single problem and its generalization.

Suppose that a manufacturer has made batteries over a long period and knows that the distribution of battery life-time (under test conditions) is normal with mean 2,000 min. and s.d. 100 min. A new process for making the batteries is proposed which, it is thought, should raise the mean life to 2,500 min. without altering the standard deviation. It is important not to undertake the expense of changing the manufacturing process if the new mean is not reached and the acceptable risk of making such a wrong decision is placed at 0·001. On the other hand, it is important not to reject the new process if the new mean is reached and the acceptable risk of that error is placed at 0·01.

A number of batteries are now made by the new process and subjected to test. The sequential procedure is designed to accept or reject the hypothesis that the mean life has increased as soon as the cumulative evidence of a succession of the new batteries justifies a decision.

Let $m_1 =$ the smaller mean (2,000 min.)
$m_2 =$ the larger mean (2,500 min.)
$\alpha =$ the risk of accepting m_2 when m_1 is true (0·001)
$\beta =$ the risk of accepting m_1 when m_2 is true (0·01)
$\sigma =$ the s.d. (100 min.)
$x =$ a reading of the variate (life in min.)
$n =$ the number of readings.

On a lattice diagram, we could plot Σx against n, that is a sequence of sample points

$$(1, x_1), (2, \overline{x_1 + x_2}), (3, \overline{x_1 + x_2 + x_3}) \text{ and so on.}$$

As in sampling for defectives, it is now required to determine parallel boundaries the crossing of which will lead to a decision.

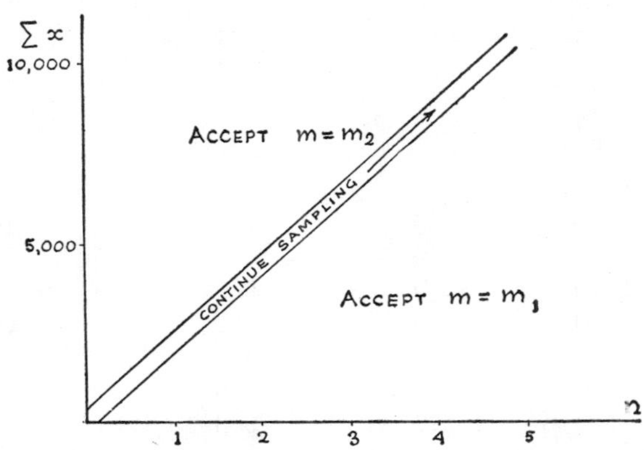

FIG. 37

The equations of the boundary lines are

$$\Sigma x = sn - a_1, \text{ (accept } m = m_1)$$

$$\Sigma x = sn + a_2, \text{ (accept } m = m_2)$$

Where

$$a_1 = \frac{\sigma^2}{m_2 - m_1} \log_e \frac{1 - a}{\beta}$$

$$a_2 = \frac{\sigma^2}{m_2 - m_1} \log_e \frac{1 - \beta}{a} \quad . \quad . \quad . \quad (10)$$

$$s = \frac{m_1 + m_2}{2}.$$

In our example

$$a_1 = \frac{100^2}{500} \log_e \frac{0{\cdot}999}{0{\cdot}01} = 92{\cdot}08$$

$$a_2 = \frac{100^2}{500} \log_e \frac{0{\cdot}99}{0{\cdot}001} = 137{\cdot}95$$

$$s = 2250$$

The boundary lines are therefore

$$\Sigma x = 2250n - 92; \text{ accept } m = 2000.$$
$$\Sigma x = 2250n + 138; \text{ accept } m = 2500.$$

Now it will be found inconvenient, on account of difficulties of scale, to record the successive sampling points graphically. We need only tabulate the values of Σx observed and compare them with the boundary values of Σx for each n. In the following table we suppose the successive observed values of x to be as in the second column; they result in the acceptance of the higher mean at the eighth observed value.

n	x	Σx	*Accept* m_1 *if* Σx *less than*	*Accept* m_2 *if* Σx *greater than*
1	2200	2200	2158	2388
2	2350	4550	4408	4638
3	2210	6760	6658	6888
4	2210	8970	8908	9138
5	2400	11370	11158	11388
6	2200	13570	13408	13638
7	2300	15870	15658	15888
8	2350	18220	17908	18138

It will be noted that if sampling continues for long the accepted mean will be m_1 or m_2 according as $\Sigma x/n$ is slightly below or slightly above $(m_1 + m_2)/2$; in other words, the best estimate of the true mean on the basis of the sample is very near $(m_1 + m_2)/2$, in fact, neither m_1 nor m_2. This arises from the form of the statistical test which only compares the probabilities of the particular values. If the object is to confirm or otherwise an acceptable increase d in the mean, it would seem reasonable to set $m_2 = m_1 + 2d$. Thus in our example we may suppose that the manufacturer will adopt the new process if there is an increase of *at least* 250 min. in the expected life.

Suggestions for further reading:

A. Wald, *Sequential Analysis* (Chapman and Hall).
H. E. Dodge and H. G. Romig, *Sampling Insection Tables* (Chapman and Hall).
Columbia University, *Sampling Inspection.*

EXAMPLES ON CHAPTER VIII

1. Which of the following two schemes is (i) more likely to accept lots containing 2% defectives, (ii) more likely to reject lots containing 20% defectives? Sketch their operating characteristics.

   Scheme A. Allow 0 or 1 defective in a random sample of 8.
   Scheme B. Allow 0, 1 or 2 defectives in a random sample of 16.

2. If a sampling scheme accepts a lot only when there are no defectives in a sample of 30, what proportion of lots containing 5% defectives will be accepted?

3. Batches of goods, each containing 1000 items, are inspected by a scheme based on taking a sample of 10 items and accepting those batches which give samples free from defectives. Find the probability that a batch will be accepted if it contains (*a*) 1% of defectives, (*b*) 20% of defectives.
   Among a large number of batches submitted for inspection, 9/10 have 1% of defectives, while the remaining 1/10 have 20% of defectives in them. Find the average percentage of defectives among all the accepted batches. (R.S.S. Cert.)

4. In a certain factory the decision to accept or reject large batches of items is based on the following sampling inspection scheme: a random sample of n items is taken from each batch and the batch is accepted if, and only if, the sample contains less than two defective items.
   Using the Poisson approximation (or otherwise), find the smallest sampling size which would ensure that when the true fraction defective in a batch is 5%, the probability of accepting the batch is 0·04. (L.U.)

5. Design a single sampling scheme in which there is a probability of 0·98 of accepting lots with 1% of defectives and a probability of about 0·02 of accepting lots with 8% of defectives.
   Calculate, for your scheme, the actual value of the second probability.

6. In a single sampling scheme the allowable risks are 1 in 100 of rejecting lots with only 2% of defectives, and 1 in 20 of accepting lots with 8% of defectives. What should be the size of sample and how many defectives should be allowed in the sample?

7. Find the operating characteristic and average sample size curve of the following scheme. Take a first sample of 20; if 0 or 1 defect occurs, accept immediately; if 4 or more defects occur, reject immediately; if 2 or 3 defects occur, test a further sample of 20 and if, in the combined sample of 40, less than 5 defects occur, accept the lot; otherwise reject the lot. Read off the chance of accepting lots with 5% defectives and the average sample size of such lots.

8. Draw a lattice diagram with the given boundary zones and determine the acceptance or rejection of the following sequences.

Boundary zones: $G = 9 + 4B$, accept;
$G = -16 + 4B$, reject.

Sequence (*a*) *G, G, G, G, B, G, G, G, G, G, G, G, G, G, B, G, G, G, G, B, G, G, G, G, G, B, G*

Sequence (*b*) *G, G, G, B, G, B, B, G, G, G, B, G, G, B, G, B, G, G, G, G, G, B, G, B, G, B, G, G, B, B, B, G, B.*

9. Draw a lattice diagram with the boundary zones:

$G = 13 + 8B$, accept;
$G = -24 + 8B$, reject.

Now plot the sequence of points and hence the acceptance or rejection given by the following trials:

(i) Toss one coin repeatedly. 'Heads' is to represent bad and 'tails' good (i.e. proportion defective $p = 1/2$).
(ii) Toss two coins repeatedly. Both heads represents bad: otherwise good. ($p = 1/4$.)
(iii) Toss three coins repeatedly. All three heads, bad: otherwise good. ($p = 1/8$.)
(iv) Toss four coins repeatedly. All four heads, bad; otherwise good. ($p = 1/16$.)

10. Determine the boundary lines and operating characteristic for the following scheme of sequential sampling; chance of accepting lots with 5% defectives is 0·10, chance of accepting lots with 1% defectives is 0·90.
What proportion of lots with 1·7% defectives will be accepted?

11. A machine is crimping cannelures in the cases of bullets. The theoretical distribution of depths of cannelure is normal with mean 0·010 in. and standard deviation 0·0005 in. It is suspected that the mean has shifted to 0·012 in. without a change in standard deviation. Taking the risks of error both to be 0·01, determine the appropriate sequential analysis to test the hypothesis, evaluating the boundaries for Σx for $n = 1, 2 \ldots 10$.
Use the scheme to accept either $m_1 = 0{\cdot}010$ in. or $m_2 = 0{\cdot}012$ in. on the basis of the following sample values:
0·0110, 0·0113, 0·0115, 0·0104, 0·0111, 0·0110, 0·0111, 0·0111, 0·0112, 0·0115, 0·0113, 0·0114, 0·0116.

CHAPTER IX

THEORY OF ERRORS

Normal distribution as law of errors. Probable error. Combination of errors. Weighted mean. Errors other than normal.

9.1 Normal distribution as 'Law of Errors'

The normal distribution is often called Gaussian on account of its derivation by Gauss as the law of the probability distribution of errors of measurement. His argument was as follows.

* Assume that the true value of a quantity is the mean of a large number of observed values. Let the deviations of the observed values from the true value be $x_1, x_2 \ldots x_n$. Then the mean of these deviations must be zero, i.e.

$$\Sigma x = 0 \quad . \quad . \quad . \quad . \quad . \quad (1)$$

Let $f(x)dx$ be the probability of an error or deviation in the range x to $x + dx$. Then the probability of the compound event that, in n observations, the errors are $x_1, x_2 \ldots x_n$ is proportional to

$$Y = f(x_1)\ f(x_2) \ldots f(x_n).$$

In other words, Y is the probability density at the point $(x_1, x_2 \ldots x_n)$ of the multi-dimensional sample space.

We seek the function $f(x)$. In addition to the assumption expressed in equation (1) we assume the appropriate function is that which will make Y, which measures the probability of the actual event, a maximum for the true value q of the measured quantity. This requires

$$\frac{d}{dq}(\log Y) = 0$$

$$\text{i.e.} \frac{f'(x_1)}{f(x_1)}\frac{dx_1}{dq} + \frac{f'(x_2)}{f(x_2)}\frac{dx_2}{dq} + \ldots = 0 \quad . \quad . \quad (2)$$

Note that the values of the x's depend on the value of the unknown q; in fact, if the observed values of the quantity measured are $s_1, s_2 \ldots$ then $x_1 = s_1 - q$, $x_2 = s_2 - q$, etc.

Hence $\dfrac{dx_1}{dq} = \dfrac{dx_2}{dq} = \ldots = -1$

so that equation (2) becomes

$$\frac{f'(x_1)}{f(x_1)} + \frac{f'(x_2)}{f(x_2)} + \ldots = 0$$

or, writing $\phi(x)$ for $f'(x)/f(x)$,

$$\Sigma\phi(x) = 0 \quad . \quad . \quad . \quad . \quad (3)$$

Now if equations (1) and (3) are simultaneously true for every large number n of errors of observation, we must have

$$\phi(x) \propto x$$

i.e.

$$\frac{f'(x)}{f(x)} = kx \text{ where } k \text{ is a constant}$$

Integrating, $\log f(x) = \frac{1}{2}kx^2 + A$

$$f(x) = Be^{\frac{1}{2}kx^2}$$

The constant B may be determined in terms of k from the fact that the total probability

$$\int_{-\infty}^{\infty} Be^{\frac{1}{2}kx^2}\,dx = 1$$

and the convergence of the integral requires that k shall be negative. We can then show that the variance of the distribution is $-1/k$ and, setting this equal to σ^2, obtain

$$dp = f(x)\,dx = \frac{1}{\sigma\sqrt{2\pi}}e^{-x^2/2\sigma^2}dx \quad . \quad . \quad . \quad (4)$$

where dp is the probability of an error in the range x to $x + dx$. *

It has been said of this 'law of errors' that physicists believe in it under the impression that mathematicians have proved it and that mathematicians believe in it under the impression that physicists have verified it. As regards mathematical proof it must be borne in mind that this would imply no more than a logical deduction from hypotheses, here of zero mean deviation and of maximum probability of observed values. To determine whether these hypotheses correspond to the operation of natural causes is a matter for experimental verification. Unfortunately, sufficiently large samples of direct measurement of a single quantity are seldom available. In order to apply the χ^2 test of goodness of fit, we should require a large enough set of observations for the frequency in each cell, after grouping, to exceed, at the very least, 5. With a sample of 20 to 30 readings it might be worth while to verify that mean (absolute) deviation is approximately 4/5 of s.d. With a small total frequency the expected frequencies in a few groups from a normal distribution are very close to corresponding frequencies in a binomial distribution with quite a low value of n; for instance, the binomial distribution 1, 4, 6, 4, 1 of total frequency 16, when $p = 1/2$, $n = 4$, compares with the normal expectations of 1·03, 3·87, 6·12, 3·87, 1·03 in corresponding intervals.

H. Jeffreys† describes some experiments planned to reproduce the conditions of certain observations in astronomy, and analyses the results with the conclusion that the deviations from the mean follow the normal law closely over the range $-2{\cdot}5\sigma$ to $+2{\cdot}5\sigma$ (this range includes 98·76% of the total probability), but show some discrepancies from the normal at the tails of the distribution.

9.2.1 Probable error

Given a moderate number of observed values of a quantity, we may accept as a working hypothesis that these constitute a sample from a normal population. The sample mean m will be the best available estimate of the true mean, and the best estimate of σ^2 is $\Sigma(x - m)^2/(n - 1)$ where the x's are the actual observed values, n in number. We distinguish between the 'error' of a reading, namely its deviation from the (unknown) true mean, and the deviation, $x - m$, from the accepted best value m, by calling the latter a *residual.*

Now the distribution of m, for all possible samples of size n, has variance σ^2/n so that $\sigma/\sqrt{n}$ is the r.m.s. error to be feared in accepting m as estimate of the true mean. It is conventional to estimate the numerical value, E, such that the probability of an error numerically less than E is 1/2, so that E is 0·6745 times the standard deviation (see §4.6.3). Then E is called the 'probable error'.

Thus E_m = probable error of the mean

$$= 0{\cdot}6745\frac{\sigma_e}{\sqrt{n}} = 0{\cdot}6745\sqrt{\frac{\Sigma(x - m)^2}{n(n - 1)}}, \quad . \quad . \quad . \quad (5)$$

where σ_e^2 is the estimate of σ^2 afforded by the n observations.

$$\begin{aligned} E_x &= \text{probable error of a single observation} \\ &= 0{\cdot}6745\sigma_e \\ &= 0{\cdot}6745\sqrt{\frac{\Sigma(x - m)^2}{n - 1}} \quad . \quad . \quad . \quad . \quad . \quad . \quad (6) \end{aligned}$$

Since the factor 0·6745 depends on the assumption of normality, we may have more confidence in equation (5) than in (6), for even if the distribution of direct readings departs from normality, the distribution of means will be normal for samples of moderate size. A large sample would be required in order to estimate the p.e. with precision (in other words, the probable error has itself a probable error) and it will usually be sufficient to replace 0·6745 by 2/3.

Most statisticians deplore the use of probable error. The name is

† *Transactions of the Royal Society*, 1936.

liable to be misunderstood by people unfamiliar with the above definition and the properties of the normal distribution. By the definition, E is the estimated semi-interquartile range of the error distribution (see §2.8). The estimated true value of a quantity, obtained as the mean, m, of a sample of readings, is often given as $m \pm E_m$. The implication is that in saying the true value lies in this range we expect to be right in 50% of cases. Other ranges may be preferred; for instance, $\pm\sigma_e/\sqrt{n}$ is the r.m.s. error to be feared in the value m, corresponding to 68% of cases. In an analysis of man-hours expended in building various types of houses, Dr. Bronowski uses a 90% range (*J.R.S.S.*, III, 1949). It is best to state clearly if the probable error or some other estimate of dispersion is being used.

9.2.2 Rejection of doubtful observations

It occasionally happens that one of a set of observed values differs notably from the others and its deviation from the mean is so large as to be highly improbable by the normal law. In such a case it is reasonable to conclude that a mistake was made or that some unusual cause operated and that the value had better be rejected. Wright and Hayford† give a rule for surveyors that if the deviation exceeds 5 times the estimated probable error, it may be rejected and that other values which show a deviation greater than $3\frac{1}{2}$ times the p.e. should be scrutinized in the light of the circumstances in which they were made. They also recommend that if large deviations are noticed at the time of observation, notes should be made of any special conditions. In applying this condition of rejection, a doubtful value is first included in the calculation of the mean and the p.e. After its rejection, a new calculation must be made, and it may then occur that another value must be rejected, for the estimate of p.e. will have been reduced by the first rejection.

Ex. i

Repeated observation of the same quantity yielded the values 16·05, 15·95, 16·00, 16·05, 16·10, 15·85, 16·0, 16·15, 15·90 and 16·15. Find the best estimate of the true value and its probable error.

$$m = \text{mean value} = 16{\cdot}02$$
$$s^2 = \text{variance of sample} = 0{\cdot}0091$$
$$\sigma^2 = \text{estimated variance of population}$$
$$= \frac{0{\cdot}091}{9} = 0{\cdot}0101$$
$$\frac{\sigma}{\sqrt{10}} = \sqrt{0{\cdot}00101} = 0{\cdot}032$$
$$E_m = 0{\cdot}6745 \times 0{\cdot}032 = 0{\cdot}020.$$

We may give the estimated true value as $16{\cdot}02 \pm 0{\cdot}020$.

In this example the largest numerical deviation from the mean is 0·17. The estimated p.e. of a single observation is $0{\cdot}6745\sigma = 0{\cdot}068$, so that the largest deviation is $2\frac{1}{2}$ times the estimated p.e.

† *Adjustment of Observations*, 2nd ed.

9.3 Combination of errors

It will often happen that a quantity must be calculated from two or more quantities whose observed or estimated values are liable to error and it is important to estimate the precision of the calculated quantity.

Suppose $z = f(x, y)$.

For small changes,

$$\delta z = \frac{\partial f}{\partial x}\,\delta x + \frac{\partial f}{\partial y}\,.\,\delta y \text{ to the first order} \quad . \quad . \quad (7)$$

This equation may be used directly to calculate the change in z due to specified small changes in x and y, or to calculate the maximum numerical error in z, given the maximum numerical errors in x and y. In the latter case we choose the signs of the errors δx and δy so as to make both terms on the r.h.s. positive. From the statistical viewpoint the procedure is pessimistic; the compound probability that the errors in x and y are such as to produce the maximum possible error in z is very small. The proper procedure is to consider the probability distribution of δz, to determine which we must know the distributions of δx and δy.

In equation (7), since δx and δy are deviations from the mean values of x and y, the partial derivatives are to be given their values at $(\bar{x}, \bar{y})$. Write the equation in the form

$$\delta_3 = A\delta_1 + B\delta_2 \quad . \quad . \quad . \quad . \quad (8)$$

where A, B are constants.

Summing for all possible errors,

$$\Sigma\delta_3^2 = \Sigma\Sigma(A\delta_1 + B\delta_2)^2$$

the double summation indicating that we must include every pair of errors δ_1, δ_2 whichever occur together. For simplicity, suppose δ_1 may take any of m individual (not necessarily unequal) values, while δ_2 may take any of n individual values. Suppose further that the errors in x and y are *independent*; then any of the values of δ_1 can occur with any of those of δ_2 and there are mn terms in the summation.

$$\Sigma\delta_3^2 = nA^2\Sigma\delta_1^2 + mB^2\Sigma\delta_2^2 + 2AB\Sigma\delta_1\delta_2 \quad . \quad . \quad (9)$$

In the last term on the r.h.s. each value of δ_1 is multiplied by $\Sigma\delta_2$, which is zero since the δ's denote deviations from the means. Hence the last term vanishes and if σ_x^2, σ_y^2, σ_z^2 denote the variances of x, y and z, equation (9) may be written

$$mn\sigma_z^2 = nA^2\, m\sigma_x^2 + mB^2\, n\sigma_y^2$$

$$\sigma_z^2 = A^2\sigma_x^2 + B^2\sigma_y^2 \quad . \quad . \quad . \quad . \quad (10)$$

The reader will realize that equation (10) is a repetition of equation (11) of Chapter VI. It is repeated here to emphasize the relationship between independence of the variates and the vanishing of the product terms.

Equation (7) and consequently equation (10) can be extended to any finite number of independent variables so that if

$$u = f(x, y, z, t \ldots)$$

$$\sigma_u^2 = \left(\frac{\partial f}{\partial x}\right)^2 \sigma_x^2 + \left(\frac{\partial f}{\partial y}\right)^2 \sigma_y^2 + \ldots \qquad (11)$$

Simple examples are

$$u = x + y + z + \ldots$$

$$\sigma_u^2 = \sigma_x^2 + \sigma_y^2 + \ldots \qquad (12)$$

and the same result is true if in the equation for u any of the signs are changed, i.e. r.m.s. error is the same for subtraction as for addition.

If $u = xy$, equation (10) gives

$\sigma_u^2 = y^2\sigma_x^2 + x^2\sigma_y^2$ or, on division by u^2,

$$\frac{\sigma_u^2}{u^2} = \frac{\sigma_x^2}{x^2} + \frac{\sigma_y^2}{y^2} \qquad (13)$$

If $u = \dfrac{x}{y}$

$\sigma_u^2 = \dfrac{1}{y^2}\sigma_x^2 + \dfrac{x^2}{y^4}\sigma_y^2$ and again on division by u^2,

$$\frac{\sigma_u^2}{u^2} = \frac{\sigma_x^2}{x^2} + \frac{\sigma_y^2}{y^2} \qquad (14)$$

These results are valid, whatever the probability distributions of the constituent errors, provided the latter are sufficiently small and are independent. If we assume further that the distribution of the constituent errors is normal, then so is that of the resultant error. For if

$$u = f(x, y, z, t \ldots)$$

$$\text{then } \delta u = A\,\delta x + B\,\delta y + C\,\delta z + \ldots \qquad (15)$$

where $A, B \ldots$ are the values of $\partial f/\partial x$, $\partial f/\partial y \ldots$ when $x, y \ldots$ take their mean values. Thus u is a linear function of normal variates δx, δy ... and so is itself a normal variate (§4.7). The range of a normal variate is, of course, theoretically infinite, while equation (15) assumes that all errors remain small. In practice, errors have not infinite range and our conclusions are valid if, say, 4 or 5 times the r.m.s. error in any of the variables $x, y \ldots$ is 'small', for the purpose of equation (15).

Equations (13) and (14) could have been derived from (12) by setting $\Delta_1 = \delta(\log x) = \delta x/\bar{x} = \delta_1/\bar{x}$, with similar equations in y and u, and then applying (12) to the relations $\log u = \log x \pm \log y$. It should be noted that if x and y are normal variates it is in general untrue that their logarithms are normal variates (see also §6.4.2 on variates which are reciprocals). But if all *errors* are small enough to justify the use of (7) and of the relation $\Delta_1 = \delta_1/\bar{x}$, then this relation is linear and if the distribution of δ_1 (and so of x) is normal, that of Δ_1 (and $\log x$) is normal also, to a sufficient degree of approximation.

With reference to equation (9), if the product terms vanish, the errors are said to be uncorrelated. We have seen that independence of the errors is sufficient to ensure this, but it is not necessary; in other words, independent errors must be uncorrelated, but uncorrelated errors are not necessarily independent. If, however, the constituent errors are normally distributed, absence of correlation does imply independence. The case of correlated errors is considered in Chapter XI.

Ex. ii

Find the p.e. in V_1, determined by the formula $V_1 = V_0\sqrt{(1 + at)}$ from the data: $V_0 = 1088$, $a = 0{\cdot}00368$, $t = 20$. The probable errors in V_0 and a are 0·5 and 0·0001 respectively, while t is supposed known exactly.

$$\frac{\partial V_1}{\partial V_0} = \sqrt{(1 + at)} = \sqrt{1{\cdot}0736}, \text{ for the given values.}$$

$$\frac{V_1}{\partial a} = \frac{V_0 t}{2\sqrt{(1 + at)}} = \frac{10880}{\sqrt{1{\cdot}0736}}$$

Since p.e. is proportional to s.d., squares of probable errors may be substituted for variances in equation (11). Hence if E denotes the required p.e. in V_1

$$E^2 = 1{\cdot}0736\,(0{\cdot}5)^2 + \frac{10880^2}{1{\cdot}0736}\,(0{\cdot}0001)^2$$

$$= 1{\cdot}37$$

The p.e. in V_1 is nearly 1·2.

9.4 Weighted mean

Consider next the problem of the best value to assign to a quantity for which we have n observed values of different precision. The observations might, for instance, be made by different observers or with different instruments.

Suppose that the observed values (not the errors) are $x_1, x_2 \ldots x_n$ made with instruments for which the law of error is in each case the normal law, the standard deviations being $\sigma_1, \sigma_2 \ldots$. If x is the true value of the quantity, the probabilities of the observed values are proportional to

$$e^{-(x-x_1)^2/2\sigma_1^2},\ e^{-(x-x_2)^2/2\sigma_2^2},\ \text{etc.}$$

and the compound probability of the set of observations is proportional to

$$e^{-\Sigma(x-x_r)^2/2\sigma_r^2}$$

Adopting the principle used in Gauss's derivation of the normal law, we choose the value of x which makes this compound probability a maximum, so that

$$\Sigma(x - x_r)^2/2\sigma_r^2 \text{ is a minimum} \quad . \quad . \quad . \quad (16)$$

Setting the differential coefficient of this quantity, with respect to x, equal to zero, we have

$$\Sigma(x - x_r)/\sigma_r^2 = 0$$

$$x = \Sigma\frac{x_r}{\sigma_r^2} \Bigg/ \Sigma\frac{1}{\sigma_r^2}$$

Writing $1/\sigma_r^2 = w_r$ we conclude that the best value of x is given by

$$x = \frac{\Sigma w_r x_r}{\Sigma w_r} \quad . \quad . \quad . \quad . \quad (17)$$

a weighted mean in which the weights are inversely proportional to the variances of the several observations.

We can now derive a measure of the precision of this best value. By equation (17) it is a linear function of the observed values $x_1, x_2 \ldots x_n$, whose variances are $\sigma_1^2, \sigma_2^2 \ldots$. If we denote by σ_m^2 the variance of the weighted mean, by equation (11) of Chap. VI

$$\sigma_m^2 = \frac{1}{(\Sigma w_r)^2}\{w_1^2\sigma_1^2 + w_2^2\sigma_2^2 + \ldots\}$$

$$= \Sigma\frac{1}{\sigma_r^2}\Bigg/\left(\Sigma\frac{1}{\sigma_r^2}\right)^2$$

$$\therefore \frac{1}{\sigma_m^2} = \Sigma\frac{1}{\sigma_r^2}$$

$$\text{or} \quad w_m = \Sigma w_r \quad . \quad . \quad . \quad . \quad . \quad . \quad (18)$$

so that the weight of the value of x given by (17) is the sum of the weights of the several observations. In particular, if n equally good observations are made of the same quantity, the best value is the arithmetic mean of the observed values (which have equal weights) and has a weight n times that of any one of the observations; its variance is $1/n$ that of a single observation.

We note in passing that equation (16) expresses the criterion of the best value in the form

$$\Sigma w_r(x - x_r)^2 \text{ is a minimum} \quad . \quad . \quad . \quad (19)$$

As in estimating the mean of a population from a sample, so here our sample of n values of differing precision affords an estimate of the true value; $x - x_r$ is a residual (the true error is not known) and the statement (19) may be put in the form: the best value of the unknown quantity, of which we have discordant observed values of different weights, is that which makes the weighted sum of the squares of the residuals a minimum. This is the Principle of Least Squares, considered more fully in the next chapter.

Ex. iii

Estimate the best value, and its p.e., of a quantity whose values, with their probable errors, from three sources are given as (i) $48{\cdot}2 \pm 0{\cdot}25$, (ii) $47{\cdot}9 \pm 0{\cdot}40$, (iii) $48{\cdot}0 \pm 0{\cdot}60$.

The given p.e.'s are in the ratios 5 : 8 : 12 and the corresponding weights 576 : 225 : 100. Shifting the origin to 48·0 we form the table:

x	w	wx
0·2	576	115·2
− 0·1	225	− 22·5
0·0	100	0·0
	901	92·7

$$\frac{\Sigma wx}{\Sigma w} = 0{\cdot}103.$$

The weight of the result is 9 times that of the 3rd value.
The p.e. of the result is 1/3 that of the 3rd value.
Hence we may give the best value as $48{\cdot}1 \pm 0{\cdot}2$.

9.5 Errors other than normal

The methods of the preceding sections are developed to deal with inevitable small discrepancies in observed values due to unknown causes whose operation is supposed to be of a random nature. A constant error of measurement would affect the calculated mean value, but would not affect deviations from the mean and so would not be reflected in the calculated p.e. Systematic errors, not necessarily constant, due to idiosyncrasies of an instrument or observer, might be revealed, if sufficient data were available, by departure from the normal law, e.g. by skewness of distribution. The avoidance of systematic error is a matter of care and planning; in some cases it may be possible to arrange that certain types of error cancel one another, by making each measurement twice in reverse directions and taking the mean of the two readings. Text-books of surveying give suitable rules and precautions.

Sometimes a number of measurements to be compounded by addition or subtraction may be given as *correct* to the nearest integer in some unit. The implication is either that it was considered sufficient to measure to this degree of precision, which was such that repetition of measurement would reveal no variation, or, if the measurements were made to a higher degree of precision, they have been rounded off. Suppose two lengths are each known to be correct to the nearest inch. We must assume that the error in each may with uniform probability have any value from $-0{\cdot}5$ to $+0{\cdot}5$ in. The error in their sum may range from -1 to $+1$ in., but not with uniform probability. Represent a pair of possible errors by a point (x, y) within the square bounded by the lines $x = \pm 0{\cdot}5$, $y = \pm 0{\cdot}5$. Over the area of the square the probability density is uniform; the whole area of the square is 1 and represents the total probability 1.

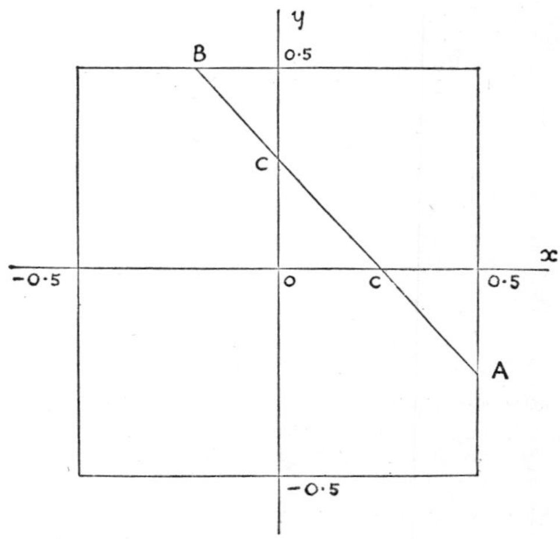

FIG. 38

The error in the sum is $x + y$ and we require the probability that this lies in a range c to $c + dc$. This element of probability will be the area, within the square, between the lines $x + y = c$, $x + y = c + dc$ and is $AB\ dc/\sqrt{2}$. The graph of $AB/\sqrt{2}$ as c varies from -1 to $+1$ gives the required distribution.

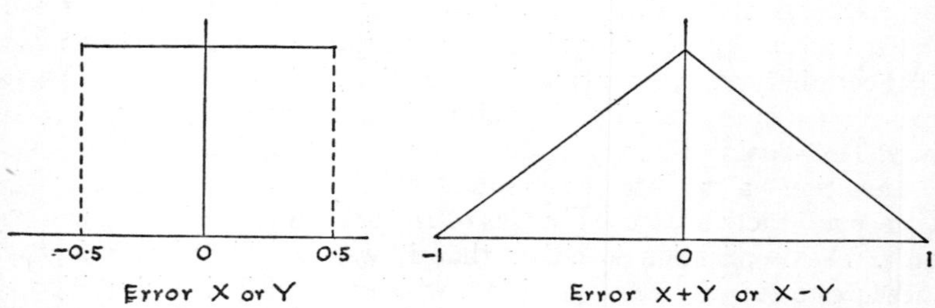

FIG. 39

Thus a rectangular distribution in x and in y gives rise to a triangular distribution in $x + y$. From consideration of the lines $x - y = c$ as c varies it is clear that $x - y$ is distributed in the same way as $x + y$.

If three lengths, whose errors x, y and z conform to the rectangular distribution, are added, the distribution of the error $x + y + z$ may be obtained by considering the area of section of a unit cube, with centre at the origin, by a plane $x + y + z = c$, as c varies from $-1{\cdot}5$ to $+1{\cdot}5$. The shape of the section changes from an equilateral triangle to a hexagon and back to an equilateral triangle. The probability curve takes the form of three parabolic arcs.

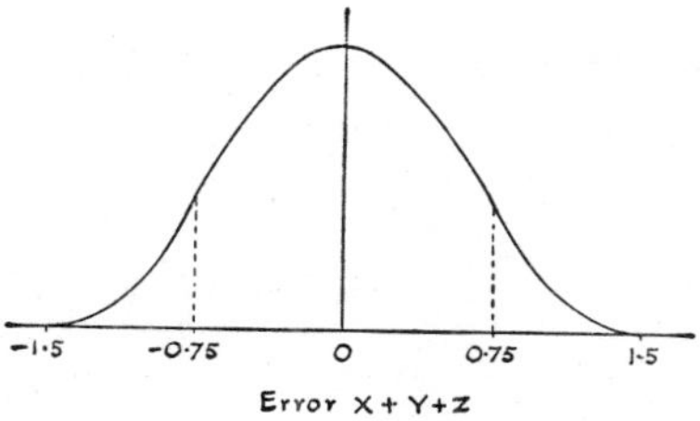

FIG. 40

If more than three such errors are added, the distribution of their sum can be shown to consist of arcs of curves of the $(n - 1)^{th}$ degree, where n is the number of errors (all having the same rectangular distribution). It can be shown that the probability curve approaches the normal form as n increases. Irrespective of the approach to normality the variance of the sum of n such errors is $n\sigma^2$ and the variance of the mean of n quantities all subject to the same law of error is σ^2/n, where σ^2 is the variance of the rectangular distribution. If the range of the latter is $-0{\cdot}5$ to $+0{\cdot}5$, $\sigma^2 = 1/12$ (see Examples on Chapter II, 13).

Ex. iv

$z = \sqrt{xy}$ and the values of x, y are known exactly. If the value of z is calculated by means of four-figure logarithm tables, show that the average numerical error to be feared is about $z/26000$, if z is not too small.

$$\log z = \tfrac{1}{2}\log x + \tfrac{1}{2}\log y$$

or

$$w = \tfrac{1}{2}u + \tfrac{1}{2}v$$
$$\delta w = \tfrac{1}{2}\delta u + \tfrac{1}{2}\delta v$$

The true logarithms are never-ending decimals which have been rounded off to the fourth place of decimals. We may assume for numbers chosen at random that the errors δu, δv in the logarithms have rectangular distributions with a range from $-0{\cdot}00005$ to $+0{\cdot}00005$. δw has the triangular distribution with the same range, since the sum of δu and δv is halved. The m.d. of the triangular distribution is 1/3 of 0·00005. Now in taking the antilogarithm we may assume a linear relationship between w and z over any small range, except when z is small. Hence the probability distribution of δz is also triangular, and is obtained from that of δw by a mere change of scale.

Since

$$w = \log_{10} z$$
$$\delta w = \frac{0{\cdot}4343}{z}\,\delta z$$
$$\delta z = \frac{z\delta w}{0{\cdot}4343}$$

In the neighbourhood of any particular z not near zero the triangular distribution of δz has a range from

$$-\frac{0{\cdot}00005z}{0{\cdot}4343} \text{ to } +\frac{0{\cdot}00005z}{0{\cdot}4343}$$

The m.d. of the error δz is $\dfrac{0{\cdot}00005z}{3(0{\cdot}4343)} = \dfrac{z}{26000}$ approximately.

Ex. v

A large set of numbers have all been rounded off to the second place of decimals. The mean of n of the numbers is calculated. What is the least value of n for which the mean can be given to three decimal places with probability 99% of being correct?

The probability distribution of the error in each number is assumed to be rectangular with range from $-$ 0·005 to $+$ 0·005.

The variance of each distribution $= \sigma^2 = (0{\cdot}005)2/12$.

The variance of the mean of n values $= \sigma^2/n$ and the distribution of the mean will be normal for moderately large values of n.

The error in the mean is to be less than 0·0005 numerically. In a normal distribution 99% of values lie within 2·57 times the s.d. from the mean.

We require

$$\frac{2{\cdot}57\sigma}{\sqrt{n}} < 0{\cdot}0005$$
$$n > \frac{2{\cdot}57^2\sigma^2}{(0{\cdot}0005)^2} = 220.$$

EXAMPLES ON CHAPTER IX

1. Eight careful measurements of a base-line yield the values 1570 + 0·41, 0·65, 0·49, 0·57, 0·39, 0·60, 0·51, 0·45 ft. Calculate the mean and its p.e.

2. If a reading is rejected when the corresponding residual is more than 5 times the estimated p.e. of a single observation, would any of the following set be rejected?

   115·8, 115·9, 116·1, 116·2, 116·3, 117·5.

3. If an angle AOC is given as α with p.e. e_1 and a part AOB of the angle as β with p.e. e_2, what is the p.e. of the value $\alpha - \beta$ of angle BOC?

4. Two observers A and B each measure the same angle 6 times. Their results differ only in the figure for seconds of angle, being

   A: 14, 17, 23, 13, 16, 19 (″)
   B: 11, 15, 17, 16, 10, 12

   Calculate each man's best estimate of the seconds figure. Show that their results should be weighted in the ratio 7 : 11 very nearly and deduce the resultant best estimate from both sets of observations.

5. In an experiment on torsion, 20 measurements were made of the diameter of a wire at different points of its length. In units of 0·0001 cm. the readings, in order, were

$$2000 + 35, 60, 45, 80, 60, 35, 65, 20, 20, 50$$
$$25, 50, 35, 50, 15, 30, 25, 15, 10, 15$$

Calculate the mean diameter and its p.e.
It is suspected that the figures afford evidence of tapering of the wire. Apply the t-test to determine whether the mean of the first 10 readings is significantly higher than the mean of the second 10.

6. The sides of a rectangle are given as 480·5 ft. and 352·0 ft. with probable errors 0·2 ft. and 0·1 ft. respectively. Find the p.e. of the calculated area.
If the given errors were maximum errors, what would be the maximum error in the calculated area?

7. If $y = x \sin \theta$ and $x = 1200 \pm 2$, $\theta = 5° 20' \pm 0·5'$, find the p.e. in the computed value of y. (1 radian = 3438′.)

8. Determine the probable error in the difference of elevation between two points in accordance with the relation

$$H = D \tan \alpha + \frac{1 - 2m}{2R} D^2$$

where the distance D = 12,425 ft. ± 2 ft.,
the angle of elevation α = 4° 13′ 20″ ± 5″,
the coefficient of refraction m = 0·065 ± 0·015,
the radius of the earth R = 3960 miles. (L.U.)

9. If a specific gravity is found from the equation

$$s = (W - W_1)/(W - W_2),$$

obtain an expression for the p.e. in s in terms of the p.e.'s in W, W_1 and W_2.

10. Citing the radii of gyration of a rectangle and a triangle, or otherwise, verify the relation $\sigma_z^2 = \sigma_x^2 + \sigma_y^2$, in the case when $z = x + y$ and x and y have the same rectangular distribution.

11. If two angles of a triangle are each given correct to the nearest minute, what is the probability that the third angle, calculated by subtraction from 180°, is also correct to the nearest minute?

12. In a topographical survey, the difference in level between two points A and B is found by three routes, via C, D, E and F, via G and H, and via J, the distances being:

Route (1)—AC = 400 ft., CD = 540 ft., DE = 1000 ft.,
EF = 860 ft., FB = 440 ft.
Route (2)—AG = 800 ft., GH = 1020 ft., HB = 1280 ft.
Route (3)—AJ = 980 ft., JB = 1540 ft.

The sections on route (1) are each levelled four times, those on route (2) eight times, and on route (3) twice; the established differences in level thus obtained being 102·7 ft., 101·9 ft. and 103·6 ft. respectively.

If the probable error in any section at each levelling is proportional to its length, and the usual laws for combination of readings hold, find the most probable value for the difference in level between A and B. (L.U.)

13. What is the s.d. of the sum of n variates, each with the rectangular distribution, of range $-a$ to $+a$? If the distribution of the sum may be treated as normal when $n = 12$, what is the semi-inter-quartile range of the sum (i.e. if the variates are errors, what is the p.e. of the sum)?

14. If, in a normal distribution $(0, \sigma)$ the probability of a deviation from the mean numerically greater than or equal to 4σ is regarded as negligible, show that we may accept as correct to $(r - 1)$ decimal places the sum of 18 numbers, each correct to r decimal places. How many such numbers may we add, with confidence that the sum is correct to $(r - 2)$ decimal places? (Use the s.d. obtained in Question 12.)

15. A student's weekly exercise is roughly assessed and allotted a mark which may be any integer from 0 to 10. Assume that in the long run errors $-2, -1, 0, 1, 2$ in the mark allotted occur with relative frequencies 2, 10, 76, 10, 2% respectively. The sum of 25 such errors may be regarded as distributed normally. Show that a student's total mark for 25 weeks will be less than 3% in error in 98·76% of cases.

16. On plain sheets of paper draw two sets of 10 lines thus: Freehand and without alteration draw a line across the sheet, from left to right or right to left, to be as nearly as you can make it by eye, 3 in. long. Cover the line and draw another under the same conditions; and so on keeping the lines covered till you have drawn 10 lines. Without uncovering these lines repeat the procedure on a second sheet, but this time draw each line up and down or, in other words, to or from yourself. Measure the lines to the nearest tenth of an inch. Interpret the results. (R.S.S. Diploma)

17. If n variates $x_1, x_2 \ldots x_n$ are each normally distributed with a common mean, show that the weighted sum $w_1x_1 + w_2x_2 + \ldots + w_nx_n$ is itself normally distributed. Find its variance in terms of those of the separate variates. Taking $w_1 + w_2 + \ldots + w_n = 1$, deduce the system of weights for which the weighted average $w_1x_1 + w_2x_2 + \ldots + w_nx_n$ has minimum variance. (R.S.S. Diploma)

CHAPTER X

THE METHOD OF LEAST SQUARES

Best straight line. Observation equations linear in the unknowns. Orthogonal polynomials. Fourier series. Approximation to a continuous function. Weighted observations. Reduction of observation equations to linear form. Error proportional to magnitude. Undetermined multipliers or correlates. Precision of the results.

10.1 Plausibility of the method

If we adopt the mean of a population as representative value, we adopt the number from which the sum of squares of deviations is least. The theory of errors assumes that the population of all possible observed values of a directly measured quantity has the true value as mean. The best available estimate of this mean afforded by a sample from the population is the mean of the sample, and in adopting it as best estimate of the true value we adopt the number from which the sum of squares of *residuals* is least. We could express this as follows. Let a be the unknown true value and let $y_1, y_2 \ldots y_n$ be the observed values. Then our best estimate of a is that which makes $\sum_1^n (y_r - a)^2$ a minimum, as a varies. Differentiating with respect to a and setting the derived function equal to zero, we obtain

$$-2\sum_1^n (y_r - a) = 0$$

$$\sum_1^n y_r - na \qquad = 0$$

$$a = \frac{1}{n} \sum_1^n y_r = \text{the mean of the } y\text{'s.}$$

We have seen further (Chap. IX, equation (16)) that when the observed values of the single quantity are of differing precision, i.e. are drawn from populations of differing variance but the same mean, the adopted best value makes the weighted sum of squares of residuals a minimum.

We now propose to extend this principle to the determination of best values of two or more unknowns, and these not necessarily observed directly. A common problem, for instance, is the determination of constants in a scientific law. Theory may suggest that the relation between certain variables is of a certain type and observed values may confirm the theory, subject to deviations which may reasonably be explained as due to errors of measurement. It remains to determine on the basis of the observed values of the variables, the most plausible values to adopt for certain constants in the relation between the variables. It is convenient for the present to suppose that all observations are of equal weight. The rules evolved are capable of simple modification (§10.6.1) when the weights are unequal.

For an account of the development of the Method of Least Squares since it was first proposed by Legendre in 1805 and of ways in which the principle may be derived from fundamental assumptions, the reader may refer to Whitaker and Robinson, *The Calculus of Observations*, Chap. IX.

10.2.1 Best straight line

Suppose we have two variables represented by Cartesian co-ordinates (x, y) of which n observed pairs of values are (x_1, y_1), (x_2, y_2) . . . (x_n, y_n). Suppose also that the corresponding n points in theory should, and in fact nearly do, lie on a straight line. The equation of this line takes the form $y = a + bx$, if we require to express the relationship between x and y in the form suitable for determining the value of y when the value of x is given. In fact, we suppose the values of x were chosen and accurately known, readings were then taken of the corresponding values of y, but these were liable to random error of measurement The unknown constants a and b are to be determined on the assumption that the sum of the squares of the residuals is a minimum, a residual being the difference between an observed value of y and its value as given by the straight line for the same value of x.

If
$$S = \sum_1^n (y_r - a - bx_r)^2$$

then for S to be a minimum

$$\frac{\partial S}{\partial a} = 0 \text{ and } \frac{\partial S}{\partial b} = 0.$$

The first of these gives

$$-2\Sigma(y_r - a - bx_r) = 0,$$

i.e.
$$\Sigma y_r = na + b\Sigma x_r \quad . \quad . \quad . \quad . \quad (1)$$

The second gives

$$-2\Sigma x_r(y_r - a - bx_r) = 0$$

i.e.
$$\Sigma x_r y_r = a\Sigma x_r + b\Sigma x_r^2 \quad . \quad . \quad . \quad (2)$$

These two equations determine the required values of a and b. There is an easy rule by which the equations may be formed. The equation for y in terms of x being $y = a + bx$, we should obtain, by substituting the observed pairs of values (x_1, y_1), $(x_2, y_2) \ldots (x_n, y_n)$, n equations

$$\begin{aligned} y_1 &= a + bx_1, \\ y_2 &= a + bx_2, \\ &\cdot \quad \cdot \quad \cdot \quad \cdot \\ y_n &= a + bx_n \end{aligned}$$

These are called the *observation equations* and, unless all the n points lie exactly on a straight line, are inconsistent, i.e. we cannot find values of a and b which will satisfy them all. The method of least squares gives equations (1) and (2) for the 'best' values of a and b. These two equations are called the *normal* equations and they may be formed from the observation equations as follows. The normal equation $\partial S/\partial a = 0$ is obtained by multiplying the observation equations by the successive coefficients of a (in this case every coefficient is 1) and adding; the normal equation $\partial S/\partial b = 0$ is obtained similarly by multiplying by the coefficients of b and adding. It is conventional to replace the sign of summation, Σ, by square brackets. Thus n observation equations of the form

$$y = a + bx$$

yield the two normal equations

$$\left.\begin{aligned} [y] &= na + b[x] \\ [xy] &= a[x] + b\,[x^2] \end{aligned}\right\} \quad . \quad . \quad . \quad (3)$$

Ex. i

Fit a straight line by the method of least squares to these data, taking y as the dependent variable:

x:	3·4	4·3	5·4	6·7	8·7	10·6
y:	4·5	5·8	6·8	8·1	10·5	12·7

(R.S.S. Cert.)

In order to obtain the sums indicated by square brackets in equations (3) it is best to tabulate the work, as follows:

x	y	xy	x^2
3·4	4·5	15·3	11·56
4·3	5·8	24·94	18·49
5·4	6·8	36·72	29·16
6·7	8·1	54·27	44·89
8·7	10·5	91·35	75·69
10·6	12·7	134·62	112·36
39·1	48·4	357·2	292·15

The normal equations are

$$\begin{aligned} 48{\cdot}4 &= 6a + 39{\cdot}1b \\ 357{\cdot}2 &= 39{\cdot}1a + 292{\cdot}15b \end{aligned}$$

These give $a = 0{\cdot}775$, $b = 1{\cdot}12$ so that the equation of the required line is $y = 0{\cdot}775 + 1{\cdot}12x$.

10.2.2 General equation of the best straight line

In numerical examples it is best to proceed as above, but for theoretical purposes a general solution is possible.

Equation (1) is $\Sigma y = na + b\Sigma x$, which may be written

$$n\bar{y} = na + nb\bar{x}$$

or

$$\bar{y} = a + b\bar{x}, \qquad (4)$$

where $\bar{x}$, $\bar{y}$ are the means of the observed values of x and y. Equation (4) shows that the required line passes through the centroid of the n points representing the observations. It remains to determine the gradient b.

$$\begin{aligned} \text{Now } \Sigma(x-\bar{x})(y-\bar{y}) &= \Sigma xy - \bar{x}\Sigma y - \bar{y}\Sigma x + n\bar{x}\bar{y} \\ &= \Sigma xy - n\bar{x}\bar{y} - n\bar{x}\bar{y} + n\bar{x}\bar{y} \\ &= \Sigma xy - n\bar{x}\bar{y} \\ \therefore \Sigma xy &= \Sigma(x-\bar{x})(y-\bar{y}) + n\bar{x}\bar{y} \qquad (5) \end{aligned}$$

(The reader may recognize this as the parallel axes theorem for products of inertia, the n points being regarded as the positions of n particles, each of unit mass.)

Also

$$\begin{aligned} n\sigma_x^2 &= \Sigma x^2 - n\bar{x}^2 \\ \therefore \Sigma x^2 &= n\sigma_x^2 + n\bar{x}^2 \end{aligned}$$

Substituting in equation (2)

$$\Sigma(x-\bar{x})(y-\bar{y}) + n\bar{x}\bar{y} = na\bar{x} + nb\sigma_x^2 + nb\bar{x}^2$$

But

$$n\bar{x}\bar{y} = na\bar{x} + nb\bar{x}^2$$

from equation (4). By subtraction,

$$\Sigma(x-\bar{x})(y-\bar{y}) = nb\sigma_x^2$$

$$b = \frac{\Sigma(x-\bar{x})(y-\bar{y})}{n\sigma_x^2} = \frac{\Sigma(x-\bar{x})(y-\bar{y})}{\Sigma(x-\bar{x})^2} \qquad (6)$$

The quantity $\frac{1}{n}\Sigma(x-\bar{x})(y-\bar{y})$ is called the covariance of x and y written cov (x, y). σ_x^2 is the variance of x, so that the equation of the best straight line may be written

$$y - \bar{y} = \frac{\text{cov}\,(x, y)}{\text{var}\,x}(x - \bar{x}) \qquad (7)$$

If the same pairs of values of (x, y) had been used to determine the best linear equation for x as a function of y, a different straight line would have been obtained, namely

$$x - \bar{x} = \frac{\text{cov}\,(x, y)}{\text{var}\,y}(y - \bar{y}) \qquad (8)$$

It is to be remembered that in proposing to determine a relation of the form $y = a + bx$, with y as dependent variable, it was stipulated that the values of x are chosen and known while those of y are readings subject to experimental error. This is often the case in scientific experiment, values of x being taken in round numbers (they may be times, loads, etc.), but those of y having to be read as accurately as possible on a continuous scale.

Both equations (7) and (8) are important in Correlation (see Chap. XI) in which we consider the case when each variable is liable to appreciable variation when the other is given. It will then be seen to be natural to expect that the lines giving the best y for given x and the best x for given y shall be different.

10.3 Observation equations linear in the unknowns

Consider the laws

$$\begin{aligned} &\text{(i)}\ y = a + bx + cx^2, \\ &\text{(ii)}\ y = ax + b/x, \\ &\text{(iii)}\ y = Ae^{kx}. \end{aligned}$$

After choosing suitable values of x and reading experimental values of y, it might be reasonable to suppose that one of these or other laws was appropriate to express y as a function of x, the constants to be determined by the method of least squares. Now in (i) and (ii) the observation equations will be linear in the unknown constants, and it is easily seen that the rule given in §10.2.1 for the formation of the normal equations is still valid. For instance, in (i) we require to make $\Sigma(y - a - bx - cx^2)^2$ a minimum. Differentiating with respect to a, b, c in turn and setting these differential coefficients equal to zero we obtain

$$\begin{aligned} -2\,\Sigma(y - a - bx - cx^2) &= 0 \\ -2\,\Sigma x(y - a - bx - cx^2) &= 0 \\ -2\,\Sigma x^2(y - a - bx - cx^2) &= 0 \end{aligned}$$

Dropping the factor -2, it is clear that the effect of the differentiations is to multiply the observation equations by the coefficients of a, b, c in turn, and we add the resulting equations to form the normal equations. It is unnecessary to transpose all terms to the L.H.S. first. For the law $y = a + bx + cx^2$ the normal equations take the form

$$\left.\begin{aligned} [y] &= na + b[x] + c[x^2] \\ [xy] &= a[x] + b[x^2] + c[x^3] \\ [x^2y] &= a[x^2] + b[x^3] + c[x^4] \end{aligned}\right\} \quad . \quad . \quad . \quad (9)$$

For the law $\quad y = ax + b/x$, they are

$$\left.\begin{aligned} [xy] &= a[x^2] + nb \\ [y/x] &= na + b[1/x^2] \end{aligned}\right\} \quad . \quad . \quad . \quad . \quad (10)$$

The rule cannot be applied to law (iii). Differentiation of $\Sigma(y - Ae^{kx})^2$ with respect to A and k produces a quite different result. We return to law (iii) in §10.7.1.

The unknowns to be determined are not restricted to the constants of a law. Observation equations represent a number, greater than that of the unknowns, of relations between the unknowns, obtained by experiment. Provided these relations are linear in the unknowns we may form the normal equations by the simple rule.

Ex. ii

A, B, C, D are four points in a straight line. A surveyor obtained the following results by measurement:

$$AB = 17{\cdot}42,\ AC = 26{\cdot}35,\ AD = 51{\cdot}70,\ BC = 8{\cdot}91,$$
$$BD = 34{\cdot}27,\ CD = 25{\cdot}38 \text{ chains.}$$

By the method of least squares obtain the best values of the lengths AB, BC, CD.

Call these unknown lengths x, y and z chains.
The observation equations are

$$\begin{aligned} x \phantom{{}+y+z} &= 17{\cdot}42 \\ x + y \phantom{{}+z} &= 26{\cdot}35 \\ x + y + z &= 51{\cdot}70 \\ y \phantom{{}+z} &= 8{\cdot}91 \\ y + z &= 34{\cdot}27 \\ z &= 25{\cdot}38 \end{aligned}$$

Multiplying by the coefficients of x (zero in the last three equations) and adding, we obtain

$$3x + 2y + z = 95{\cdot}47$$

Similarly,

$$2x + 4y + 2z = 121{\cdot}23$$
$$x + 2y + 3z = 111{\cdot}35$$

The first two equations give

$$4x = 69{\cdot}71,\ \underline{x = 17{\cdot}4275.}$$

The second and third equations give

$$4z = 101{\cdot}47,\ \underline{z = 25{\cdot}3675.}$$

From the second equation

$$\begin{aligned} 4y &= 121{\cdot}23 - 2(x + z) \\ &= 121{\cdot}23 - 85{\cdot}59 \\ &= 35{\cdot}64,\ \underline{y = 8{\cdot}91.} \end{aligned}$$

As is to be expected in such an experiment, the results do not differ much from the directly observed values. The work can be made simpler by taking as unknowns the corrections to be applied to the observed values. Here suppose

$$AB = 17{\cdot}42 + \alpha,\ BC = 8{\cdot}91 + \beta,\ CD = 25{\cdot}38 + \gamma.$$

The observation equations will now read

$$\begin{aligned} \alpha &= 0 \\ 26{\cdot}33 + \alpha + \beta &= 26{\cdot}35 \qquad \text{i.e. } \alpha + \beta = 0{\cdot}02 \\ 51{\cdot}71 + \alpha + \beta + \gamma &= 51{\cdot}70 \qquad \alpha + \beta + \gamma = 0{\cdot}01 \\ \beta &= 0 \\ 34{\cdot}29 + \beta + \gamma &= 34{\cdot}27 \qquad \beta + \gamma = -\ 0{\cdot}02 \\ \gamma &= 0 \end{aligned}$$

The normal equations reduce to

$$\begin{aligned} 3\alpha + 2\beta + \gamma &= 0{\cdot}01 \\ 2\alpha + 4\beta + 2\gamma &= -\ 0{\cdot}01 \\ \alpha + 2\beta + 3\gamma &= -\ 0{\cdot}03 \end{aligned}$$

giving $\alpha = 0{\cdot}0075$, $\beta = 0$, $\gamma = -\ 0{\cdot}0125$ as the corrections to be applied to the direct readings of the lengths AB, BC, CD.

Ex. iii

It is known that the readings for t and y given below should follow a law of the form $y = a/t + bt$, where a and b are constants. Use the method of least squares to find the best values of a and b.

t:	1	2	3	4	5	6	7	8
y:	5·43	6·28	8·23	10·32	12·63	14·86	17·27	19·51

The normal equations take the form

$$[y/t] = a[1/t^2] + 8b$$
$$[yt] = 8a + b[t^2].$$

The required calculations are as follows:

y/t	yt	$1/t^2$	t^2
5·43	5·43	1·000	1
3·14	12·56	0·250	4
2·74	24·69	0·111	9
2·58	41·28	0·063	16
2·53	63·15	0·040	25
2·48	89·16	0·028	36
2·47	120·89	0·020	49
2·44	156·08	0·016	64
23·8	513	1·53	204

$$23{\cdot}8 = 1{\cdot}53a + 8b$$
$$513 = 8a + 204b \qquad \underline{a = 3{\cdot}03} \qquad \underline{b = 2{\cdot}40}.$$

10.4.1 Orthogonal polynomials

When the assumed law is of the form

$$y = a_0 + a_1x + a_2x^2 + \ldots + a_mx^m \quad . \quad . \quad (11)$$

the solution of the normal equations becomes very arduous, if m is greater than, say, 2. The work is much simplified by taking the readings of y at equal intervals in x; then by shift of origin and change of scale of x, if necessary, we may suppose, if there are n readings, that the values of x are $0, 1, 2 \ldots (n - 1)$. Suppose it is desired to fit a polynomial of degree m: then there are $m + 1$ coefficients, so that n must be at least equal to $m + 1$, which would be just sufficient to determine the coefficients, or n must be greater than $m + 1$, in which case the method of least squares will be used. Instead of assuming equation (11) let

$$y = c_0P_0 + c_1P_1 + c_2P_2 + \ldots + c_mP_m \ldots \quad . \quad (12)$$

where P_r is a polynomial in x of degree r, having the property

$$\sum_{x=0}^{n-1} P_{r_1} P_{r_2} = 0, \text{ if } r_1 \neq r_2 \quad . \quad . \quad . \quad (13)$$

Such polynomials are called orthogonal. Forming the normal equation corresponding to c_r by multiplying the n observation equations by the coefficients of c_r, namely the values taken by P_r when $x = 0, 1 \ldots n - 1$, and adding, we obtain

$$\sum_{x=0}^{n-1} yP_r = c_r \sum_{x=0}^{n-1} P_r^2 \quad . \quad . \quad . \quad . \quad (14)$$

all other coefficients vanishing by virtue of equation (13). Suitable polynomials for different values of m and n have been tabulated by Fisher and Yates. It is not necessary to know the P's explicitly in terms of x; they can be found when required from the tabulated values by the method used in the following example.

Ex. iv

Express y as a polynomial in x of the third degree from the following readings:

x:	0	1	2	3	4	5
y:	18	80	122	151	201	267

P_0 is always to be taken as 1 so that by equation (14), $\Sigma y = c_0 \Sigma 1 = 6c_0$; in fact, c_0 is always the mean of the recorded values of y. From the tables the values of the appropriate polynomials P_1, P_2, P_3 for the 6 values of x are obtained.

x	y	P_1	P_2	P_3	yP_1	yP_2	yP_3
0	18	− 5	+ 5	− 5	− 90	+ 90	− 90
1	80	− 3	− 1	+ 7	− 240	− 80	+ 560
2	122	− 1	− 4	+ 4	− 122	− 488	+ 488
3	151	+ 1	− 4	− 4	+ 151	− 604	− 604
4	201	+ 3	− 1	− 7	+ 603	− 201	− 1407
5	267	+ 5	+ 5	+ 5	+ 1335	+ 1335	+ 1335
Sums:	839				1637	52	282
Sums of squares:		70	84	180			

Applying equation (14), we obtain

$$c_0 = 839/6 = 139{\cdot}8, \qquad c_2 = 52/84 = 0{\cdot}62$$
$$c_1 = 1637/70 = 23{\cdot}4, \qquad c_0 = 282/180 = 1{\cdot}57.$$

To identify the polynomials, we may use the tabulated values as follows. Form the differences between successive values, then the differences of the differences and so on. For an rth degree polynomial the rth set of differences will be constant. In our example, we obtain

P_1		P_2			P_3			
− 5		5			− 5			
	2		− 6			12		
− 3		− 1		3	7		− 15	
	2		− 3			− 3		10
− 1		− 4		3	4		− 5	
	2		0			− 8		10
1		− 4		3	− 4		5	
	2		3			− 3		10
3		− 1		3	− 7		15	
	2		6			12		
5		5			5			

We need only the top diagonal line of figures in each case.

$$P_1 = -5 + 2x$$
$$P_2 = 5 - 6x + 3x(x-1)/2!$$
$$P_3 = -5 + 12x - 15x(x-1)/2! + 10x(x-1)(x-2)/3!$$

The top diagonal line of figures may be denoted by y_0, $\Delta(0)$, $\Delta^2(0)$, $\Delta^3(0)$, etc. The general formula for a polynomial y, tabulated in this manner at unit intervals starting with $x = 0$ is

$$y_0 + x\Delta(0) + \frac{x(x-1)}{2!}\Delta^2(0) + \frac{x(x-1)(x-2)}{3!}\Delta^3(0) + . \quad (15)$$

This is a well-known result in the Calculus of Finite Differences.

A valuable property of the method of orthogonal polynomials is that the least squares approximation to any *lower* degree is obtained by stopping at that degree.

$y = c_0 + c_1P_1$ is the best straight line

$y = c_0 + c_1P_1 + c_2P_2$ is the best curve of the second degree.

In the present example it remains to substitute in

$$y = c_0 + c_1P_1 + c_2P_2 + c_3P_3 \text{ and reduce this to}$$

$$\begin{aligned} y = {} & 139{\cdot}8 \\ & + 23{\cdot}4\,(-5 + 2x) \\ & + 0{\cdot}62\,(5 - 7{\cdot}5x + 1{\cdot}5x^2) \\ & + 1{\cdot}57\,(-5 + 22{\cdot}83x - 12{\cdot}5x^2 + 1{\cdot}67x^3) \\ = {} & 18 + 78x - 18{\cdot}7x^2 + 2{\cdot}6x^3. \end{aligned}$$

10.4.2 Fourier series

If a function of x is periodic with period 2π, it may be expanded in a Fourier series, usually infinite, in the form

$$\begin{aligned} y = \tfrac{1}{2}\,a_0 + a_1 \cos x + a_2 \cos 2x + \ldots + a_r \cos rx + \ldots \\ + b_1 \sin x + b_2 \sin 2x + \ldots + b_r \sin rx + \ldots \end{aligned}$$

If the higher harmonics are of negligible amplitude, a sufficiently close approximation may be obtained by terminating the series at the nth harmonic, namely $a_n \cos nx + b_n \sin nx$. There are then $2n + 1$ constants to be determined. Given the values of y for the $2m$ values 0, π/m, $2\pi/m \ldots (2m - 1)\pi/m$ of x, we obtain $2m$ observation equations, linear in the unknown constants. It is, of course, supposed that $2m > 2n + 1$. When the observation equations are multiplied by the coefficients of any one of the unknowns, say a_r, and added, all other coefficients vanish on the R.H.S. by virtue of the properties of the trigonometrical series involved. (The trigonometrical functions are, in fact, orthogonal functions for summation over the range.) We obtain the formulae

$$\tfrac{1}{2}a_0 = \frac{1}{2m}\Sigma y$$

$$a_r = \frac{1}{m}\Sigma y \cos rx,\ b_r = \frac{1}{m}\Sigma y \sin rx.$$

These formulae are independent of n, so that if the series is terminated at any earlier harmonic, the constants, as given by the method of least squares, are unaltered up to the last included harmonic.

With $2m$ ordinates, the observation equations have a unique solution if the series is taken as far as the $(m - 1)^{th}$ harmonic, with the additional term $a_m \cos mx$. The formula for a_m then differs from that for a_r, being

$$a_m = \frac{1}{2m}\Sigma y \cos mx$$

b_m is in any case indeterminate, since $b_m \sin mx$ vanishes for all the chosen values of x; in other words, b_m could be given any value whatsoever without altering the observed ordinates. The inclusion of a term $b_m \sin mx$ would, however, produce in the continuous graph an oscillation about the smooth curve through the tops of the chosen ordinates. It is therefore reasonable to assume that b_m is negligible if the ordinates were read from a graph; for if the oscillation existed, it would be realized that intermediate ordinates must be included, by increasing the value of m, in order to get a good approximation to the graph.

We do not include examples of harmonic analysis, which is dealt with in text-books of Mathematics. It has its importance in the analysis of time-series in Statistics, but these also are outside our scope.

10.5.1 Approximation to a continuous function

The method of least squares may be applied so as to obtain an approximation of given form to a given continuous function over a given range. The method is to minimize the sum, in this case an integral, of the squared difference between the true value of the function and the value given by the approximation.

10.5.2

For instance, a periodic function may be defined by a simple algebraical formula over the range $0 \leqslant x \leqslant 2\pi$ and by the relation $f(x + 2\pi) = f(x)$ elsewhere. To find the constant coefficients of its Fourier series up to any desired harmonic, we minimize

$$S = \int_0^{2\pi} (y - \tfrac{1}{2}a_0 - a_1 \cos x - a_2 \cos 2x \ldots - b_1 \sin x - b_2 \sin 2x \ldots)^2 \, dx$$

The normal equation $\partial s/\partial a_r = 0$ is obtained by differentiating under the integral sign, giving

$$-2\int_0^{2\pi} \cos rx(y - \tfrac{1}{2}a_0 - a_1 \cos x \ldots - b_1 \sin x \ldots)\, dx = 0$$

All terms in the integral vanish except the integrals of $y \cos rx$ and $a_r \cos^2 rx$ and we obtain

$$a_r = \frac{1}{\pi}\int_0^{2\pi} y \cos rx \, dx.$$

Similarly,
$$b_r = \frac{1}{\pi}\int_0^{2\pi} y \sin rx \, dx.$$

The reader familiar with Fourier series will know that these are the coefficients of the infinite series for the periodic function. It follows that the least squares approximation to the function up to any harmonic is obtained by terminating its complete Fourier series at that harmonic.

10.5.3

It is sometimes convenient to approximate to a known continuous function over a given range by means of a polynomial.

Ex. v

Let it be required to find the best quadratic function $a + bx + cx^2$, to be used as an approximation to $\sin x$ over the range $0 \leqslant x \leqslant \pi/2$. The coefficients a, b, c are chosen so as to minimize

$$\int_0^{\pi/2} (a + bx + cx^2 - \sin x)^2 dx$$

Differentiating with respect to a, b and c we obtain the normal equations

$$a\int_0^{\pi/2} dx \quad + b\int_0^{\pi/2} xdx \quad + c\int_0^{\pi/2} x^2dx = \int_0^{\pi/2} \sin x \, dx$$

$$a\int_0^{\pi/2} xdx \quad + b\int_0^{\pi/2} x^2dx + c\int_0^{\pi/2} x^3dx = \int_0^{\pi/2} x \sin x \, dx$$

$$a\int_0^{\pi/2} x^2dx + b\int_0^{\pi/2} x^3dx + c\int_0^{\pi/2} x^4dx = \int_0^{\pi/2} x^2 \sin x \, dx$$

These equations determine a, b and c.

The solution of such simultaneous equations can again be avoided by the use of orthogonal functions. By change of scale the range can be made $0 \leqslant x \leqslant 1$. An approximation to the given function $f(x)$ can be found in the form

$$f(x) = c_0 + c_1P_1 + c_2P_2 + \ldots$$

where P_r is a polynomial in x of degree r such that

$$\int_0^1 P_mP_n \, dx = 0, \text{ if } m \neq n.$$

10.6.1 Weighted observations

In §9.4 the weighted mean of n observations of a single quantity was adopted as best estimate of the true value, the weights being inversely proportional to the variances of the several observations. Moreover, the adoption of the weighted mean was equivalent to making the weighted sum of squares of residuals a minimum. Similarly for quantities not directly observed, but derived from observation equations of unequal weight, the method of least squares consists in minimizing the weighted sum of squares of residuals, a residual being the difference between the L.H.S. and R.H.S. of an observation equation when the unknowns are given their estimated best values. Thus if the observation equations, linear in the unknowns a, b, c, etc., are

$$y_1 = h_1a + k_1b + \ldots \qquad \text{weight } w_1$$

$$y_2 = h_2a + k_2b + \ldots \qquad \text{weight } w_2$$

$$y_n = h_na + k_nb + \ldots \qquad \text{weight } w_n$$

the resulting normal equations are $\partial S/\partial a = 0$, etc., where

$$S = \Sigma w(y - ha - kb - \ldots)^2. \text{ They become}$$

$$\left.\begin{aligned}[why] &= [wh^2]a + [whk]b + \ldots \\ [wky] &= [whk]a + [wk^2]b + \ldots\end{aligned}\right\} \quad . \quad . \quad (16)$$

and so on.

The same equations may be obtained by first multiplying each observation by the *square root* of its weight and then applying the rule of §10.2, i.e. by treating the modified equations as of equal weight.

10.6.2 Examples from surveying

Observations in surveying can be made to a high degree of precision, but many problems arise of adjusting small discordances. The sum of the angles at a point in a plane must be exactly 360°; if the whole angle is observed as the sum of a number of parts it is practically inevitable that the sum of the observed angles differs from 360° by a few seconds of angle. In levelling, the heights (positive or negative) of n points above a chosen point could be determined by n direct observations; if further observations are made of the differences of level of pairs chosen from the n points, the results are unlikely to be exactly consistent with the first n measurements. The sum of the angles of a polygon is fixed by the number of its sides, and if observed values do not have the correct sum they must be adjusted. If the area surveyed is large the curvature of the earth becomes important; the true sum of the angles of a triangle will now exceed 180° by a quantity called the spherical excess. Difficult problems of adjustment arise in triangulation; besides the exact angle sums, there are exact relations which must be satisfied, expressing the fact that a particular length occurs as the side of more than one triangle.

There are two methods of incorporating exact relations. The first is to reduce the number of unknowns to the minimum and express each observation in terms of such unknowns. At present we shall use only this method. The second method is given later, in §10.8.

Ex. vi

The angles of a plane triangle are measured. Show that the error in the sum should be distributed in ratios inversely proportional to the weights of the three observed angles.

Suppose the observation equations are

$$\begin{aligned} A &= \alpha \quad \text{weight } w_1 \\ B &= \beta \quad \text{weight } w_2 \\ C &= \gamma \quad \text{weight } w_3 \end{aligned}$$

and that $\alpha + \beta + \gamma = 180° + e''$.

The minimum number of unknowns is two. Suppose x'' and y'' are the corrections to be applied to angles A and B. The observation equations are

$$\begin{aligned} x &= 0 \quad \text{weight } w_1 \\ y &= 0 \quad \text{weight } w_2 \\ x + y &= -e \quad \text{weight } w_3 \end{aligned}$$

The observation equation C = γ takes the form

$$180° - (\alpha + x'') - (\beta + y'') = \gamma$$

i.e.

$$x'' + y'' = 180° - \alpha - \beta - \gamma$$

The normal equations are

$$(w_1 + w_3)x + w_3 y = - e w_3$$
$$w_3 x + (w_2 + w_3)y = - e w_3$$

Subtracting

$$w_1 x - w_2 y = 0$$

i.e.

$$x : y = \frac{1}{w_1} : \frac{1}{w_2}$$

The first normal equation gives

$$(w_1 + w_3)x + \frac{w_3 w_1 x}{w_2} = - e w_3$$

$$x = - \frac{e w_2 w_3}{w_1 w_2 + w_2 w_3 + w_3 w_1} = - e \frac{1/w_1}{1/w_1 + 1/w_2 + 1/w_3}$$

Ex. vii

Observed differences of level between four points A, B, C, D, and the distances between the points are as follows:

A to B	Rise	83·45 ft.	Distance 5 miles
B to C	Rise	24·07 ft.	Distance 2 miles
C to D	Fall	136·82 ft.	Distance 4 miles
D to A	Rise	29·18 ft.	Distance 4 miles
B to D	Fall	112·61 ft.	Distance 3 miles

Find the heights of B, C, D relative to A.

Here the observations are to be given weights inversely proportional to the corresponding distances. The reason for this is that each recorded rise or fall is the sum of a number of differences of level. The number of terms in each sum is the number of individual observations and may be assumed proportional to the distance. The variance of a sum of n terms is n times the variance of a single reading; hence the weight of a sum is proportional to $1/n$.

The first three observations give heights of B, C and D above A as + 83·45, + 107·52 and − 29·30 ft. Let x, y and z ft. be the corrections required. The observation equations are

$$\begin{aligned} x \qquad\qquad &= 0 \quad &&\text{weight } 1/5 \\ -x + y \qquad &= 0 &&\text{weight } 1/2 \\ -y + z &= 0 &&\text{weight } 1/4 \\ z &= 0{\cdot}12 &&\text{weight } 1/4 \\ -x \qquad + z &= 0{\cdot}14 &&\text{weight } 1/3 \end{aligned}$$

We may take the weights as 12, 30, 15, 15, 20 and form the normal equations

$$62x - 30y - 20z = - 2{\cdot}8$$
$$- 30x + 45y - 15z = 0$$
$$- 20x - 15y + 50z = 4{\cdot}6$$

These give $x = 0{\cdot}01(2)$, $y = 0{\cdot}04(5)$, $z = 0{\cdot}11(0)$. The corrected heights are: *B*, 83·46 ft.; C, 107·56 ft.; D, − 29·19 ft.

10.7.1. Reduction of observation equations to linear form

It has so far been assumed that the observation equations are linear in the unknowns to be determined by the method of least squares. When they are not so, it is necessary to obtain first approximations to the values of the unknowns, by graphical or other methods, and then to express the observation equations in terms of the unknown small corrections to be applied to these first approximations. On the assumption that the corrections are sufficiently small for their squares and higher powers to be neglected, the observation equations will now be linear.

Consider the law $y = Ae^{Kx}$, values of y being directly observed and subject to experimental error. It is required to find the best values of A and K. A test of conformity to this law is to plot values of log y against x; exact conformity would yield a straight line graph since the law may be written

$$Y = \log_e y = \log_e A + Kx$$
$$\text{or } Y = \log_{10} y = \log_{10} A + 0{\cdot}4343\, Kx.$$

We could, of course, find a 'best' straight line expressing the relation between the observed values of Y and x, deriving 'best' values of log A (and hence of A) and K. If, however, the basis of the method of least squares is the assumption of the normal law of errors in the directly observed quantity, in this case y, the procedure is incorrect. A straight line drawn by eye will usually be sufficient to yield first approximations A_0 and K_0 to the values of the required constants. Substitute $A = A_0 + \alpha$, $K = K_0 + \beta$ in the assumed law, which is also the form of the observation equations, and expand the R.H.S. as far as first-degree terms in α and β. We obtain

$$\begin{aligned} y &= (A_0 + \alpha)\, e^{(K_0+\beta)x} \\ &= (A_0 + \alpha)\, e^{K_0x}e^{\beta x} \\ &= (A_0 + \alpha)\, e^{K_0x}(1 + \beta x), \text{ neglecting } \beta^2 \text{ and higher powers} \\ &= A_0e^{K_0x} + e^{K_0x}\alpha + A_0e^{K_0x}\beta x, \qquad \qquad (17) \end{aligned}$$

neglecting the product $\alpha\beta$.

This is the form of the observation equations in terms of α and β as the unknowns. When the observed values are substituted it will be necessary to calculate the term $A_0e^{K_0x}$ with a high degree of accuracy, for it will be fairly nearly equal to y in each case. The method is illustrated in the following example. An alternative method is given in §10.7.2.

Ex. viii

The following values of x and y, in which y is directly observed, very nearly satisfy a law of the form $y = Ae^{Kx}$. Obtain the best values of A and K by the method of least squares.

x:	0	1	2	3
y:	1·05	2·10	3·85	8·30

By plotting $\log_e y$ against x, it will be found that suitable first approximations are $A_0 = 1$, $K_0 = 0{\cdot}7$. Substituting the observed values in equation (17) and transposing the term $A_0e^{K_0x}$ to the L.H.S. we obtain the observation equations:

$$\begin{aligned} 0{\cdot}05 &= \alpha \\ 0{\cdot}086 &= 2{\cdot}014\alpha + 2{\cdot}014\beta \\ -\,0{\cdot}205 &= 4{\cdot}055\alpha + 8{\cdot}11\beta \\ 0{\cdot}134 &= 8{\cdot}166\alpha + 24{\cdot}5\beta \end{aligned}$$

From these the normal equations are

$$0{\cdot}303 = 88\alpha + 237\beta$$
$$1{\cdot}427 = 237\alpha + 670\beta$$
$$\text{giving } \alpha = -\,0{\cdot}028,\ \beta = 0{\cdot}013$$
$$A = 1 - 0{\cdot}028 = \underline{0{\cdot}972}$$
$$K = 0{\cdot}7 + 0{\cdot}013 = \underline{0{\cdot}713}.$$

10.7.2 Error proportional to magnitude

It is sometimes reasonable to assume that the error to be feared in a directly measured quantity increases proportionally with the magnitude of the reading. There are two ways of allowing for this. If σ_y^2 denotes the mean square error to be feared when the magnitude is y, we are supposing that $\sigma_y^2 \propto y^2$. Each reading in y should be given a weight inversely proportional to its variance (§9.4) and we can do this by multiplying each observation equation by $1/y$, the square root of the weight, before forming the normal equations by the usual rule.

Again, to suppose that the error to be feared is proportional to y is to suppose a constant distribution of $\delta y/y$ where δy is a small error in y. But $\delta y/y = \delta(\log y)$, so that we suppose a constant liability to error in $\log y$. Hence if we make $\log y$ a new variable, Y say, we shall have no need of weighting. This will be very convenient when the law is of a form which becomes linear in the unknown constants when Y is the variable; two such laws are $y = Ae^{Kx}$ and $y = Ax^n$. Thus if we express the first of these in the form $Y = \log_e y = \log_e A + Kx$ and proceed to determine the best linear relation between Y and x, on the assumption of constant liability to error in Y, then the resulting values of A and K are appropriate to the assumption of error in y proportional to magnitude; they will give the best fit to the straight-line graph of Y against x, not to the curve of y against x.

If now we wish to fit the best curve $y = Ae^{Kx}$ to the observed values, i.e. we assume constant liability to error in y, we have two procedures available. As shown in §10.7.1 we may use suitable first approximations to A and K to put the equation in a form which is linear in the required small corrections. Alternatively, since the determination of the linear law for Y in terms of x is equivalent to giving every observed value of y a weight $1/y^2$, we may reverse this weighting, giving each value of Y a weight y^2. This may be achieved by multiplying each observation equation in Y by y before forming the normal equations. Hence if there is constant liability to error in y, the weighted observation equations, linear in the unknown constants, are of the form

$$y \log y = y \log A + Kxy, \qquad (18)$$

and the normal equations are

$$\left.\begin{aligned}[y^2\log y] &= [y^2]\log A + [xy^2]K \\ [xy^2\log y] &= [xy^2]\log A + [x^2y^2]K\end{aligned}\right\} \qquad (19)$$

If common logarithms are used, K must be replaced by $0{\cdot}4343K$. We will illustrate this method with the data of Ex. viii, contrasting the results with those obtained on the assumption of error proportional to magnitude.

Ex. ix

From the values

x:	0	1	2	3
y:	1·05	2·10	3·85	8·30

find the best values of the constants A, K in the law $y = Ae^{Kx}$, assuming (a) constant liability to error in y, (b) error proportional to magnitude.

(a) To form the normal equations (19) we require the table:

x	y	$\log y$	y^2	xy^2	x^2y^2	$y^2\log y$	$xy^2\log y$
0	1·05	0·0488	1·0525	0	0	0·05	0
1	2·10	0·7419	4·4100	4·41	4·41	3·27	3·27
2	3·85	1·3481	14·8225	29·65	59·29	19·98	39·96
3	8·30	2·1163	68·8900	206·67	620·01	145·79	437·37
			89·17	240·71	683·71	169·09	480·60

Thus

$$89{\cdot}17 \log A + 240{\cdot}71K = 169{\cdot}09$$
$$240{\cdot}71 \log A + 683{\cdot}71K = 480{\cdot}60$$

$\log A = -0{\cdot}025$ $\qquad$ $\underline{A = 0{\cdot}975}$

$\underline{K = 0{\cdot}712}$

These are in close agreement with the values obtained in Ex. viii.

(b) We now require the best linear relation

$$Y = \log y = \log A + Kx,$$

the observation equations being of equal weight. The normal equations take the forms

$$[\log y] = 4 \log A + [x]K$$
$$[x \log y] = [x] \log A + [x^2]K.$$

These are

$$4{\cdot}2511 = 4 \log A + 6K$$
$$9{\cdot}7870 = 6 \log A + 14K$$

and give

$$\underline{A = 1{\cdot}044}, \ \underline{K = 0{\cdot}682}.$$

10.7.3

As a further example of a law which is not linear in the unknown constants, consider the relation $y = a/(x + b)$. A test of conformity to the law is to plot values of $1/y$ against x, for

$$Y = \frac{1}{y} = \frac{x + b}{a}.$$

If y is the directly observed quantity supposed subject to the normal law of errors, there are again two ways of determining the best values of a and b.

Firstly, if the graphical test has yielded first approximations a_0 and b_0,

$$y = \frac{a_0 + \alpha}{x + b_0 + \beta}$$

where α, β are required small corrections. Expanding to the first order in α and β, the observation equations take the form

$$y = \frac{a_0 + \alpha}{x + b_0}\left(1 - \frac{\beta}{x + b_0}\right)$$
$$= \frac{a_0 + \alpha}{x + b_0} - \frac{a_0\beta}{(x + b_0)^2}$$

This form is linear in the unknowns.

Secondly, if $Y = \frac{1}{y}$ then $\delta Y = -\frac{1}{y^2}\delta y$ and $\sigma_Y^2 = \frac{1}{y^4}\sigma_y^2$.

The linear relation $Y = \frac{1}{a}x + \frac{b}{a}$ would give weight $\frac{1}{y^4}$ to readings in y which should have equal weight. The equal weight may be obtained by giving the equations in Y a weight y^4 and this may be done by multiplying each equation by y^2 before forming the normal equations.

Write $1/y = cx + d$ where $c = 1/a$, $d = b/a$.

Multiplying by y^2,

$$y = cxy^2 + dy^2$$

Normal equations which give equal weight to readings in y are

$$[xy^3] = c[x^2y^4] + d[xy^4]$$
$$[y^3] = c[xy^4] + d[y^4].$$

10.8 Undetermined multipliers or correlates

The conditions for a stationary value of a function of several variables which are not independent but satisfy given equations, may be obtained by Lagrange's method of undetermined multipliers,† as follows.

Let $f(x_1, x_2 \ldots x_m)$ be a function of m variables which satisfy k equations

$$\phi_r(x_1, x_2 \ldots x_m) = 0 \ (r = 1, 2 \ldots k) \quad . \quad . \quad (20)$$

Then we can find k numbers $\lambda_1, \lambda_2 \ldots \lambda_k$ as well as m values of the variables $x_1, x_2 \ldots x_m$ to satisfy

(i) the k equations (20) and also

(ii) m equations

$$\frac{\partial f}{\partial x_s} + \lambda_1\frac{\partial \phi_1}{\partial x_s} + \lambda_2\frac{\partial \phi_2}{\partial x_s} + \ldots = 0 \ (s = 1, 2 \ldots m) \quad . \quad (21)$$

and for the m values of the variables, the value of $f(x_1, x_2 \ldots x_m)$ is a stationary one.

The effect of this theorem is that although we cannot say, when the variables are not independent, that the conditions for a stationary value are

$$\frac{\partial f}{\partial x_s} = 0, \ (s = 1, 2 \ldots m)$$

† See, for instance, Dockeray, *Elementary Treatise on Pure Mathematics*, §225

we can say that multipliers $\lambda_1, \lambda_2 \ldots \lambda_k$ exist such that

$$\frac{\partial}{\partial x_s}(f + \lambda_1\phi_1 + \lambda_2\phi_2 + \ldots) = 0.$$

In the method of least squares we seek to make Σv^2, the sum of squares of residuals, a minimum. Σv^2 is a function of several unknowns and if we take $\Sigma\frac{1}{2}v^2$ as $f(x_1, \ldots x_m)$ in the above theorem then $\partial f/\partial x_s = 0$, $(s = 1, 2 \ldots m)$ represents the normal equations. (The factor 1/2 makes this form of the normal equations consistent with the rule in §10.3 for the case when the observation equations are linear in the unknowns.) If the unknowns must satisfy exactly certain rigorous equations, these will be the equations (20). They also can be made linear in the unknowns, by the method of §10.7.1, making the unknowns small corrections to be applied to suitable first approximations to the required values of the variables. Then equations (21) will take the form

$$N_s + \lambda_1 a_s + \lambda_2 b_s + \ldots = 0 \ (s = 1, 2 \ldots m) \quad . \quad . \quad (22)$$

where $N_s = 0$ represents the normal equations as formed by the usual rule and a_s, b_s, etc., are the coefficients of x_s in the rigorous equations $\phi_1 = 0$, $\phi_2 = 0$, etc., reduced to linear form.

In many problems of adjustment by least squares in surveying, each observation represents a directly observed value of an angle or a length, but there are rigorous equations connecting the true values of the observed quantities. If we regard every observed value as the first approximation to the corresponding true value, the unknowns to be determined may be made equal in number to the observation equations, being the corrections to be applied to all the observations. If these corrections are x_1, x_2, etc., the observation equations take the form

$$\begin{aligned} x_1 \qquad\qquad &= 0 \quad \text{weight } w_1 \\ x_2 \qquad &= 0 \quad \text{weight } w_2 \\ x_3 &= 0 \quad \text{weight } w_3 \end{aligned}$$

and so on.

Suppose now the rigorous equations become

$$\begin{aligned} a_1x_1 + a_2x_2 + \ldots &= e_1 \\ b_1x_1 + b_2x_2 + \ldots &= e_2, \text{ etc.} \quad . \quad . \quad (23) \end{aligned}$$

Take $-\lambda_1$, $-\lambda_2$, etc., as Lagrange's multipliers (the λ's are now called correlates). Then equations (22) take the form

$$\left.\begin{aligned} w_1x_1 - \lambda_1a_1 - \lambda_2b_1 - \ldots = 0 \\ w_2x_2 - \lambda_1a_2 - \lambda_2b_2 - \ldots = 0 \end{aligned}\right\} \quad . \quad . \quad (24)$$

and so on.

Using equation (24) we may substitute for x_1, x_2, etc., in equation (23) obtaining linear equations for the λ's; solve these and use equation (24) to determine the x's.

The most arduous part of the solution of a problem in least squares is the solution of the normal equations. When there is a choice between (i) reduction to a minimum number of unknowns and (ii) the use of correlates, we may with advantage choose the latter if the number of correlates will be smaller than the minimum number of unknowns. In Ex. vii we had 5 observations in 3 independent unknowns. We will now solve the same problem, treating every observation as requiring correction but with two equations to be exactly satisfied by the corrected values.

Ex. x

Solution of Ex. vii, using correlates.

Let x_1, x_2, x_3, x_4, x_5 be the required corrections to the observed differences of level, in the order given.

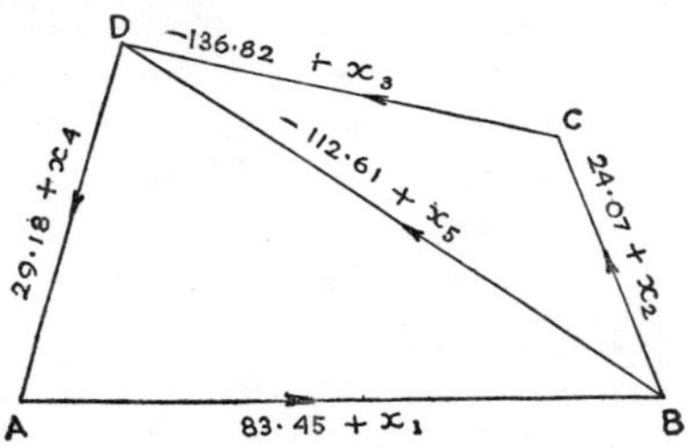

FIG. 41

The diagram shows the relations between the points in respect of difference of height. The observation equations are

		weight
x_1	$= 0$	1/5
x_2	$= 0$	1/2
x_3	$= 0$	1/4
x_4	$= 0$	1/4
x_5	$= 0$	1/3

Since there are three independent unknowns (the heights of B, C and D above A) there must be two rigorous equations. These are obtained by using the fact that the total rise in height in a closed circuit must be zero. For the circuits ABD, BCD

$$83{\cdot}45 + x_1 - 112{\cdot}61 + x_5 + 29{\cdot}18 + x_4 = 0$$
$$24{\cdot}07 + x_2 - 136{\cdot}82 + x_3 = -112{\cdot}61 + x_5$$

These reduce to

Correlate

$$\left.\begin{array}{ll} x_1 + x_4 + x_5 = -0{\cdot}02 & \lambda_1 \\ x_2 + x_3 - x_5 = 0{\cdot}14 & \lambda_2 \end{array}\right\} \quad . \quad . \quad (25)$$

The conditions for the minimum sum of squares are

$$\begin{array}{ll} \tfrac{1}{5}x_1 - \lambda_1 = 0 & x_1 = 5\lambda_1 \\ \tfrac{1}{2}x_2 - \lambda_2 = 0 & x_2 = 2\lambda_2 \\ \tfrac{1}{4}x_3 - \lambda_2 = 0 & x_3 = 4\lambda_2 \\ \tfrac{1}{4}x_4 - \lambda_1 = 0 & x_4 = 4\lambda_1 \\ \tfrac{1}{3}x_5 - \lambda_1 + \lambda_2 = 0 & x_5 = 3\lambda_1 - 3\lambda_2 \end{array}$$

Substituting in equations (25), we have

$$12\lambda_1 - 3\lambda_2 = -\ 0{\cdot}02$$
$$-\ 3\lambda_1 + 9\lambda_2 = \quad 0{\cdot}14$$

giving $\lambda_1 = 0{\cdot}0024$, $\lambda_2 = 0{\cdot}0164$

$$x_1 = 0{\cdot}012,\quad x_2 = 0{\cdot}033,\quad x_4 = 0{\cdot}010.$$

Reference to the diagram gives the corrected heights above A as

B, 83·46(2) ft.; C, 107·56(5) ft.; D, − 29·19(0) ft.

10.9.1 Precision of the results

In the case of direct observations of a single quantity, the best estimate of the true value, on the basis of a number of observed values of differing precision, is the weighted mean of the observed values, and this mean has a weight equal to the sum of the weights of the observations. Thus the probable error of the mean can be estimated.

When the unknowns are not directly measured but are deduced from observation equations by the method of least squares, the problem of estimating the precision of the results is more difficult. The rules, for the proof of which the reader is referred to Whitaker and Robinson, *Calculus of Observations*, Chap. IX, are as follows.

First reduce the observation equations to linear form and equal weight (*see* §§10.6.1 and 10.7.1). After solution by the method of least squares, substitute in the observation equations the values obtained for the unknowns and so determine the residual, v, in each equation, namely the numerical difference between the two sides of the equation. If there were n observation equations and k normal equations, the variance of an observation of unit weight is

$$\sum_1^n v^2/(n-k) \text{ or } [v^2]/(n-k) \quad . \quad . \quad . \quad (26)$$

Now suppose that the unknowns are $x_1, x_2 \ldots x_k$. The observation equations will be of the form

$$p_j x_1 + q_j x_2 + \ldots = l_j,\ (j = 1, 2 \ldots n)$$

and the normal equations

$$a_{11}\,x_1 + a_{12}\,x_2 + \ldots + a_{1k}\,x_k = [pl]$$
$$a_{21}\,x_1 + a_{22}\,x_2 + \ldots + a_{2k}\,x_k = [ql], \text{ etc.}$$

If Δ denotes the determinant of the coefficients on the L.H.S. and A_{rr} the minor determinant of a_{rr}, that is of the coefficient of x_r in the rth equation, then the calculated value of x_r has weight Δ/A_{rr}. Since variance is inversely proportional to weight, and equation (26) gives the variance associated with unit weight, it follows that the variance of x_r is

$$\frac{A_{rr}[v^2]}{\Delta(n-k)}$$

This is best expressed by saying that the r.m.s. error to be feared in the calculated value of x_r is

$$\sqrt{\frac{A_{rr}[v^2]}{\Delta(n-k)}} \quad . \quad . \quad . \quad . \quad (27)$$

The factor A_{rr}/Δ may be obtained by the following method. Replace the terms $[pl]$, $[ql]$, etc., on the R.H.S. of the normal equations by zeroes, except in the rth equation, where the R.H.S. is made 1. Now solve for x_r, the value of which will be A_{rr}/Δ.

10.9.2 Methods of solving equations

There are three recognized methods of solving simple simultaneous equations: by means of determinants, by elimination, and by successive approximations. The determinant of the coefficients in a set of normal equations is symmetrical, that is, the determinant is unaltered if rows and columns are interchanged. This property simplifies the solution of the equations by the use of determinants (see Whitaker and Robinson, *Calculus of Observations*). Of the methods using successive approxima-tions, that known as the relaxation method is particularly suitable if the coefficients a_{11}, a_{22}, a_{33} . . . are relatively large. The best form of the method of elimination is known as the Gauss-Doolittle method and is illustrated in the following example. The equations are represented by their coefficients, all (including the constant term) on the L.H.S. The sum of the coefficients in each line appears in the column S, which is used as a check on the computation. Equation A_1 is derived from A by making the leading coefficient equal to -1; A_2, A_3 and A_4 are all derived from A_1 and similarly for other letters. The process consists of the successive elimination of the unknowns x, y and z. Bracketed equa-tions are added.

Ex. xi

Gauss-Doolittle method of solution of simple simultaneous equations.

To solve	A	$2x + 6y + 9z + 4u = 22$
	B	$6x + 30y - 2z + 17u = 95$
	C	$9x - 2y + 11z - 2u = -11$
	D	$4x + 17y - 2z + 12u = 64$

	x	y	z	u	c	S
A_1	− 1	− 3	− 4·5	− 2	11	0·5
B	6	30	− 2	17	− 95	− 44
A_2	− 6	− 18	− 27	− 12	66	3
E		12	− 29	5	− 29	− 41
E_1		− 1	2·417	− 0·417	2·417	3·417
C	9	− 2	11	− 2	11	27
A_3	− 9	− 27	− 40·5	− 18	99	4·5
E_2		29	− 70·09	12·09	− 70·09	− 99·09
F			− 99·59	− 7·91	39·91	− 67·59
F_1			− 1	− 0·079	0·400	− 0·679
D	4	17	− 2	12	− 64	− 33
A_4	− 4	− 12	− 18	− 8	44	2
E_3		− 5	12·08	− 2·08	12·08	17·08
F_2			7·92	0·63	− 3·17	5·38
G				2·55	− 11·09	− 8·54
G_1				− 1	4·349	3·349

$u = 4{\cdot}34(9)$

$F_1 \quad z = -0{\cdot}079\,(4{\cdot}349) + 0{\cdot}400 = \underline{0{\cdot}05(6)}$

$E_1 \quad y = 2{\cdot}417\,(0{\cdot}056) - 0{\cdot}417\,(4{\cdot}349) + 2{\cdot}417 = \underline{0{\cdot}73(9)}$

$A_1 \quad x = -3(0{\cdot}739) - 4{\cdot}5\,(0{\cdot}056) - 2(4{\cdot}349) + 11 = \underline{-\,0{\cdot}16(7)}$

Suggestions for further reading:

E. T. Whitaker and G. Robinson, *Calculus of Observations*, Chapter 9.

M. G. Kendall, *Advanced Theory of Statistics*, Chapter 22 (Griffin).

EXAMPLES ON CHAPTER X

1. Determine the relation $y = a + bx$ which best fits the following observed values:

x:	0	1	2	3	4	5	6
y:	0·4	0·7	1·1	1·6	1·9	2·3	2·6

2. From the following values find the best linear relations expressing (i) y as a function of x, (ii) x as a function of y.

x:	0	100	200	300	400
y:	50	40	25	15	0

3. The number of grams of a given salt which will dissolve in 100 gms. of water at different temperatures is shown in the table below:

Temperature: $x°$ C.	0	10	20	30	40	50
Weight of salt: y gms.	53·5	59·5	65·2	70·6	75·5	80·2
Temperature: $x°$ C.	60	70	80	90	100	
Weight of salt: y gms.	85·5	90·0	95·0	99·2	104·0	

Use the method of least squares to find the linear formula, $y = a + bx$, which best fits these observations. (L.U.)

4. Determine the best values of the constants a, b, c in the law

$$R = a + bv + cv^2$$

from the following observed values.

v:	10	20	30	40	50	60
R:	157	179	210	252	302	361

Use your results to estimate the true values of R, correct to one place of decimals, for the given values of v.

5. Complete the calculations of Ex. v, page 181, obtaining the best function $a + bx + cx^2$ to represent the function sin x over the range $x = 0$ to $x = \pi/2$.

6. Obtain by the method of least squares the function $A + Bx + Cx^2$ which best represents the function $\sqrt{(1 - x^2)}$ over the range $x = 0$ to $x = 1$.
By change of scales deduce the best quadratic function representing the ordinates, in the positive quadrant, of the ellipse $x^2/a^2 + y^2/b^2 = 1$.

7. The following lengths were observed in standardizing a 300-ft. invar tape on a plane surface under the field tension:

0–100	ft.	99·9972
0–200	,,	199·9958
0–300	,,	299·9936
100–200	,,	99·9967
100–300	,,	199·9952
200–300	,,	99·9970

Determine the most probable values of the individual 100-ft. lengths and the corresponding total length. Also state the probable error of the adjusted length. (L.U.)

8. ABCD is a quadrilateral in a plane. The following are observed values, of equal weight: BAD = 70° 14′ 07″, ABD = 43° 22′ 11″, ADB = 66° 23′ 46″, DBC = 51° 44′ 24″, BDC = 80° 31′ 14″, BCD = 47° 44′ 24″, ABC = 95° 06′ 30″, ADC = 146° 55′ 05″.
Taking as independent unknowns the angles ABD, ADB, DBC and BDC, find their most probable values.

9. Variables y and x are known to satisfy the law $y = Kx$. Observed values of y are liable to error and are of unequal weight. If w denotes the weight of an observed value, obtain a formula for the best value of K.

10. A base-line was set out in three separate lengths: AB = p, BC = q and CD = r. The average of three measurements of AB was 5·82 chains and the averages of BC and CD (both measured twice) 9·74 and 4·59 chains respectively. In addition a single measurement of BD was taken, giving a length of 14·28 chains.
Determine the most probable values of p, q and r. (Note that an average of n measurements has n times the weight of a single measurement.) (L.U.)

11. Use the method of correlates (one undetermined multiplier or correlate is required) to establish the result of Ex. vi, page 182.
Show that the rule applies to the error in the sum of any number of angles of known sum (e.g. the angles of a polygon or n adjacent angles at a point together forming the whole angle 360°).

12. ABC is a triangle on the earth's surface. It is known (from the area of the triangle) that the spherical excess is 5·3″. If observed values of the angles A, B, C are 58° 17′ 30·5″, 41° 20′ 17″ and 80° 22′ 21·5″ with weights 2, 1, 2 respectively, obtain their corrected values.

13. Derive, by the method of least squares, an expression for the correction of a triangle when the weights of the observations are unequal.
The following mean values were recorded for a triangle ABC, the individual measurements being uniformly precise:

A = 60° 10′ 16″	6 observations
B = 58° 40′ 36″	4 observations
C = 61° 08′ 56″	2 observations

Give the corrected value of the angles. (L.U.)

14. The mean observed values of the angles of a traverse ABCDEA, together with the number of times each angle was read, are given below.

Angle	*Mean Observed Value*	*Number of Readings*
EAB	73° 21′ 18″	2
ABC	115° 16′ 24″	2
BCD	123° 07′ 52″	1
CDE	89° 32′ 42″	3
DEA	138° 40′ 46″	1

Prepare the normal equations and determine the most probable values of the angles. Check the results by an alternative method. (L.U.)

15. The following are the recorded differences of level between four bench marks A, B, C and D and the lengths of the route levelled.

A to B	Fall	66·22 ft.	3 miles
B to C	Rise	121·68 ft.	4 miles
C to D	Rise	11·97 ft.	3 miles
D to A	Fall	67·25 ft.	4 miles
A to C	Rise	55·34 ft.	6 miles

Each difference is the mean of the levellings in opposite directions with sight lengths equalized. The reduced level of A is 120·32. Determine the most probable levels of the other three bench marks. (L.U.)

16. The relation between x and t is of the form $1/x = a + bt$. Assuming uniform precision in the measurement of x, find the best values of a and b from the following observed values of x and t.

t:	0	1	2·5	5	10
x:	1·02	0·08	0·62	0·46	0·32

17. The following table gives corresponding values of two variables d and A

d:	1·1	2·0	3·2	4·0	5·5	6·3
A:	5·3	14·2	30·1	43·8	77·3	97·8

It is thought that they may be connected by a law of the form $A = nd + md^2$, where m and n are constants.
Test this assumption and find the best values for m and n by the method of least squares.

18. By means of a straight-line graph verify that the following values of x and t satisfy very nearly an equation of the form $x = Ae^{Kt}$. Assuming that error in x may be expected to be proportional to the magnitude of x, find the most plausible values of A and K.

t:	0	5	8	12	20
x:	3·0	1·5	1·0	0·55	0·18

19. With the data of Question 18, find the best values of A and K on the assumption of constant liability to error in x.

CHAPTER XI

CORRELATION

Regression lines. Linear regression. Coefficient of correlation. Correlation table. Correlation-ratios. Bivariate normal distribution. Precision and significance of r. Accuracy of prediction. Correlation by ranks. Trivariate correlation.

11.1.1 Scatter diagram. Regression lines

We have used the method of least squares to determine the values of unknown constants in a relation between two variables, when the deviations of the observed values of the dependent variable from those given by the assumed relation are small, being such as might reasonably be expected as inevitable errors of measurement. The relationship between the variables is then so well defined that repeated measurements of the dependent variable, for a fixed value of the independent variable, will all yield values very near one another.

Now when statistical data take the form of paired values of two variables, all things are possible from a well-defined relationship to complete statistical independence. In many cases, something between these two extremes will be apparent, namely that it is possible to say that 'on the average' as one variable, chosen as independent variable, increases, the other changes in a particular way.

Suppose each pair of observed values of two variables x and y represented by a dot at the point (x, y). The dots form a *scatter diagram* and the three possibilities: (i) marked relationship, (ii) a degree of interdependence, (iii) independence, will be reflected in the scatter diagram, as in Fig. 42.

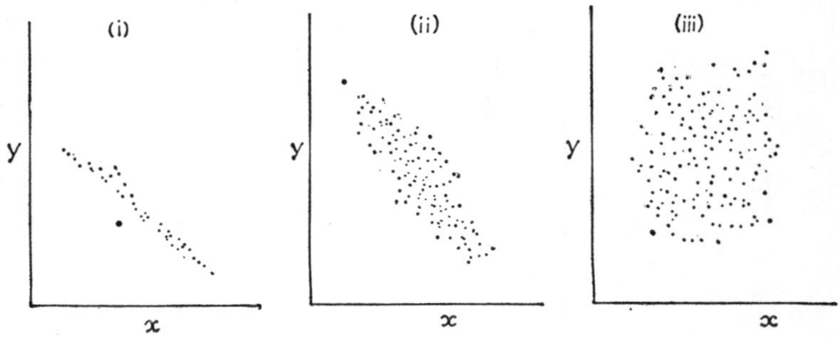

SCATTER DIAGRAMS

FIG. 42

The main purpose of an empirical law expressing the relation between two variables is to enable us to predict or estimate the value of one when the other is given. It is clear that if, for a given value of x, the value of y can vary appreciably, we can at best attempt to estimate the most probable value of y. This is usually taken to be the mean of the observed values. Denoting by y_M the mean value of y for given x, the locus of (x, y_M) is called *the line of regression* of y on x. Similarly the line of regression of x on y is a second locus, that of (x_M, y) where x_M is the mean value of x for given y.

11.1.2 Linear regression

The regression lines will in general be curved, but the simplest and most important case is that in which each locus of means approximates to a straight line. The method of least squares is used to determine the equation of either line, the coefficients in an equation so obtained being 'best' estimates on the basis of the observed values of (x, y). (Even if either locus is far from straight, the method will yield a "best" straight line; its value, however, is small unless the corresponding locus of means of observed values is nearly linear.)

Suppose there are N pairs of observed values. Let y_{Mi} be the mean of n_i values of y which occur with the value x_i of x, so that $\Sigma n_i = N$. In applying the method of least squares we give the pair (x_i, y_{Mi}) a weight n_i, since the corresponding point represents the mean of n_i observations. The best straight line passes through the centroid of the weighted observations, namely through $(\bar{x}, \bar{y})$ where

$$\bar{x} = \frac{\Sigma n_i x_i}{\Sigma n_i} = \frac{\sum_1^N x}{N}$$

$$\bar{y} = \frac{\Sigma n_i y_{Mi}}{\Sigma n_i} = \frac{\sum_1^N y}{N}$$

so that $(\bar{x}, \bar{y})$ is simply the centroid of the original observed set of points, given equal weights.

To find the gradient of the line of regression of y on x, suppose the origin shifted to $(\bar{x}, \bar{y})$ and use capital letters for co-ordinates referred to the new origin. Applying equation (6) of Chapter X, with the appropriate weights, we have

$$b = \frac{\Sigma n_i X_i Y_{Mi}}{\Sigma n_i X_i^2} = \frac{\sum_1^N XY}{\sum_1^N X^2} = \frac{\Sigma XY}{N\sigma_1^2} = \frac{\text{cov}(x, y)}{\text{var } x}$$

where σ_1^2 is the variance of the x's.

The line of regression of y on x is thus

$$y_M - \bar{y} = \frac{\text{cov }(x, y)}{\text{var } x}(x - \bar{x}) \quad . \quad . \quad . \quad (1)$$

exactly as in equation (7) of Chapter X.

The line of regression of x on y is, similarly,

$$x_M - \bar{x} = \frac{\text{cov }(x, y)}{\text{var } y}(y - \bar{y}) \quad . \quad . \quad . \quad (2)$$

The coefficients on the R.H.S. of these equations are called the *regression coefficients* and may be denoted by β_2 and β_1 respectively, the Greek letters being applicable to the true values for the real or hypothetical complete 'population'. Values derived from a sample are to be denoted by b_2 and b_1; var x and var y will be replaced by the sample variances s_1^2 and s_2^2.

11.1.3 Calculation of regression coefficients

It is convenient to make use of trial means or shift of origin in the calculation of x, y, s_1^2, s_2^2 and also of cov $(x, y) = \frac{1}{N}\Sigma(x - \bar{x})(y - \bar{y})$.

If a, b are chosen trial means of x, y respectively and

$$X = x - a,\ Y = y - b$$
$$\text{then } \bar{X} = \bar{x} - a,\ \bar{Y} = \bar{y} - b$$
$$\begin{aligned}\text{so that } \Sigma(x - \bar{x})(y - \bar{y}) &= \Sigma(X - \bar{X})(Y - \bar{Y}) \\ &= \Sigma XY - \bar{X}\Sigma Y - \bar{Y}\Sigma X + N\bar{X}\bar{Y} \\ &= \Sigma XY - \bar{X}N\bar{Y} - \bar{Y}N\bar{X} + N\bar{X}\bar{Y} \\ &= \Sigma XY - N\bar{X}\bar{Y} \quad . \quad . \quad . \quad (3)\end{aligned}$$

This equation may be compared with equation (5) of Chapter X; it shows that covariance, like variance, is independent of origin, being a function of deviations from means. Equation (3) is in terms of summations over all individual values; if a sample is large it will be convenient to group the values of both variates and form a frequency table. (See §11.2.3.)

Ex. i

Efficiency of Smoke Collection

x: mean particle size ($m\mu$)	28	29	34	34	34	35	36	36	38	39
y: efficiency (%)	29	44	39	40	48	28	44	59	67	54

x:	40	40	40	41	41	42	42	42	43	43	43	44	45	45	45
y:	51	52	83	41	51	56	65	70	57	61	67	55	62	69	71

x:	45	45	46	47	47
y:	76	76	69	66	81

Obtain the equation of the regression line of y on x.

Put $X = x - 40$, $Y = y - 55$ and tabulate the values of X, Y, X^2 and XY, obtaining

$$\Sigma X = 9,\ \Sigma Y = 91,\ \Sigma X^2 = 751,\ \Sigma XY = 1633.$$

Since $N = 30$, $\bar{x} = 40 + 9/30 = 40{\cdot}3$, $\bar{y} = 55 + 91/30 = 58{\cdot}0$

$$Ns_1^2 = 751 - 30\,(0{\cdot}3)^2 = 748{\cdot}3$$

$$b_2 = \frac{\Sigma XY - 30\bar{X}\bar{Y}}{Ns_1^2} = \frac{1630}{748} = 2{\cdot}18.$$

The regression line is $y - 58 = 2{\cdot}18\,(x - 40{\cdot}3)$
or $\underline{y = 2{\cdot}18x - 30.}$

The reader will appreciate that we have merely found the best straight line through the points, with y as dependent variable, by the method of least squares, the only differences from the procedure of Chapter X being that we have used the general equation for the best line, that we could equally well have made x the dependent variable and that a wider scatter of points about the regression line is envisaged than was the case when the equation $y = a + bx$ was regarded as a 'law'. The reader is recommended to plot the given values and draw the regression line. It represents the *general trend* of the efficiency of collection with increasing particle size.

11.2.1 The coefficient of correlation

If, as in §11.1.2, X and Y denote deviations from the mean values of x and y, the equations of the regression lines may be written

$$Y = \beta_2 X \text{ and } X = \beta_1 Y \,. \qquad (4)$$

where $$\beta_2 = \sum_{r=1}^{N} X_r Y_r \Big/ N\sigma_1^2, \quad \beta_1 = \sum_{r=1}^{N} X_r Y_r \Big/ N\sigma_2^2$$

Thes two gradients are unequal, but the equations (4) may be written

$$\frac{Y}{\sigma_2} = \varrho \frac{X}{\sigma_1} \text{ and } \frac{X}{\sigma_1} = \varrho \frac{Y}{\sigma_2} \,. \qquad (5)$$

where $$\varrho = \frac{\sigma_1 \beta_2}{\sigma_2} = \frac{\sigma_2 \beta_1}{\sigma_1} = \sum_{r=1}^{N} X_r Y_r \Big/ N\sigma_1\sigma_2 \,. \qquad (6)$$

Equations (5) show that if the deviations from the means are standardized, by dividing by the corresponding standard deviations, the regression lines with the standardized deviations as variables, make equal angles, the one with the horizontal and the other with the vertical axis. Numerically this means that, considering x as the independent variable, the mean value of y for given x, in standardized units, increases by ϱ units for unit increase in x, and a precisely similar statement is valid with y as independent variable. The coefficient ϱ is thus a measure of the tendency of *either* variable, expressed in standardized units, to increase when the other increases. ϱ is called the *coefficient* of (product-moment) *correlation*. The use of the Greek letter implies that the value relates to a complete bivariate population; the corresponding quantity derived from a sample is denoted by r.

Note that by virtue of equations (6) $\varrho^2 = \beta_1\beta_2$ and similarly r is the geometric mean of the regression coefficients derived from a sample.

Ex. ii

From the data of Ex. i find the coefficient of correlation and the equation of the second line of regression.

It is only necessary to calculate s_2^2, the sample variance of the y's. This is found to be given by $30s_2^2 = 5969$.

Then
$$r = \frac{1630}{\sqrt{(748 \times 5969)}} = 0{\cdot}77$$
$$b_1 = \frac{1630}{5969} = 0{\cdot}273$$

The line of regression of x on y is
$$x - 40{\cdot}3 = 0{\cdot}273\ (y - 58{\cdot}0)$$
$$\underline{x = 0{\cdot}273y + 24{\cdot}5.}$$

11.2.2 Positive and negative correlation

Assuming approximately linear regression, the trend of either variable may be either to increase or to decrease as the other increases. In Exs. i and ii each variable increases, on the average, when the other increases; the value of r is positive. Data showing, at a sequence of times, the marriage rate and numbers of unemployed would exhibit the opposite condition, that on the average as either variable increases the other decreases, the value of r being negative.

Numerically, ϱ (or r) cannot exceed 1. If X, Y denote deviations from the means of x and y,
$$\varrho^2 = \frac{(\Sigma XY)^2}{N^2\sigma_1^2\sigma_2^2} = \frac{(\Sigma XY)^2}{\Sigma X^2 . \Sigma Y^2}$$

Now
$$\begin{aligned} &\Sigma X^2 . \Sigma Y^2 - (\Sigma XY)^2 \\ &= (X_1^2 + X_2^2 + \ldots + X_N^2)(Y_1^2 + \ldots + Y_N^2) \\ &\qquad - (X_1Y_1 + X_2Y_2 + \ldots + X_NY_N)^2 \\ &= \Sigma(X_rY_s - X_sY_r)^2 \end{aligned}$$

Since a sum of squares cannot be negative, the denominator of $\varrho^2 \geqslant$ the numerator, i.e. $\varrho^2 \leqslant 1$.

(a) $\varrho^2 = 1$ if and only if $X_rY_s - X_sY_r = 0$ for all values r, s, i.e. if $X_r/Y_r = X_s/Y_s$. In this case all points (X_r, Y_r) lie on a straight line through the origin (which is the centroid). Correlation is perfect and the two regression lines coincide.

(b) In general $-1 < \varrho < 1$. With positive correlation the lines of regression lie in the first and third quadrants (the centroid being the origin); with negative correlation the lines are in the second and fourth quadrants.

(c) If $\varrho = 0$, the variables are said to be uncorrelated. This implies that $\Sigma XY = 0$ and occurs in particular if the variables are independent, but it is not conversely true that if $\varrho = 0$, the variables are independent.

The regression lines for uncorrelated variables are parallel to the axes; in other words, one variable takes on the average the same value, whatever the value of the other.

We consider in §11.5 the significance of a computed value of r.

11.2.3 Computation of *r*. Correlation table

(a) If a calculating machine is available, the shift of origin used in Exs. i and ii is hardly worth while. A convenient form of r^2 is

$$r^2 = \frac{\{N\Sigma xy - (\Sigma x)(\Sigma y)\}^2}{\{N\Sigma x^2 - (\Sigma x)^2\}\{N\Sigma y^2 - (\Sigma y)^2\}} \quad . \quad . \quad (7)$$

If the values of x and y are all positive and not too large, all the sums required can be obtained simultaneously on the machine. Set the number $x \ldots y$, the dots denoting missing or zero digits between those of x and y, and square it, obtaining in the product register $x^2 \ldots 2xy \ldots y^2$. Repeat this for each pair of values, clearing only the setting register in between. Then Σx, Σy will be obtained in the multiplier register and Σx^2, $2\Sigma xy$ and Σy^2 in the product register.

(b) Using tables of squares only, the value of Σxy may be obtained from either of the identities

$$\Sigma(x - y)^2 = \Sigma x^2 - 2\Sigma xy + \Sigma y^2$$
$$\Sigma(x + y)^2 = \Sigma x^2 + 2\Sigma xy + \Sigma y^2.$$

The first is the more convenient if the correlation is positive, the second if the correlation is negative, since the values of $x - y$, $x + y$ respectively will tend to be small.

(c) *Large sample. Correlation table*

Given a large sample, it will be convenient, as in the case of a single variable, to operate with grouped data, dividing the ranges in both x and y into suitable equal class-intervals and so the area in the (x, y) plane into class-rectangles. We can then tabulate the frequencies in the rectangles, the resulting table constituting a correlation table.

Ex. iii

In the following table the variates are the marks obtained by the same candidate in an examination, in the two subjects Theory of Machines (mark x) and Mathematics (mark y). The numbers in the body of the table are frequencies with which certain marks were associated; e.g. 24 candidates scored a mark in the range 22–28 for Theory of Machines and a mark in the range 31–40 for Mathematics.

		x							
		1–7	8–14	15–21	22–28	29–35	36–42	43–49	50–56
	1–10		1	1					
	11–20	2	1	2	2	1			
	21–30		3	13	12	10	3		
y	31–40	1	2	5	24	20	19	3	1
	41–50			5	10	22	29	8	3
	51–60			2	1	13	14	13	7
	61–70				1	5	2	20	5
	71–80					1	1	6	6

Before proceeding to calculate r, we may note that the quantity $\Sigma XY/N\sigma_1\sigma_2$ is invariant to change of origin or scale, for ΣXY, $N\sigma_1^2$ and $N\sigma_2^2$ are product-moment and second moments with respect to axes through the centroid and are therefore independent of origin, while a change of scale in x affects X and σ_1 in the same proportion and a change of scale in y affects Y and σ_2 in the same proportion. Hence we may denote the centres of class-intervals by consecutive integers, placing the zero of each variate at about its mean value. By so doing we shall have smaller numbers to deal with. The work may be tabulated as follows.

	− 4	− 3	− 2	− 1	0		2	3	f	fv	fy^2	Σx	$y\Sigma x$
− 4		1	1						2	− 8	32	− 5	20
− 3	2	1	2	2	1				8	− 24	72	− 17	51
− 2		3	13	12	10	3			41	− 82	164	− 44	88
− 1	1	2	5	24	20	19	3	1	75	− 75	75	− 16	16
0			5	10	22	29	8	3	77	0	0	34	0
1			2	1	13	14	13	7	50	50	50	56	56
2				1	5	2	20	5	33	66	132	56	112
3					1	1	6	6	14	42	126	31	93
f	3	7	28	50	72	68	50	22	300	− 31	651	95	436
fx	− 12	− 21	− 56	− 50	0	68	100	66	95				
fx^2	48	63	112	50	0	68	200	198	739				
Σy	− 7	− 15	− 39	− 51	− 17	− 4	68	34	− 31				
$x\Sigma y$	28	45	78	51	0	− 4	136	102	436				

$\bar{x} = \frac{95}{300} = 0{\cdot}317$

$\bar{y} = -\frac{31}{300} = -0{\cdot}103$

$$s_1^2 = \frac{739}{300} - (0{\cdot}317)^2 = 2{\cdot}362$$

$$s_2^2 = \frac{651}{300} - (0{\cdot}103)^2 = 2{\cdot}159$$

$$\text{cov}\,(x, y) = \frac{\Sigma XY}{N} = \frac{436}{300} - (0{\cdot}317)\,(-\,0{\cdot}103) = 1{\cdot}486$$

$$r = \frac{1{\cdot}486}{\sqrt{2{\cdot}362 \times 2{\cdot}159}} = \underline{0{\cdot}66.}$$

Note that the method of tabulation affords checks on three of the required totals. The numbers labelled Σx and Σy require mental allowance for frequencies, e.g. (second row)

$$-\,17 = 2(-\,4) + 1(-\,3) + 2(-\,2) + 2(-\,1) + 1(0).$$

An alternative way of obtaining the sum of products is to tabulate the values of $x - y$ and their frequencies, the latter being the sums of the frequencies in the diagonals corresponding to the possible values of $x - y$.

$x - y$:	4	3	2	1	0	-1	-2	-3	Totals
f:	1	10	47	82	100	41	14	5	300
$f(x - y)^2$:	16	90	188	82	0	41	56	45	518

$$\Sigma(x - y)^2 = 518 = \Sigma x^2 + \Sigma y^2 - 2\Sigma xy$$
$$= 739 + 651 - 2\Sigma xy$$
$$\therefore \Sigma xy = \tfrac{1}{2}(739 + 651 - 518) = 436.$$

With these variables (x, y) the lines of regression are

$$y + 0{\cdot}103 = b_2\,(x - 0{\cdot}317)$$
$$x - 0{\cdot}317 = b_1\,(y + 0{\cdot}103)$$

where $$b_2 = \frac{\Sigma XY}{Ns_1^2} = \frac{1{\cdot}486}{2{\cdot}362} = 0{\cdot}63,\ b_1 = \frac{1{\cdot}486}{2{\cdot}159} = 0{\cdot}69$$

We can now plot the observed values of y against x and of x against y, as in the figure below, for comparison with the regression lines and so test the assumption of linear regression. The deviations are not unduly large.

Replacing x by $(x - 32)/7$ and y by $(y - 45{\cdot}5)/10$, we obtain the equations of the regression lines in terms of the original marks. They become

$$y = 0{\cdot}9\ x + 13{\cdot}7$$
$$x = 0{\cdot}48y + 12{\cdot}7$$

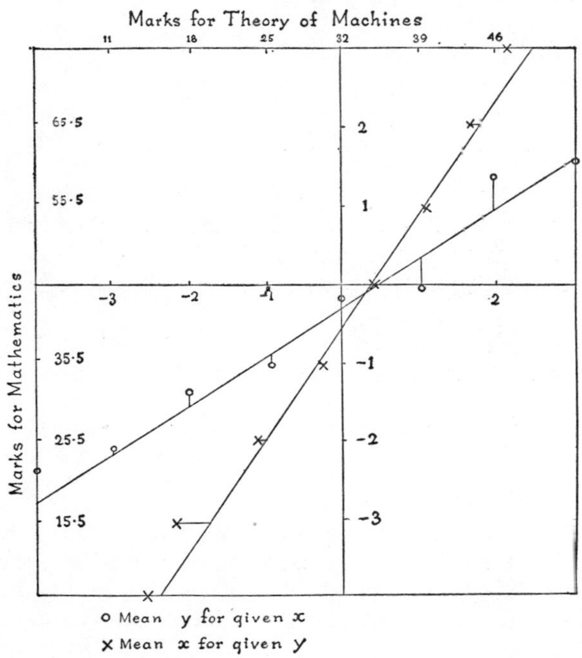

FIG. 43

11.2.4 Correlated errors

If $z = f(x, y)$ and δ_1, δ_2, δ_3 denote small errors in x, y and z,

$$\delta_3 = \frac{\partial z}{\partial x}\delta_1 + \frac{\partial z}{\partial y}\delta_2$$

from which we deduced (§9.3) that if δ_1 and δ_2 are independent, the relation

$$\sigma_z^2 = \left(\frac{\partial z}{\partial x}\right)^2 \sigma_x^2 + \left(\frac{\partial z}{\partial y}\right)^2 \sigma_y^2$$

holds between the variances, the partial differential coefficients taking their values at $(\bar{x}, \bar{y})$.

If the errors are not independent the sum $\Sigma\delta_1\delta_2$ cannot be assumed to vanish; in fact,

$$\Sigma\delta_1\delta_2 = N\sigma_x\sigma_y\varrho$$

where N is the total number of paired errors and ϱ is their coefficient of correlation. The relation between the variances now becomes

$$\sigma_z^2 = \left(\frac{\partial z}{\partial x}\right)^2 \sigma_x^2 + 2\varrho\frac{\partial z}{\partial x}\frac{\partial z}{\partial y}\sigma_x\sigma_y + \left(\frac{\partial z}{\partial y}\right)^2 \sigma_y^2 \quad . \quad . \quad (8)$$

In particular, if $z = x + y$

$$\sigma_z^2 = \sigma_x^2 + 2\varrho\sigma_x\sigma_y + \sigma_y^2$$

while if

$$z = x - y$$

$$\sigma_z^2 = \sigma_x^2 - 2\varrho\sigma_x\sigma_y + \sigma_y^2$$

so that the r.m.s. errors in a sum and difference are no longer equal, if the errors in the two terms are correlated.

If any variates x, y, z satisfy a relation

$$z = Ax + By \quad . \quad . \quad . \quad . \quad . \quad . \quad (9)$$

where A and B are constants, equation (8) takes the form

$$\sigma_z^2 = A^2\sigma_x^2 + 2\varrho AB\sigma_x\sigma_y + B^2\sigma_y^2 . \quad . \quad . \quad . \quad (10)$$

11.3 Correlation ratios

We have so far made the large assumption that the regressions are linear, i.e. that the loci of y_M against x and of x_M against y are straight lines. If the dot diagram revealed a well-marked curve for y_M against x, it might be possible to choose a suitable function of x to represent y, determining the coefficients in the function by the method of least squares.

Without obtaining an equation for the regression curve we can still measure the degree of correlation by Pearson's *correlation-ratio*. The criterion of high correlation is that the observed values are closely clustered round the curve of means. When this is so, the r.m.s. deviation, σ_2, of all y's from $\bar{y}$ will not differ much from the r.m.s. deviation of the y_M's from their (weighted) mean which is also $\bar{y}$. Calling this latter r.m.s. deviation σ_{M_2}, K. Pearson's correlation-ratio, η_{yx}, of y on x is defined by

$$\eta_{yx} = \frac{\sigma_{M_2}}{\sigma_2}$$

Thus if y_{Mi} is the mean of the n_i values of y occurring with the value x_i of x

$$\sigma_{M_2}^2 = \frac{1}{N} \Sigma n_i (y_{Mi} - \bar{y})^2$$

so that

$$\eta_{yx}^2 = \frac{\Sigma n_i (y_{Mi} - \bar{y})^2}{N \sigma_2^2} \quad . \quad . \quad . \quad . \quad (11)$$

There is a similar correlation ratio η_{xy} of x on y

given by

$$\eta_{xy}{}^2 = \frac{\Sigma n_j (x_{Mj} - \bar{x})^2}{N \sigma_1^2} \quad . \quad . \quad . \quad . \quad (12)$$

It can be shown that each $\eta^2 \geqslant \varrho^2$ and that regression is linear if, and only if, each $\eta = \varrho$ numerically. It is useful therefore to calculate the η's, or rather their estimates from sample, for comparison with the estimate of ϱ, as a test of linear regression. The table of calculated values in §11.2.3 may be enlarged by the inclusion of four additional columns and four additional rows, which, for the example given, would be as follows.

x_M	$x_M - \bar{x}$	$(x_M - \bar{x})^2$	$f(x_M - \bar{x})^2$
− 2·500	− 2·817	7·935	15·87
− 2·125	− 2·442	5·963	47·70
− 1·073	− 1·390	1·932	79·21
− 0·213	− 0·530	0·281	21·08
0·442	0·125	0·016	0·12
1·120	0·803	0·645	32·25
1·697	1·380	1·905	62·86
2·214	1·897	3·600	50·40
			309·5

y_M:

− 2·333	− 2·143	− 1·393	− 1·020	− 0·059	1·360	1·545	

$y_M - \bar{y}$:

− 2·230	− 2·040	− 1·290	− 0·917	0·044	1·463	1·648	

$(y_M - \bar{y})^2$:

4·973	4·162	1·664	0·841	0·002	2·140	2·717	

$f(y_M - \bar{y})^2$:

14·92	29·13	46·59	42·05	0·14	107·00	59·77	300·8

Estimated $\eta_{xy} = \sqrt{\dfrac{309{\cdot}5}{300 \times 2{\cdot}362}} = 0{\cdot}66$

$$\eta_{yx} = \sqrt{\frac{300{\cdot}8}{300 \times 2{\cdot}159}} = 0{\cdot}68$$

These are very near our estimate for ϱ, so that the assumption of linear regression is well justified.

11.4 Bivariate normal distribution

In view of the importance of the normal distribution for a single variate, it is natural to investigate the possibility of its extension to two variables. This may be done and the resulting distribution throws light on the association between linear regression and the product-moment coefficient of correlation. The reader is referred to Whitaker and Robinson, *The Calculus of Observations*, Chap. XII, where the bivariate normal distribution is obtained by two methods. The second method is analogous to Gauss's derivation of the law of errors; the first is perhaps apposite here, since it uses hypotheses which express the idea of partial dependence or correlation. Suppose that two marksmen firing at a target are, in the absence of wind, independently liable to error, each in accordance with his own normal law. When there is a wind each is affected, though not necessarily equally. The errors x, y of the two marksmen are supposed represented by

$$x = u + aw, y = v + bw$$

where u, v, w are independent normal variates and a, b are constants, the terms in w being the errors due to the wind. It is then shown that the probability distribution of (x, y) is $dp = \phi(x, y)dx\,dy$ where

$$\phi\,(x, y) = \frac{1}{2\pi\sigma_1\sigma_2\sqrt{1-\varrho^2}}\,e^{-\frac{1}{2(1-\rho^2)}\left\{\frac{x^2}{\sigma_1^2} - \frac{2\rho xy}{\sigma_1\sigma_2} + \frac{y^2}{\sigma_2^2}\right\}} \quad . \quad (13)$$

The distribution is represented geometrically by the surface $z = \phi(x, y)$; $\phi(x, y)$ is the probability density at the point (x, y) of the plane XOY;

the volume under the surface, standing on any area in the plane XOY, represents the probability that a pair of values (x, y) falls within the area. The origin has been taken at the mean of x and of y.

Since $$\frac{x^2}{\sigma_1^2} - \frac{2\varrho xy}{\sigma_1\sigma_2} + \frac{y^2}{\sigma_2^2} = \left(\frac{x}{\sigma_1} - \frac{\varrho y}{\sigma_2}\right)^2 + \frac{(1 - \varrho^2)y^2}{\sigma_2^2}$$

it is evident from equation (13) that, if y is constant, x is distributed normally about a mean given by

$$\frac{x}{\sigma_1} = \frac{\varrho y}{\sigma_2}$$

and this is the equation we obtained for the line of regression of x on y; in other words, if (x, y) is distributed in accordance with the bivariate normal law, the means of x's lie exactly on this straight line. Similarly the means of y's lie on the line

$$\frac{y}{\sigma_2} = \frac{\varrho x}{\sigma_1}$$

If $\varrho = 0$, $\phi(x, y)$ can be expressed as a product $\phi_1(x)\phi_2(y)$ showing that x and y are then statistically independent. This is a special property of the normal distribution; it is not true in general that uncorrelated variables are necessarily independent.

A convenient way of representing the surface $z = \phi(x, y)$ is by means of its contours on the (x, y) plane. These will be the loci $z =$ constant, so that

$$\frac{x^2}{\sigma_1^2} - \frac{2\varrho xy}{\sigma_1\sigma_2} + \frac{y^2}{\sigma_2^2} = \text{constant.}$$

If the scales are changed so as to standardize the variates, a contour has the equation

$$x^2 - 2\varrho xy + y^2 = \text{constant.}$$

The contours now form a family of similar ellipses with centre at the origin and with the lines $x = y$ and $x = -y$ as principal axes.

The regression line $y = \varrho x$ is the diameter bisecting all chords parallel to the y-axis; similarly $x = \varrho y$ is the diameter conjugate to the x-axis. A plane perpendicular to OX, through N, cuts the surface $z = \phi(x, y)$ in a normal curve; M is the projection of the vertex of the curve on to the plane XOY. Thus, for a given x, the distribution of y is normal and MN is the mean value of y; as x varies, $y = \varrho x$ is the locus of means of y. Similarly all sections of the surface by planes perpendicular to OY are normal curves and the locus of the means of the corresponding distribution of x is the line $x = \varrho y$.

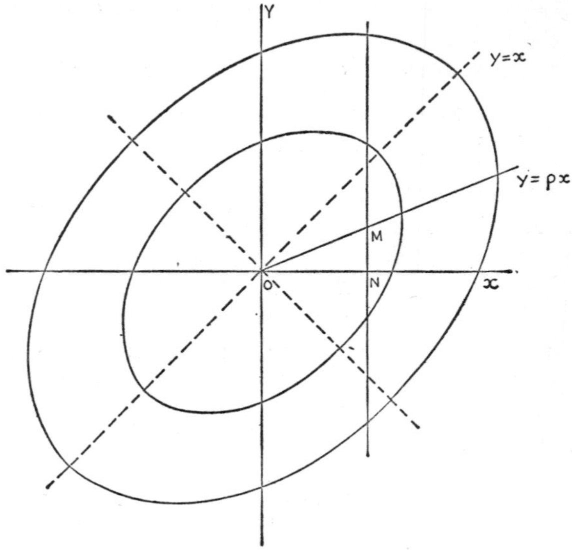

FIG. 44

A correlation table of grouped frequencies can be represented in three dimensions by a bivariate histogram. Each class-rectangle in the plane XOY is made the base of a rectangular block, of volume proportional to the frequency in the class-rectangle. Allowing for the effects of sampling and grouping, the top surface of the resulting solid figure will be more or less indicative of the probability surface of the parent distribution from which the sample is drawn.

11.5 Precision and significance of the correlation coefficient

When an estimate of ϱ has been obtained from a sample it is important to estimate further the liability to error of the result. The usual criterion, viz. the standard error, or sampling variance, of the calculated parameter is of little value unless the sampling distribution is approximately normal. In the present case it is necessary to determine the distribution of r, estimate of ϱ, when all possible samples of size n are drawn from a bivariate normal population, having given $\bar{x}$, $\bar{y}$, σ_1, σ_2 and ϱ.

It is found † that, supposing ϱ given, the only estimates of the other parameters being their sample values, the distribution of r is given by

$$dp \propto (1 - r^2)^{(n-4)/2} \frac{d^{n-2}}{d(r\varrho)^{n-2}} \left\{ \frac{\cos^{-1}(-r\varrho)}{\sqrt{1 - r^2\varrho^2}} \right\} dr \quad . \quad (14)$$

† M. G. Kendall, *Advanced Theory of Statistics*, Vol. I, pp. 337–46 (Griffin).

To make use of this formidable result it is now necessary to calculate the probabilities of values of r less than or equal to specified values, as has been done by F. N. David. It is possible, however, to convert the distribution (14) into a nearly normal form and then the standard error affords a sufficient criterion of the precision. R. A. Fisher showed that, if $z = \tanh^{-1}r = \frac{1}{2}\log_e\frac{1+r}{1-r}$ and $\zeta = \tanh^{-1}\varrho$ and n is sufficiently large,† then $z - \zeta$ may be taken as distributed normally about zero mean with variance $1/(n-3)$. A more accurate approximation is that $z - \zeta$ is distributed normally about mean $\varrho/2(n-1)$ with variance $1/(n-1) + (4-\varrho^2)/2(n-1)^2$. The correction to the mean is more important if ϱ is relatively large and n small.

In the example of §11.2.3, $n = 300$ so that the variance of $z - \zeta$ is $1/297 = 0{\cdot}00337$. To three decimal places $r = 0{\cdot}658$, giving $z = \tanh^{-1}r = 0{\cdot}7893$. We have no other estimate of ϱ or ζ, but we could say on the basis of the normal distribution that it is highly improbable that our z differs from the true ζ by more than three times the standard error. If the true value of ϱ were $0{\cdot}7893$ we should not expect a value of z outside the range $z = 0{\cdot}7893 \pm 3\sqrt{0{\cdot}00337} = 0{\cdot}7342$ to $0{\cdot}8444$. The corresponding range in r is from $r = \tanh\,(0{\cdot}7342) = 0{\cdot}626$ to $r = \tanh\,(0{\cdot}8444) = 0{\cdot}682$.

Significance

If the number n in sample is small, it becomes quite probable that a small value of r, suggesting some degree of correlation, is obtained from the sample, even if the latter is drawn from a normal distribution for which $\varrho = 0$; in other words the value of r has no significance if it might well have arisen as the result of random sampling from an uncorrelated distribution. The extreme example is afforded by $n = 2$, which is bound to give $r = 1$, suggesting perfect correlation; for we have only two points in our dot diagram and the line joining them represents both regression lines. It follows that for small samples it is important to test the significance of the calculated value of r by determining the probability that such a value (or a larger one) could have arisen as the result of random sampling from an uncorrelated population. If this probability is low the value of r is significant. When $\varrho = 0$, the distribution (14) of r becomes $dp \propto (1 - r^2)^{(n-4)/2}dr$ and the substitutions $\nu = n - 2$, $t^2(1 - r^2) = \nu r^2$ reduce the distribution to that of Student's t with ν degrees of freedom, so that the significance of a value of r could be tested by reference to tables of t. For convenience, however, tables have been prepared showing the values of r at various levels of significance for various values of ν. Suppose the number of paired observations is 20 and the computed value of r is $0{\cdot}35$. The tables give $P = 0{\cdot}10$ with

† Kendall gives $n > 50$; Aitken, $n > 20$.

$\nu = 18$ and $r = 0{\cdot}378$, which means that there is a probability of 0·1 that $r \geqslant 0{\cdot}378$ as the result of random sampling from an uncorrelated normal population; the apparent correlation is therefore not significant. If, however, we had found $r = 0{\cdot}53$ it appears from the table that the probability of so large a value from an uncorrelated population is less than 0·02; we may with considerable confidence infer that the apparent correlation is real.

Ex. iv

Evaluate the correlation coefficient and test its significance for the following data.

Distance between plots (ft.):	100	200	200	250	300	340	360	400
Crossing of varieties of wheat (%):	0·84	0·44	0·52	0·66	0·12	0·14	0·14	0·36

Shifting the origin to 250, 0·44 and multiplying the variables by 1/10 and 100 respectively, we have

									Totals
x:	− 15	− 5	− 5	0	5	9	11	15	15
y:	40	0	8	22	− 32	− 30	− 30	− 8	− 30
x^2:	225	25	25	0	25	81	121	225	727
y^2:	1600	0	64	484	1024	900	900	64	5036
xy:	− 600	0	− 40	0	− 160	− 270	− 330	− 120	− 1520

Using equation (7), $r = \dfrac{8(-1520) - 15(-30)}{\sqrt{\{8(727) - 15^2\}\{8(5036) - 30^2\}}}$

$= -0{\cdot}73$

With $\nu = 6$, the probability of so large a coefficient from an uncorrelated population is between 0·02 and 0·05. The apparent negative correlation may be accepted as significant, but very little importance is to be attached to the numerical value of r, computed from so small a sample.

11.6 Variance of departure from regression line. Accuracy of prediction

A practical use of regression equations is to forecast the value of one variable from an observed value of the other. If the correlation is sufficiently high it may be possible in this way to use an easily measured variable to estimate the value of one which it is difficult to measure. Examples are: degree of corrosion from electrical resistance, hardness of metal from electrical resistance.

On the assumption of linear regression we obtain first-degree equations from which to calculate the mean y for given x and vice versa. Part of the variation in y is thus explained as due to the general trend, but an observed value of y is still liable to differ from the mean or expected value for given x. If X, Y denote deviations from the means, the equation $Y_M = \beta_2 X$ gives the mean Y for given X and $Y - \beta_2 X$ is the deviation of any Y from Y_M.

$$\begin{aligned}\text{Now } \Sigma(Y - \beta_2 X)^2 &= \Sigma Y^2 - 2\beta_2 \Sigma XY + \beta_2^2 \Sigma X^2 \\ &= N\sigma_2^2 - \frac{2\varrho\sigma_2}{\sigma_1}\varrho N\sigma_1\sigma_2 + \frac{\varrho^2\sigma_2^2}{\sigma_1^2}N\sigma_1^2 \\ &= N\sigma_2^2(1 - 2\varrho^2 + \varrho^2) \\ &= N\sigma_2^2(1 - \varrho^2)\end{aligned}$$

The r.m.s. value of $Y - \beta_2 X$ is therefore

$$\sigma_2\sqrt{(1 - \varrho^2)}, \quad . \quad . \quad . \quad . \quad . \quad (15)$$

and this is the s.d. of the departure from the regression line. Similarly $\sigma_1\sqrt{(1 - \varrho^2)}$ is the r.m.s. value of $X - \beta_1 Y$. In estimating (15) from a small sample, although we use the sample variances s_1^2 and s_2^2 in the computation of r, the factor σ_2 is replaced by its best estimate from the data. There is a loss of two degrees of freedom in fitting the best straight line and the estimate of σ_2^2 to be adopted is $\Sigma(y - \bar{y})^2/(N - 2)$.

Ex. v

From the data of Exs. i and ii find the r.m.s. error to be feared in estimating the efficiency from the mean particle size, and vice versa, from the regression equations.

We had $r = 0{\cdot}77 \quad \therefore 1 - r^2 = 0{\cdot}407$
Estimated $\sigma_2^2 = 5969/28 = 213$
In estimating the efficiency from the mean particle size, the r.m.s. error is

$$\sqrt{(213)\ (0{\cdot}407)} = 9{\cdot}3.$$

Estimated $\sigma_1^2 = 748{\cdot}3/28 = 26{\cdot}7.$
The r.m.s. error in mean particle size derived from efficiency is

$$\sqrt{(26{\cdot}7)\ (0{\cdot}407)} = \underline{3{\cdot}3.}$$

We may say that the whole variance of y can be expressed as the sum of two terms, viz.

$$\sigma_2^2 = \sigma_2^2\varrho^2 + \sigma_2^2(1 - \varrho^2)$$

the first representing the variance due to the general trend and the second the residual variance about the repression line. This is a particular example of the Analysis of Variance (Chapter XII).

When the regression is non-linear the correlation ratios η_{xy} and η_{xy} enter into the corresponding partition of the variance. The equations

$$\begin{aligned}\sigma_1^2 &= \sigma_1^2\eta_{xy}^2 + \sigma_1^2(1 - \eta_{xy}^2) \\ \sigma_2^2 &= \sigma_2^2\eta_{yx}^2 + \sigma_2^2(1 - \eta_{yx}^2)\end{aligned}$$

similarly represent the variances, of x and y respectively, as the sum of the variances (i) of the means (of x for given y, or y for given x) and (ii) about the means.

11.7 Correlation by ranks

Individuals may be classified, in respect of some chosen characteristic in an order of merit. It may or may not be possible to measure the characteristic on a calibrated scale. For instance, two judges may each allot an order of merit to the competitors in a fancy dress parade or a baby show. There is no objective standard of measurement and no correct order with which the orders awarded by the two judges can be compared; but these two orders can be compared with one another to test how well the standards of the two judges agree. On the other hand, suppose a person called upon to place in order, from light to dark, various shades of grey. Here it would be possible to determine the true order by scientific measurement; the person's goodness of judgment by eye could be assessed by comparing his order with the true order.

In either type of problem, if there are n individuals, let their placings in the two orders be $x_1, x_2 \ldots x_n; y_1, y_2 \ldots y_n$ where the x's (and the y's) are simply the numbers 1, 2 . . . n in some order. They have equal means, namely

$$\bar{x} = \bar{y} = \tfrac{1}{2}(n + 1)$$

and equal variances, namely

$$\sigma_1^2 = \sigma_2^2 = \frac{1}{n}\Sigma x^2 - \bar{x}^2 \text{ where } \Sigma x^2 = \frac{1}{6}n(n + 1)(2n + 1).$$

Suppose now we calculate ϱ by the formula for product-moment correlation; in other words we allot equally spaced marks to the individuals in agreement with either order.

$$\varrho = \frac{\Sigma xy - n\bar{x}\bar{y}}{n\sigma_1^2}$$

$$= \frac{\frac{1}{2}\Sigma x^2 + \frac{1}{2}\Sigma y^2 - \frac{1}{2}\Sigma(x - y)^2 - n\bar{x}\bar{y}}{n\sigma_1^2}$$

$$= \frac{\Sigma x^2 - n\bar{x}^2 - \frac{1}{2}\Sigma d^2}{\Sigma x^2 - n\bar{x}^2} \text{ where } d = x - y$$

$$= 1 - \frac{3\Sigma d^2}{n(n + 1)(2n + 1) - 6n(n + 1)^2/4}$$

$$= 1 - \frac{6\Sigma d^2}{n^3 - n} \quad \ldots \ldots \ldots \quad (16)$$

This quantity is known as Spearman's coefficient of rank correlation. It is easy to evaluate: we have only to tabulate the individuals in any order with their placings x, y and the differences $x - y$. It may be noted that $\varrho = 1$ if and only if every d is zero, i.e. if the two orders are identical; it can also be shown that $\varrho = -1$ when Σd^2 is a maximum, which occurs when one order is the reverse of the other. Of course Spearman's ϱ is distinct from the product-moment coefficient of correlation between measured variates: it is merely a suggested measure of correlation when no calibrated scale of the variates is available. Other measures have been suggested, including Kendall's τ; the reader is referred to Kendall, *Advanced Theory of Statistics*, Vol. I, Chapter 16 (Griffin), for these.

Even when scales for the variates exist, we may obtain a first estimate of the degree of correlation by means of Spearman's ϱ. As an illustration, consider the data of Ex. i and Ex. ii in which the product-moment coefficient was found to be 0·77. Instead, we now replace the values of x and y by their ranks or placings in order of magnitude. Where two or more values of x or y are equal, they are each allotted the average of the corresponding ranks.

Ex. vi

Mean particle size and efficiency of smoke collection.

m = order in particle size.
n = order in efficiency.

m:	1	2	4	4	4	6	$7\frac{1}{2}$	$7\frac{1}{2}$	9	10	12	12	12	$14\frac{1}{2}$	$14\frac{1}{2}$
n:	2	$6\frac{1}{2}$	3	4	8	1	$6\frac{1}{2}$	16	$21\frac{1}{2}$	12	$9\frac{1}{2}$	11	30	5	$9\frac{1}{2}$
d:	1	$4\frac{1}{2}$	1	0	4	5	1	$8\frac{1}{2}$	$12\frac{1}{2}$	2	$2\frac{1}{2}$	1	18	$9\frac{1}{2}$	5

m:	17	17	17	20	20	20	22	25	25	25	25	25	28	$29\frac{1}{2}$	$29\frac{1}{2}$
n:	14	19	25	15	17	$21\frac{1}{2}$	13	18	$23\frac{1}{2}$	26	$27\frac{1}{2}$	$27\frac{1}{2}$	$23\frac{1}{2}$	20	29
d:	3	2	8	5	3	$1\frac{1}{2}$	9	7	$1\frac{1}{2}$	1	$2\frac{1}{2}$	$2\frac{1}{2}$	$4\frac{1}{2}$	$9\frac{1}{2}$	$\frac{1}{2}$

It will be found that $\Sigma d^2 = 1113$

By ranks
$$\varrho = 1 - \frac{6(1113)}{(29)\,(30)\,(31)}$$
$$= \underline{0{\cdot}754}$$

This agrees remarkably well with the value, 0·77, of r.

11.8.1 Trivariate correlation

The theory of correlation and linear regression is extended to more than two variables. We shall merely state without proof the results for three variables.

The assumption of linear regression implies that the best value of any one of the three variables, when the other two are given, is obtained by an equation of the first degree expressing the first variable in terms of the others. There are three such equations, representing three planes in a three-dimensional space. If the method of least squares is used to determine the constants of the equations, a first result is that the three planes all pass through the mean centre of the points representing the observed triads of values. We therefore suppose the origin shifted to the mean centre, so that if x, y and z are the variates in the form of deviations from means, the regression equations are

$$\left.\begin{aligned}\frac{x}{\sigma_1} &= \beta_{12}\frac{y}{\sigma_2} + \beta_{13}\frac{z}{\sigma_3}\\ \frac{y}{\sigma_2} &= \beta_{23}\frac{z}{\sigma_3} + \beta_{21}\frac{x}{\sigma_1}\\ \frac{z}{\sigma_3} &= \beta_{31}\frac{x}{\sigma_1} + \beta_{32}\frac{y}{\sigma_2}\end{aligned}\right\} \quad . \quad . \quad . \quad . \quad (17)$$

where σ_1^2, σ_2^2, σ_3^2 are the variances of x, y and z.

Further it is found that the best values of the β's are

$$\beta_{12} = \frac{\varrho_{12} - \varrho_{13}\varrho_{23}}{1 - \varrho_{23}^2}, \; \beta_{13} = \frac{\varrho_{13} - \varrho_{12}\varrho_{23}}{1 - \varrho_{23}^2} \quad . \quad . \quad (18)$$

and two similar pairs, where the ϱ's are the product-moment coefficients of correlation as already defined:

$$\text{i.e. } \varrho_{12} = \frac{\Sigma xy}{n\sigma_1\sigma_2}, \; \varrho_{23} = \frac{\Sigma yz}{n\sigma_2\sigma_3}, \; \varrho_{31} = \frac{\Sigma zx}{n\sigma_3\sigma_1}.$$

Note that $\varrho_{12} = \varrho_{21}$, etc., but $\beta_{12} \neq \beta_{21}$, etc.

Analogously to the case of two variables, it is found that equations (17) accurately represent the most probable values of the dependent variables if x, y and z are the variates in a trivariate *normal* distribution. In practice, if a set of values of (x, y, z) may be regarded as a sample from a nearly normal distribution, equations corresponding to (17) and (18), but with b's for β's and r's for ϱ's, may be used to estimate the regression coefficients.

If there is correlation between y and z, this affects the correlation between x and y; for if the value of y is to some extent dependent on that of z, then the apparent dependence of x on y may be regarded as partly representing dependence of x on z. To eliminate the effect of z we need to measure the correlation between x and y, *when z is constant.* For this purpose a *partial correlation* coefficient $\varrho_{12.3}$ is defined, the suffix after the dot denoting that the third variable is constant.

Now if we set $z = c$ in the first two of equations (17) we have equations for the most probable x, given y, and for the most probable y, given x. After shifting the origin to the point of intersection of the regression lines which these equations represent, the equations can be arranged as

$$x = \frac{\beta_{12}\sigma_1}{\sigma_2}y, \quad y = \frac{\beta_{21}\sigma_2}{\sigma_1}x \quad . \quad . \quad . \quad (19)$$

The comparable equations, when there is no question of a third variable, are (5) and (4), arranged as

$$x = \frac{\varrho\sigma_1}{\sigma_2}y, \quad y = \frac{\varrho\sigma_2}{\sigma_1}x$$

Here ϱ has the property that it is the geometric mean of the gradients. We define $\varrho_{12\cdot3}$ to have the same relation to the gradients of the lines (19), so that

$$\varrho_{12.3}^2 = \beta_{12}\,\beta_{21} \quad . \quad . \quad . \quad . \quad . \quad . \quad . \quad (20)$$

with the sign of $\varrho_{12.3}$ still to be determined. Using equations (18) we obtain $\varrho_{12.3}$ in terms of the previous ϱ's as

$$\varrho_{12.3} = \frac{\varrho_{12} - \varrho_{13}\,\varrho_{23}}{\sqrt{1 - \varrho_{13}^2}\,\sqrt{1 - \varrho_{23}^2}} \quad . \quad . \quad . \quad (21)$$

the sign being now fixed as that of the common numerator of β_{12} and β_{21}.

For the ideal case of a trivariate normal distribution it is found that $\varrho_{12.3}$ so defined is the correlation coefficient between x and y when z is constant and is the same for every value of z. In a practical case $\varrho_{12.3}$ must be regarded as measuring the average dependence of x on y, or of y on x, at constant z.

An important feature of equation (21) is that it determines the partial correlation coefficients from the coefficients ϱ_{12}, ϱ_{23}, ϱ_{31}, which are called total correlation coefficients. In other words we have been able to sort out, from experimental data relating to three variables, a measure of the relation between two of them if the third were held constant. The scientist in the laboratory is sometimes able to control the factors which may be expected to affect a chosen variable; if he can succeed in allowing only one factor to vary, he may be able to determine a fairly precise relationship between the chosen variable and the single variable factor. But often such control is impossible, and it is one of the chief objects of statistical method to cope with data which represent the combined effect of several causes. A widely used technique for this purpose is the analysis of variance, considered in the next chapter.

It is necessary to beware of confusing correlation with causation. An apparent relationship between two variables might merely express the fact that both depend on a third. Examples are sometimes given of what is called 'nonsense correlation' in which a marked degree of correlation is exhibited between paired values of two characteristics which can obviously not stand in the relation of cause and effect. One suggested by W. Hope-Jones† is as follows:

"If you capture a large number of people and measure for each the

† *A Plea for Teaching Probability in Schools*, Mathematical Gazette, Vol. XII.

number of aunts he possesses and the percentage of Calcium in his bones, I think you will find a high negative correlation between the two, much Calcium going with few aunts. . . . Aunts do not repel Calcium; nor does Calcium murder aunts. What happens is that as you grow older the percentage of Calcium in your bones increases, while aunts have a tendency to die in greater quantity than new ones are born. The two are correlated only because they both depend on age."

Ex. vii

Hardness, tensile strength and abrasion loss were measured for each of 30 samples of rubber. The correlation coefficients between hardness and tensile strength, hardness and abrasion loss, and between tensile strength and abrasion loss were respectively $-0{\cdot}30$, $-0{\cdot}74$ and $-0{\cdot}39$. Find the partial correlation between abrasion loss and tensile strength and discuss briefly its practical meaning. (L.U.)

Using the suffixes 1, 2, 3 to correspond to abrasion loss, tensile strength and hardness respectively, we require $\varrho_{12\cdot3}$, given that

$$\varrho_{12} = -0{\cdot}39,\ \varrho_{23} = -0{\cdot}30,\ \varrho_{13} = -0{\cdot}74.$$

By equation (21), $\varrho_{12\cdot3} = \underline{-0{\cdot}965.}$

Thus if hardness could be kept constant, abrasion loss would be very highly correlated with tensile strength, decreasing much faster with increased tensile strength. In practice, increased tensile strength goes with decreased hardness, which tends to increase abrasion loss.

11.8.2 Coefficient of multiple correlation

The regression equations (17) may be used to predict values of one variable, say x, when the other two are given. The precision of an estimate so obtained is measured by the variance of the departure of observed values from predicted values. This variance is denoted by $\sigma^2_{1.\,{}^3_2}$ and is given by

$$\begin{aligned}\sigma^2_{1\cdot23} &= \sigma_1^2(1-\varrho_{12}^2)(1-\varrho_{13.2}^2)\\ &= \sigma_1^2(1-\varrho_{13}^2)(1-\varrho_{1\cdot23}^2)\\ &= \sigma_1^2\Delta/(1-\varrho_{23}^2), \qquad . \quad . \quad . \quad . \quad (22)\end{aligned}$$

where

$$\Delta = \begin{vmatrix} 1 & \varrho_{12} & \varrho_{13} \\ \varrho_{12} & 1 & \varrho_{23} \\ \varrho_{13} & \varrho_{23} & 1 \end{vmatrix}$$

Another measure of the closeness of observed to predicted values is obtained by determining the coefficient of correlation between them. This coefficient is called the multiple correlation coefficient and denoted by $\underline{R}_{1.23}$. Its value is given by

$$\underline{R}_{1\cdot23} = \sqrt{(1-\sigma^2_{1.23}/\sigma_1^2)}, \qquad . \quad . \quad . \quad . \quad (23)$$

the positive root being taken.

For the proofs of these results and their extension to four or more variables, the reader may consult Kendall, *Advanced Theory of Statistics*, Vol. I, Chapter 15 (Griffin).

Suggestions for further reading:

R. A. Fisher, *Statistical Methods for Research Workers* (Oliver & Boyd).
W. P. Elderton, *Frequency Curves and Correlation* (Griffin).
F. N. David, *Tables of the Correlation Coefficient* (Cambridge University Press).
R. A. Fisher, *The Frequency Distribution of the Values of the Correlation Coefficient in Samples from an Indefinitely Large Population*, Biometrika, 10, 507.

EXAMPLES ON CHAPTER XI

1. Obtain the equation of the regression line of y on x for the following observed values.

x (Temp. °C.):	17	19	19	20	22	22	23	23	25	25
y (Invert sugar %):	0·2	0·25	0·3	0·35	0·4	0·6	0·5	0·6	0·55	0·65

What is the r.m.s. error to be feared in using the equation of the regression line to predict the value of y for given x?

2. Soil Temperature and Germination Interval (interval between sowing and appearance above ground) for Winter Wheat, 1926–7 in 12 places.

Mean soil temperature at 4 in. (° F.):	57	42	38	42	45	42	44	40	46	44	43	40
Interval (days):	10	26	41	29	27	27	19	18	19	31	29	33

Calculate to two decimal places the coefficient of correlation between soil temperature and germination interval and interpret the result.

(R.S.S. Cert.)

3. Prove that if the regressions of y on x and x on y are both linear, then the product of the two regression coefficients is equal to the square of the product moment correlation coefficient. In the table below, verify that the means of the x-arrays are collinear, and also those of the y-arrays, and deduce that the correlation coefficient is $-0{\cdot}535$.

	x: 0	1	2	3	4
y: 0				4	3
1			18	36	9
2		12	54	36	3
3	1	12	18	4	

(R.S.S. Cert.

4. For certain data, $y = 1{\cdot}2x$ and $x = 0{\cdot}6y$ are the regression lines. Compute r_{xy} and σ_x/σ_y.
If $y = x + u$, compute r_{xu}, r_{yu} and σ_u/σ_y. (R.S.S. Cert.)

5. Personal Expenditure on Consumers' Goods and Services at Current Market Prices (£ million), (quarterly).

Year:	1945		1946		1947	
Quarter	*Fuel and Light*	*Travel*	*Fuel and Light*	*Travel*	*Fuel and Light*	*Travel*
1	78	52	84	56	92	58
2	62	62	64	69	70	77
3	56	73	61	82	63	88
4	71	57	82	63	85	73

—(Monthly Digest)

Compute the product-moment correlation coefficient for the two series. Explain the meaning of this result. (R.S.S. Cert.)

6. A factory welfare service collects information concerning, amongst other things, weight, height, education and type of work of the employees. This information is summarized on an index card for each employee. The following table contains information from 30 of the cards selected at random, each column of the table being obtained from one card. The letter S denotes the person was on the staff, the letter I that he was an industrial worker. The latter E denotes that the person did attend a secondary school, the letter N that he did not.
Prepare a scatter diagram to show the connection between height and weight in the sample and calculate the correlation coefficient. Carry out a test to see whether there is any evidence of association between education and type of work.

Weight (to the nearest 5 lb.):	145	155	160	130	145	150	125	165
Height (to the nearest in.):	67	71	70	68	69	69	63	71
Type of work:	S	S	S	S	I	S	I	S
Education:	N	E	E	E	N	E	E	N

Weight:	130	145	135	170	140	150	115	145	145	165	135
Height:	71	70	68	75	69	68	66	67	71	66	70
Type of work:	S	I	I	S	I	S	I	S	I	S	I
Education:	E	N	N	E	N	N	N	E	E	E	E

Weight:	120	135	135	140	160	120	145	180	160	135	130
Height:	67	69	64	71	74	70	69	70	71	66	72
Type of work:	I	I	I	S	I	S	I	I	I	I	I
Education:	N	N	E	E	E	N	N	N	N	N	N

(R.S.S. Cert.)

7. What do you understand by a 'test of significance'? Explain without proof how you would test the significance of a correlation coefficient in simple sampling.
The correlation for the annual figures over the past 30 years of (1) the number of horse-drawn vehicles on the roads of Great Britain and (2) the maternal mortality rate is positive, and is significantly different from zero in the sense of the ordinary theory of simple sampling. How far would you accept this as evidence of genuine relationship?
(R.S.S. Cert.)

8. Test the significance of the values of r obtained in questions 2, 5 and 6.

9. Show that the linear regression coefficient of y on x is given by

$$b_{y.x} = \Sigma(y - \bar{y})(x - \bar{x})/\Sigma(x - \bar{x})^2,$$

the summations being over all pairs of values in the data. Defining the correlation coefficient to be

$$r = \sqrt{\{b_{y.x} \times b_{x.y}\}} \text{ show that}$$
$$\Sigma(y - Y)^2 = \Sigma(y - \bar{y})^2(1 - r^2)$$

where
$$Y = \bar{y} + b(x - \bar{x})$$

and hence derive limits for the possible values of r.
An ordinary pack of 52 cards is dealt to 4 whist players (13 each). If one player has r hearts, what is the average number held by his partner? Deduce that the correlation coefficient between the number of hearts in the two hands is $-1/3$. (R.S.S. Cert.)

10. X_1 and X_2 are two variates whose correlation coefficient is ϱ and whose variances are σ_1^2, σ_2^2 respectively. Show that if a is a constant, the variance of $X_1 + aX_2$ is

$$\sigma^2 = \sigma_1^2 + a^2\sigma_2^2 + 2a\varrho\sigma_1\sigma_2$$

If $Y_1 = X_1 + aX_2$ and $Y_2 = X_1 + bX_2$, determine values of a and b which are independent of ϱ and such that Y_1 and Y_2 are uncorrelated.
(L.U.)

11. A manufacturer makes shoes whose widths and lengths vary. The widths W increase in steps of 1/8 in. while the lengths L increase in steps of 1/4 in. He proposes to market his shoes among a population in which the required values of L and W are jointly normally distributed, L having mean 9 in. and standard deviation 1 in., W having mean 3 in. and standard deviation 1/4 in.
The correlation coefficient between L and W is $+ 0{\cdot}5$. Roughly what proportion of his shoes should the manufacturer make in the size $L = 10$ in., $W = 4$ in.? Roughly what proportion of his shoes should the manufacturer make with $L = 9$ in.? (With the origin at the mean values, the probability density function for the normal bivariate distribution is

$$\frac{1}{2\pi\sigma_1\sigma_2\sqrt{(1 - \varrho^2)}} \exp. - \left\{\left(\frac{x}{\sigma_1}\right)^2 - 2\varrho\left(\frac{x}{\sigma_1}\right)\left(\frac{y}{\sigma_2}\right) + \left(\frac{y}{\sigma_2}\right)^2\right\} \Big/ 2(1 - \varrho^2).)$$

(R.S.S. Cert.)

12. Compute the coefficient of rank correlation for the data of worked Ex. iv, page 210.

13. Compute the coefficient of rank correlation for the following two orders.

A:	1	2	3	4	5	6	7	8	9	10	11	12
B:	12	9	6	10	3	5	4	7	8	2	11	1

14. *50 Varieties of Wheat*

		Days to Mature				
		88 and under 91	91–	94–	97 and under 100	
Days to Head	52 and under 55	4	2	—	—	6
	55–	—	8	12	—	20
	58–	—	1	14	7	22
	61 and under 64	—	—	—	2	2
		4	11	26	9	50

Both variates give the time from appearance above ground. Calculate the Coefficient of Correlation between 'Days to Head' and 'Days to Mature' and discuss the meaning of the result. (R.S.S. Cert.)

15. The values obtained in measurements of characters X and Y on each of 500 individuals led to the table below, in which each row corresponds to an array of values of Y, the values of X falling within a fixed interval.

Central Value of X	*No. of Observations*	*Average Y*
10	40	5·7
12	90	6·3
14	130	6·7
16	120	7·2
18	80	7·5
20	40	7·9

Obtain the linear regression equation of Y on X.
What additional information would you require in order to calculate the sample coefficient of correlation between X and Y? How would you test whether the population correlation coefficient was equal to zero? (R.S.S. Cert.)

16. Counts of antibodies, before and after vaccination, yield powers of 2, 2^x being the count before and 2^y the count after vaccination. Values of x and y occurred with frequencies as in the table below and the regres-

sions appear to be approximately linear. Calculate the coefficient of correlation and the two correlation ratios.

y \ x	4	5	6	7	8	9
6	2					
7	2	1	2	1		
8		3	6	2	1	
9	2	1	3	5		1
10		2	2	3	1	1
11			1		3	

17. Classify the following 50 pairs of readings in a correlation table and find the coefficient of correlation.

x:	0·33	0·57	0·76	0·89	0·86	0·90	0·94	0·99	1·05	0·50
y:	0·73	1·03	1·60	2·33	3·01	2·73	2·14	2·67	2·93	0·80
x:	0·32	0·87	0·78	0·90	0·87	1·09	1·15	1·23	1·17	0·79
y:	0·78	1·55	1·53	2·27	1·55	2·39	2·84	3·34	2·92	2·90
x:	0·98	1·00	0·32	0·94	0·99	0·72	0·93	1·09	1·15	0·89
y:	3·71	3·25	0·74	1·82	2·24	1·05	2·59	3·09	3·39	1·58
x:	0·73	1·00	0·92	0·93	0·77	0·81	0·84	0·65	0·75	0·76
y:	1·45	2·96	2·77	2·16	1·54	1·77	2·57	1·02	1·79	2·25
x:	0·62	0·64	0·75	0·85	0·61	0·62	0·72	0·81	0·82	0·88
y:	1·27	2·13	2·78	2·54	0·94	1·59	1·67	2·27	1·22	2·28

18. Explain what is meant by partial correlation and state without proof the formula for $r_{12\cdot3}$ in terms of r_{12}, r_{13} and r_{23}.
A large number of boys are tested at the high jump and long jump. Their heights are correlated with their performances, giving correlations of $+ 0{\cdot}60$ (high jump) and $+ 0{\cdot}28$ (long jump), and the total correlation between the two events is $+ 0{\cdot}56$. Find the partial coefficient of correlation between them. (A.I.S.)

19. x, y, z measure characteristics of an individual. y and z are uncorrelated. The coefficients of correlation between x and y, x and z are ϱ_1 and ϱ_2 respectively. What is the partial correlation coefficient between x and y?
The mean winter temperature in a large number of European localities is negatively correlated with latitude (coefficient $- 0{\cdot}75$) and with altitude above sea-level (coefficient $- 0{\cdot}30$). Latitude and altitude are uncorrelated. Find the coefficient of correlation between temperature and altitude at constant latitude.

CHAPTER XII

THE ANALYSIS OF VARIANCE

Single-factor analysis. Two factors. Three factors. Two factors with replications

12.1 Purpose of the method

The statistical methods grouped under the heading 'The Analysis of Variance' were introduced by Prof. R. A. Fisher to deal with problems in agricultural research, but are applicable in many other fields. Consider, for instance, an industrial problem such as the following. A factory produces components, many machines being at work on the same operation. The process is not purely mechanical, the machine operators having an influence on the quality of the work. Moreover, it is thought that on certain days of the week (e.g. Monday) the work is apt to be of poorer quality than on other days (e.g. Friday). The quality therefore depends on at least three factors: the machine, the operator and the day of the week. There may be other factors in operation and some of the factors mentioned may have no significant effect. It would be possible, by an analysis of variance, to estimate whether machines, operators, days of the week, or some combination of these, has an appreciable effect on the quality and also to estimate the contribution made by each factor to the overall variability in the product. Thus the purpose of the analysis is to establish relations of 'cause' and 'effect'. In this connection, when we are investigating the effect of one or more causes, it must usually be supposed that, apart from these, there are minor unknowable causes of an inherent variability to be assumed normal. Thus if the three factors in our example all had no effect on the product we should still expect an inevitable variation, expressed by the normal distribution, in the product.

12.2.1 Single-factor analysis

The simplest case is that in which we are trying to assess the effect of a single cause. Suppose, for example, that holes of depth x are bored in the faces of blocks, the operation being purely mechanical, by a number of machines. The mean bore depth should be constant for all machines, but it is feared that this is not so. We wish to investigate the effect of a single factor, namely the machine, on the depth of the hole.

Let the number of machines be r and select a random sample of n items from the output of each machine. Measure the depth of bore for each of the rn items.

Denote the set of measurements by x_{ij}, where x_{ij} is the jth value of x taken from the ith machine. i will run from 1 to r and j from 1 to n. In table form:

Table of Bore Depth (x_{ij})

Machine number						
1	x_{11}	x_{12}	x_{13}	.	.	. x_{1n}
2	x_{21}	x_{22}	x_{23}	.	.	. x_{2n}
3	x_{31}	x_{32}	x_{33}	.	.	. x_{3n}
. .	.	.	.	.	.	.
r	x_{r1}	x_{r2}	x_{r3}	.	.	. x_{rn}

Let $x_{1.}$ denote the mean of the n values from the first machine
$x_{2.}$ denote the mean of the n values from the second machine
$x_{r.}$ denote the mean of the n values from the rth machine.

Then $x_{1.} = \frac{\sum_{j=1}^{n} x_{1j}}{n}$, $x_{2.} = \frac{\sum_{j=1}^{n} x_{2j}}{n}$ . . . $x_{r.} = \frac{\sum_{j=1}^{n} x_{rj}}{n}$

It is required to test whether the true machine means are constant. The working hypothesis (or null hypothesis) will be:

"The x's from each machine are normally distributed; the means of the r normal distributions are all equal as are the variances. The total population of the x's is thus normal (μ, σ), independent of machine."

It should be noted that even if this hypothesis is true the $x_{i\cdot}$ (machine sample means) will not all be equal since we have here only a sample from the whole (supposed infinite) population of the sample means. The method of attack will be to test whether the variability noted in the $x_{i\cdot}$ is more than one would expect under the null hypothesis that the machine (population) means are constant. If the machine means are not constant then we shall expect a greater variability to occur in the $x_{i\cdot}$.

Let $x_{..}$ be the mean of the rn values of x, i.e. $x_{..} = \frac{\sum_{ij} x_{ij}}{rn}$. Then, by the null hypothesis, $\frac{\sum_{ij}(x_{ij} - x_{..})^2}{rn - 1}$ is an unbiased estimate of σ^2 (see Chapter VI, equation (9)).

But $\sum_{ij}(x_{ij} - x_{..})^2 = \sum_{ij}(x_{ij} - x_{i\cdot} + x_{i\cdot} - x_{..})^2$

$$= \sum_{ij}(x_{ij} - x_{i\cdot})^2 + \sum_{ij}(x_{i\cdot} - x_{..})^2 + 2\sum_{ij}(x_{ij} - x_{i\cdot})(x_{i\cdot} - x_{..}) \quad . \quad (1)$$

$$\sum_{ij}(x_{ij} - x_{i\cdot})(x_{i\cdot} - x_{..}) = \sum_{ij} x_{ij}\, x_{i\cdot} + \sum_{ij} x_{i\cdot} x_{..} - \sum_{ij} x_{i\cdot}^2 - \sum_{ij} x_{ij} x_{..}$$
$$= \sum_i (x_{i\cdot} \sum_j x_{ij}) + x_{..}\, n \sum_i x_{i\cdot} - n \sum_i x_{i\cdot}^2 - rnx_{..}^2$$
$$= n\sum_i x_{i\cdot}^2 + rnx_{..}^2 - n\sum_i x_{i\cdot}^2 - rnx_{..}^2$$
$$= 0. \quad . \quad . \quad . \quad . \quad . \quad . \quad . \quad (2)$$

Equations (1) and (2) give:

$$\sum_{ij}(x_{ij} - x_{..})^2 = \sum_{ij}(x_{ij} - x_{i\cdot})^2 + \sum_{ij}(x_{i\cdot} - x_{..})^2 \quad . \quad . \quad . \quad (3)$$

By §6.7

$$\frac{1}{\sigma^2}\sum_{ij}(x_{ij} - x_{..})^2 \text{ is distributed as } \chi^2 \text{ with } (rn - 1) \text{ d.f.}$$

$$\frac{1}{\sigma^2}\sum_{ij}(x_{ij} - x_{i.})^2 \text{ is distributed as } \chi^2 \text{ with } (rn - r) \text{ d.f.}$$

Using equation (3), Cochran's theorem gives:

$$\frac{1}{\sigma^2}\sum_{ij}(x_{i.} - x_{..})^2 \text{ is distributed as } \chi^2 \text{ with } (r - 1) \text{ d.f.}$$

In each case the mean of the distribution is the number of d.f. Moreover $\sum_{ij}(x_{i.} - x_{..})^2$ and $\sum_{ij}(x_{ij} - x_{..})^2$ are independent and hence $\frac{\sum_{ij}(x_i - x_{..})^2}{r - 1} \Big/ \frac{\sum_{ij}(x_{ij} - x_{..})^2}{rn - 1}$ is distributed as Fisher's F (see Chapter VI) and this will give a test of the constancy of machine means.

The whole process can be summarized thus:

$$\frac{1}{rn - 1}\sum_{ij}(x_{ij} - x_{..})^2; \quad \frac{1}{rn - r}\sum_{ij}(x_{ij} - x_{i.})^2; \quad \frac{1}{r - 1}\sum_{ij}(x_{i.} - x_{..})^2$$

each estimate σ^2. The first is based on the total variation occurring in the whole sample of rn, independent of machine. The second is based on the variation occurring *within* each machine sample. The third is based on the variation of the machine means from the whole sample mean. These variations are called, respectively, the 'TOTAL', the 'WITHIN', the 'BETWEEN' variations and equation (3) shows that the total variability as measured by sums of squares has two components, a within machine and a between machine variability.

Completed Analysis of Variance

Cause	*Sum of Squares*	*Degrees of Freedom*	*Mean Square (Ratio of Sum of Squares to d.f.)*
Between machines	$\sum_{ij}(x_{i.} - x_{..})^2$	$r - 1$	$\frac{1}{r - 1}\sum_{ij}(x_{i.} - x_{..})^2$
Within machines .	$\sum_{ij}(x_{ij} - x_{i.})^2$	$rn - r$	$\frac{1}{rn - r}\sum_{ij}(x_{ij} - x_{i.})^2$
Total . .	$\sum_{ij}(x_{ij} - x_{..})^2$	$rn - 1$	$\frac{1}{rn - 1}\sum_{ij}(x_{ij} - x_{..})^2$

The total sum of squares is the sum of the component sums of squares and the total degrees of freedom is the sum of the component degrees of freedom.

Ex. i

A sample of ten bore holes was taken from each of five machines. The following are the data in inches. Assuming that the distribution of bore holes from each machine is normal with constant variance, investigate whether the means are constant.

Machine											Machine mean
1:	1·105	1·108	1·104	1·105	1·107	1·105	1·105	1·106	1·105	1·103	$x_{1.} = 1{\cdot}1054$
2:	1·109	1·105	1·104	1·103	1·106	1·102	1·104	1·103	1·105	1·105	$x_{2.} = 1{\cdot}1046$
3:	1·101	1·104	1·105	1·106	1·104	1·103	1·103	1·104	1·106	1·102	$x_{3.} = 1{\cdot}1038$
4:	1·106	1·105	1·102	1·108	1·103	1·104	1·105	1·104	1·104	1·104	$x_{4.} = 1{\cdot}1045$
5:	1·108	1·104	1·103	1·105	1·101	1·104	1·103	1·106	1·105	1·104	$x_{5.} = 1{\cdot}1043$

Grand mean $x_{..} = 1{\cdot}1045$

Then
$$\sum_{ij}(x_{ij} - x_{..})^2 = 0{\cdot}0001465$$
$$\sum_{ij}(x_{i.} - x_{..})^2 = 0{\cdot}0000131$$
$$\sum_{ij}(x_{ij} - x_{i.})^2 = 0{\cdot}0001334$$

We show in §12.2.2 how these sums may be more conveniently calculated, without using the machine means.

Completed Analysis of Variance

Cause	*Sum of Squares*	*Degrees of Freedom*	*Mean Square*
Between machines .	0·0000131	4	0·0000033
Within machines .	0·0001334	45	0·0000030
Total . . .	0·0001465	49	0·0000030

Testing the mean square between machines against the mean square within machines:

$$\frac{\text{Mean square between machines}}{\text{Mean square within machines}} = 1{\cdot}07 = F$$

The corresponding degrees of freedom are 4, 45. Reference to the F table shows no significance and so the hypothesis of constancy of machine means has not been disproved.

12.2.2 Method of computation

At this stage it may be helpful to give a method which simplifies the computation of the sums of squares. The formulae obtained are analogous to that often used in computing a single sample variance, namely

$$ns^2 = \Sigma(x - \bar{x})^2 = \Sigma x^2 - \frac{1}{n}(\Sigma x)^2$$

Consider first,
$$\begin{aligned}\sum_{ij}(x_{ij} - x_{..})^2 &= \sum_{ij}x_{ij}^2 - 2\sum_{ij}x_{ij}x_{..} + \sum_{ij}x_{..}^2 \\ &= \sum_{ij}x_{ij}^2 - 2x_{..}\sum_{ij}x_{ij} + rnx_{..}^2 \\ &= \sum_{ij}x_{ij}^2 - 2x_{..}(rnx_{..}) + rnx_{..}^2 \\ &= \sum_{ij}x_{ij}^2 - rnx_{..}^2 \\ &= \sum_{ij}x_{ij}^2 - \frac{(\sum_{ij}x_{ij})^2}{rn} \qquad (4)\end{aligned}$$

$$\sum_{ij}(x_{i\cdot} - x_{\cdot\cdot})^2 = n\sum_i(x_{i.} - x_{\cdot\cdot})^2$$
$$= n\{\sum_i x_{i.}^2 - 2\sum_i x_{i.}\, x_{\cdot\cdot} + \sum_i x_{\cdot\cdot}^2\}$$
$$= n\{\sum_i x_{i.}^2 - r x_{\cdot\cdot}^2\}$$
$$= n\left\{\sum_i \frac{(\sum_j x_{ij})^2}{n^2} - r\frac{(\sum_{ij} x_{ij})^2}{r^2 n^2}\right\}$$
$$= \frac{\sum_i(\sum_j x_{ij})^2}{n} - \frac{(\sum_{ij} x_{ij})^2}{rn} \quad . \quad . \quad . \quad . \quad (5)$$
$$\sum_{ij}(x_{ij} - x_{i.})^2 = \sum_{ij} x_{ij}^2 - 2\sum_{ij} x_{ij}\, x_{i\cdot} + \sum_{ij} x_{i.}^2$$
$$= \sum_{ij} x_{ij}^2 - \sum_i(\sum_j x_{ij})^2/_n \quad . \quad . \quad . \quad . \quad (6)$$

From the table of data, x_{ij}, we need only calculate

(i) A = sum of all x's = $\sum_{ij} x_{ij}$,
(ii) S = sum of all squares = $\sum_{ij} x_{ij}^2$,
and (iii) the r row totals, $R_i = \sum_j x_{ij}$.

Then the 'between rows' sum of squares $= \frac{1}{n}\Sigma R^2 - \frac{1}{rn}A^2$

the 'within rows' sum of squares $= S - \frac{1}{n}\Sigma R^2$

the 'total' sum of squares $= S - \frac{1}{rn}A^2$.

The sum of the column totals $= \Sigma C_j = \sum_{ij} x_{ij}$ is calculated as a check on A. There is also no loss of generality in referring the data to a working origin.

Take 1·100 as origin and let 0·001 = 1 unit. In this way all variances and sums of squares will be 10^6 times their real value. The numbers in parentheses are the corresponding squares.

Machine											$\sum_j x$	$(\sum_j x)^2$	$\sum_j x^2$
1	5(25)	8(64)	4(16)	6(36)	7(49)	5(25)	5(25)	6(36)	5(25)	3(9)	54	2916	310
2	9(81)	5(25)	4(16)	3(9)	6(36)	2(4)	4(16)	3(9)	5(25)	5(25)	46	2116	246
3	1(1)	4(16)	5(25)	6(36)	4(16)	3(9)	3(9)	4(16)	6(36)	2(4)	38	1444	168
4	6(36)	5(25)	2(4)	8(64)	3(9)	4(16)	5(25)	4(16)	4(16)	4(16)	45	2025	227
5	8(64)	4(16)	3(9)	5(25)	1(1)	4(16)	3(9)	6(36)	5(25)	4(16)	43	1849	217
$\sum_i x$	29	26	18	28	21	18	20	23	25	18	226	10346	
$\sum_i x^2$	207	146	70	170	111	70	84	113	127	70			1168

$$\frac{(\sum_{ij} x_{ij})^2}{rn} = \frac{51076}{50} = 1021{\cdot}5 \qquad \frac{\sum_i(\sum_j x_{ij})^2}{n} = \frac{10346}{10} = 1034{\cdot}6$$

Between machine sum of squares = 1034·6 − 1021·5 = 13·1
Within machine sum of squares = 1168 − 1034·6 = 133·4
Total sum of squares = 1168 − 1021·5 = 146·5

It is seen that these results are equivalent to the previous values.

Ex. ii

Six technicians take hardness figures on metal strip. It is required to test whether there is any difference in mean level (i.e. any bias) between them. Using the same hardness-testing machine each technician makes eight readings on the strip. Assuming that the strip is homogeneous, do the following data give any indication of differences between technicians?

Hardness Figures (V.D.H.)

Technician								
A:	729	728	726	727	725	725	723	727
B:	727	726	724	725	727	724	725	726
C:	729	726	727	726	723	725	726	724
D:	727	722	723	726	723	725	723	722
E:	726	721	722	725	723	724	723	723
F:	725	727	724	723	726	724	724	728

Reducing the data to a working mean of 720, the following is the calculation:

									Σx	$(\Sigma x)^2$	Σx^2
A:	9(81)	8(64)	6(36)	7(49)	5(25)	5(25)	6(36)	7(49)	53	2809	365
B:	7(49)	6(36)	4(16)	5(25)	7(49)	4(16)	5(25)	6(36)	44	1936	252
C:	9(81)	6(36)	7(49)	6(36)	3(9)	5(25)	6(36)	4(16)	46	2116	288
D:	7(49)	2(4)	3(9)	6(36)	3(9)	5(25)	3(9)	2(4)	31	961	145
E:	6(36)	1(1)	2(4)	5(25)	3(9)	4(16)	3(9)	3(9)	27	729	109
F:	5(25)	7(49)	4(16)	3(9)	6(36)	4(16)	4(16)	8(64)	41	1681	231
Σx	43	30	26	32	27	27	27	30	242		
Σx^2	321	190	130	180	137	123	131	178			1390

$$\sum_{ij} x_{ij}^2 = 1390 \qquad \sum_i \Big(\sum_j x_{ij}\Big)^2 = 10232 \qquad \sum_{ij} x_{ij} = 242$$

$$\frac{\sum_i \Big(\sum_j x_{ij}\Big)^2}{8} = 1279{\cdot}0 \qquad \frac{\Big(\sum_{ij} x_{ij}\Big)^2}{48} = 1220{\cdot}1$$

Components of Variance

Cause	*Sum of Squares*	*d.f.*	*Mean Square*
Between technicians . .	58·9	5	11·78
Within technicians . . .	111·0	42	2·64
Total	169·9	47	3·61

Testing the between technicians effect against the within technicians effect gives $F = \dfrac{11{\cdot}78}{2{\cdot}64} = 4{\cdot}46$, with 5 and 42 d.f. The level of significance is less than 1% and so the null hypothesis is abandoned.

12.2.3 Rejection of the null hypothesis

The null hypothesis is virtually divided into three parts:

(a) The group populations are normal;
(b) The group means are the same;
(c) The group variances are the same.

The reasons for assuming (a) are that a large number of distributions that occur in practice are either normal or very close to it and, unfortunately, the ratio F has only been tabulated for normal populations. This normality assumption should always be tested, although in practice, it must be confessed, this is rarely done. It would be difficult to assess the normality of group populations at all effectively when the number sampled from each group is small.

Suppose therefore that (a) is true and that the F test does not give a confirmation of the null hypothesis (see Ex. ii). In this case either (b), or (c), or both, must be false.

Assume first that (c) is true and that (b) is false; that is, we have abandoned the hypothesis of constancy of group means. The first problem is to estimate the common variance, σ^2 (say). Previously there were three estimates of σ^2, but, as we shall see, we now have only one.

Let the number of groups (or machines) be r and suppose we have taken a sample of n from each group (or machine).

For any given i, $\Sigma_j \frac{(x_{ij} - x_{i\cdot})^2}{n-1}$ is an unbiased estimate of σ^2.

Hence $\Sigma_{ij}(x_{ij} - x_{i\cdot})^2$ estimates $\Sigma_i(n-1)\sigma^2 = (rn - r)\sigma^2$ . . (7)

and $\Sigma_{ij} \frac{(x_{ij} - x_{i\cdot})^2}{rn - r}$ estimates σ^2.

Since we now assume that the true group means are not all equal, let the true mean of the ith group be μ_i, i.e. $E(x_{i\cdot}) = \mu_i$ and let $\mu_\cdot$ be the true mean of all the x (independent of group).

Write $x_{ij} = \mu_i + y_{ij}$. The y_{ij} for each group are therefore normal, by the null hypothesis, with zero mean (since $E(x_{ij}) = \mu_i$) and with variance σ^2, by the null hypothesis.

To estimate σ^2 we proceed as follows:

$$x_{i\cdot} = \mu_i + y_{i\cdot}\,;\; x_{\cdot\cdot} = \mu_\cdot + y_{\cdot\cdot}$$

$$\begin{aligned}\Sigma_i(x_{i\cdot} - x_{\cdot\cdot})^2 &= \Sigma_i(x_{i\cdot} - y_{i\cdot} + y_{i\cdot} - y_{\cdot\cdot} + y_{\cdot\cdot} - x_{\cdot\cdot})^2 \\ &= \Sigma_i(\mu_i - \mu_\cdot + y_{i\cdot} - y_{\cdot\cdot})^2 \\ &= \Sigma_i(\mu_i - \mu_\cdot)^2 + 2\Sigma_i(\mu_i - \mu_\cdot)(y_{i\cdot} - y_{\cdot\cdot}) + \Sigma_i(y_{i\cdot} - y_{\cdot\cdot})^2\end{aligned}$$

But $E\Sigma_i(\mu_i - \mu_\cdot)(y_{i\cdot} - y_{\cdot\cdot}) = 0$ since the two brackets are independent and $E(y_{i\cdot} - y_{\cdot\cdot}) = 0$.

$$E\Sigma_i(x_{i\cdot} - x_{\cdot\cdot})^2 = E\Sigma_i(\mu_i - \mu_\cdot)^2 + E\Sigma_i(y_{i\cdot} - y_{\cdot\cdot})^2$$

$$E\Sigma_{ij}(x_{i\cdot} - x_{\cdot\cdot})^2 = E\Sigma_{ij}(\mu_i - \mu_\cdot)^2 + E\Sigma_{ij}(y_{i\cdot} - y_{\cdot\cdot})^2$$

$(y_{i\cdot} - y_{\cdot\cdot})$ is normal and $\Sigma_{ij}(y_{i\cdot} - y_{\cdot\cdot})^2 = n\Sigma_i(y_{i\cdot} - y_{\cdot\cdot})^2$

$\therefore \sum_{ij}(y_{i\cdot} - y_{\cdot\cdot})^2$ is distributed as $\sigma^2\chi^2$ with $(r - 1)$ d.f.

Hence $E\sum_{ij}(x_{i\cdot} - x_{\cdot\cdot})^2 = E\sum_{ij}(\mu_i - \mu_{\cdot})^2 + (r - 1)\sigma^2$ . . . (8)

Again,

$$\sum_{ij}(x_{ij} - x_{\cdot\cdot})^2 = \sum_{ij}(x_{ij} - x_{i\cdot} + x_{i\cdot} - x_{\cdot\cdot})^2$$
$$= \sum_{ij}(y_{ij} - y_{i\cdot} + y_{i\cdot} - y_{\cdot\cdot} + \mu_i - \mu_{\cdot})^2$$
$$= \sum_{ij}(y_{ij} - y_{\cdot\cdot})^2 + \sum_{ij}(\mu_i - \mu_{\cdot})^2 + 2\sum_{ij}(\mu_i - \mu_{\cdot})(y_{ij} - y_{\cdot\cdot})$$

$\sum_{ij}(y_{ij} - y_{\cdot\cdot})^2$ is distributed as $\sigma^2\chi^2$ with $(rn - 1)$ d.f. and

$E\sum_{ij}(\mu_i - \mu_{\cdot})(y_{ij} - y_{\cdot\cdot}) = 0$, by argument as above.

Hence $E\sum_{ij}(x_{ij} - x_{\cdot\cdot})^2 = E\sum_{ij}(\mu_i - \mu_{\cdot})^2 + (rn - 1)\sigma^2$. (9)

Also $E\sum_{ij}(x_{ij} - x_{i\cdot})^2 = (rn - r)\sigma^2$ . . . . (7)

Now in equations (8) and (9),

$$E\sum_{ij}(\mu_i - \mu_{\cdot})^2 = n\, E\sum_{i}(\mu_i - \mu_{\cdot})^2$$

since in the summation over i and j each μ_i occurs n times, once for each value of j.

Moreover $E\sum_{i}(\mu_i - \mu_{\cdot})^2/(r - 1)$ is the variance of the group means on the assumption that our r groups are a sample from a large population of groups.

Denoting the variance of group means by σ_μ^2, the analysis of variance may be tabulated as follows, the last column showing the quantity estimated by the computed mean squares.

Cause	*Sum of Squares*	*d.f.*	*Mean Square Estimates*
Between groups	$\sum_{ij}(x_{i\cdot} - x_{\cdot\cdot})^2$	$r - 1$	$\sigma^2 + n\sigma_\mu^2$
Within groups	$\sum_{ij}(x_{ij} - x_{i\cdot})^2$	$rn - r$	σ^2
Total .	$\sum_{ij}(x_{ij} - x_{\cdot\cdot})^2$	$rn - 1$	$\sigma^2 + n\,E\sum_{i}(\mu_i - \mu_{\cdot})^2/(rn - 1)$

For instance, in Ex. ii above our estimates are

$$\sigma^2 = 2{\cdot}64$$
$$\sigma^2 + 8\sigma_\mu^2 = 11{\cdot}78, \text{ giving } \sigma_\mu^2 = 1{\cdot}18.$$

Thus, regarding the six technicians as a random sample of technicians, we estimate as 1·18 the variance to be expected as the result of measurement by different technicians, over and above the inherent variability in measured hardness represented by the variance 2·64.

It should be noted that the groups cannot always be regarded as a sample from a large population of groups; for instance the factor being investigated might be the day of the week. If the r groups include all possible groups we should use $\sum_i (\mu_i - \mu_.)^2/r$ to estimate σ_μ^2. In either case as n (and where possible, r) tends to infinity the *total* mean square tends to $\sigma^2 + \sigma_\mu^2$, in accordance with Chapter VI, equation (11) for the variance of the sum of independent variates.

If r is fixed, being the total number of possible groups, our last mean square estimates

$$\sigma^2 + \frac{nr}{nr-1}\sigma_\mu^2 \to \sigma^2 + \sigma_\mu^2 \text{ as } n \to \infty$$

If the r groups are a sample from an infinity of possible groups, the last mean square estimates

$$\sigma^2 + \frac{n(r-1)}{nr-1}\sigma_\mu^2 \to \sigma^2 + \sigma_\mu^2 \text{ as } n, r \to \infty$$

Before leaving the case of single-factor analysis, the reader should observe that the analysis of variance is here used as a test of homogeneity of *means*, yielding, when the null hypothesis (of homogeneity) is rejected, estimates of the components of variance due to (i) variation of group means and (ii) all other causes.

The analogous test of homogeneity of group *variances* is Bartlett's (§6.6.2).

12.3.1 Two-factor analysis

Consider next the problem of estimating the effects of two contributory causes. In Ex. i, for instance, suppose now that the operation of boring is not purely mechanical but that operators as well as machines may affect the quality of the output, namely the depth of bore. It is required to estimate whether the variability in depth and mean depth may be attributed to differences between operators or between machines, and also whether particular operators on particular machines give significantly different results—this last effect is called an operator × machine interaction.

Let the number of machines be r and the number of operators c. Let each operator work each of the r machines. From the output of each of the rc combinations we might suppose a sample of n values taken but we shall restrict ourselves at present to the case when $n = 1$, so that one bore depth is observed for each combination of operator and machine. The case when n is greater than one is considered in §12.4.4 and Ex. ix as a special form of three-factor analysis.

The data will consist of values x_{ij}, where i refers to the machine and runs from 1 to r, and j refers to operator and runs from 1 to c.

Table of Bore Depths

	Operator 1	2	3	.	.	c
Machine						
1:	x_{11}	x_{12}	x_{13}	—	—	x_{1c}
2:	x_{21}	x_{22}	x_{23}	—	—	x_{2c}
3:	x_{31}	x_{32}	x_{33}	—	—	x_{3c}
.	.	.	.	.	.	.
r:	x_{r1}	x_{r2}	x_{r3}	—	—	x_{rc}

The null hypothesis will be that the distribution of bore depths is normal with constant mean μ and constant variance σ^2 for each of the rc combinations of operator and machine, and this hypothesis will be tested by a random sample of rc selected one from each of the combinations.

By properties of mean values, used in establishing equations (1), (2) and (3), we now express the total sum of squares as the sum of three components. We denote by $x_{\cdot j}$ the mean of the jth column, namely $\frac{1}{r}\Sigma_i x_{ij}$.

$$\Sigma_{ij}(x_{ij} - x_{\cdot\cdot})^2$$
$$= \Sigma_{ij}(x_{i\cdot} - x_{\cdot\cdot})^2 + \Sigma_{ij}(x_{\cdot j} - x_{\cdot\cdot})^2 + \Sigma_{ij}(x_{ij} - x_{i\cdot} - x_{\cdot j} + x_{\cdot\cdot})^2$$
$$= c\Sigma_i(x_{i\cdot} - x_{\cdot\cdot})^2 + r\Sigma_j(x_{\cdot j} - x_{\cdot\cdot})^2 + \Sigma_{ij}(x_{ij} - x_{i\cdot} - x_{\cdot j} + x_{\cdot\cdot})^2 \quad . \quad (10)$$

But $\Sigma_{ij}(x_{ij} - x_{\cdot\cdot})^2$ is distributed as $\sigma^2\chi^2$ with $(rc - 1)$ d.f. . . (11)

$c\Sigma_i(x_{i\cdot} - x_{\cdot\cdot})^2$ is distributed as $c \, . \, \frac{c^2}{c}\chi^2$ with $(r - 1)$ d.f. . . (12)

$r\Sigma_j(x_{\cdot j} - x_{\cdot\cdot})^2$ is distributed as $r \, . \, \frac{\sigma^2}{r}\chi^2$ with $(c - 1)$ d.f. . (13)

Hence by equation (10) and Cochran's theorem,

$\Sigma_{ij}(x_{ij} - x_{i\cdot} - x_{\cdot j} + x_{\cdot\cdot})^2$ is distributed as $\sigma^2\chi^2$ with

$(rc - 1) - (r - 1) - (c - 1) = (r - 1)(c - 1)$ d.f. . (14)

and the three sums on the R.H.S. of equation (10) are independent χ^2 variates. Equations (11), (12), (13) and (14) give:

$$E\Sigma_{ij}(x_{ij} - x_{\cdot\cdot})^2 = (rc - 1)\sigma^2 \quad . \quad . \quad (15)$$
$$Ec\Sigma_i(x_{i\cdot} - x_{\cdot\cdot})^2 = (r - 1)\sigma^2 \quad . \quad . \quad . \quad (16)$$
$$Er\Sigma_j(x_{\cdot j} - x_{\cdot\cdot})^2 = (c - 1)\sigma^2 \quad . \quad . \quad . \quad (17)$$
$$E\Sigma_{ij}(x_{ij} - x_{i\cdot} - x_{\cdot} + x_{\cdot\cdot})^2 = (r - 1)(c - 1)\sigma^2 \quad . \quad (18)$$

All these four sums of squares can be used to estimate σ^2. The variance from equation (15), $\dfrac{\Sigma_{ij}(x_{ij} - x_{..})^2}{rc - 1}$ arises from the total variability. The variance from equation (16), $\dfrac{\Sigma_i(x_{i.} - x_{..})^2}{r - 1}$ arises from the variability between $x_{i.}$, i.e. the 'between machine' variability. The variance from equation (17), $\dfrac{\Sigma_j(x_{.j} - x_{..})^2}{c - 1}$ arises from the variability between $x_{.j}$, i.e. the 'between operator' variability. From equation (10), in this context, we can refer to $\dfrac{\Sigma_{ij}(x_{ij} - x_{i.} - x_{.j} + x_{..})^2}{(r - 1)(c - 1)}$ as a residual variability.

In table form:

Cause	*Sum of Squares*	*d.f.*	*Mean square Estimates*
Between machines .	$c\Sigma_i(x_{i.} - x_{..})^2$	$r - 1$	σ^2
Between operators .	$r\Sigma_j(x_{.j} - x_{..})^2$	$c - 1$	σ^2
Residual .	$\Sigma_{ij}(x_{ij} - x_{i.} - x_{.j} + x_{..})^2$	$(r - 1)(c - 1)$	σ^2
Total .	$\Sigma_{ij}(x_{ij} - x_{..})^2$	$rc - 1$	σ^2

The total sum of squares is, by equation (10), the sum of the component sums of squares, and the total degrees of freedom is the sum of the component degrees of freedom.

The assumption of homogeneity of means is tested by comparing the between machines mean square with the residual mean square and the between operators mean square with the residual mean square using Fisher's F test. If either of these, say the former, gives a significant result it would mean that the corresponding effect, i.e. the between machines, is of significance and the hypothesis of homogeneity of machine means would have to be abandoned. In this case the between machines mean square would, of course, no longer be a valid estimate of σ^2.

If both comparisons on the F test give a significant result, then the hypothesis of homogeneity of means would have to be abandoned for machines and operators. In this case we would use the residual mean square to estimate σ^2.

If no significant result occurs in either comparison we have four estimators of σ^2 and we may take the total mean square, $\dfrac{\sum'_{ij}(x_{ij} - x_{..})^2}{rc - 1}$ as the best estimate as this is based on the greatest number of degrees of freedom.

Ex. iii

The following data are values of the weight (in oz.) of a pellet selected at random from the work of ten operators on each of five pellet machines. Is there any evidence of a difference in workmanship of the operators or in the working of the machines?

Weight (oz.)

Operator

	A	B	C	D	E	F	G	H	I	J
a:	83	84	84	86	85	84	87	88	86	86
b:	84	83	85	84	84	88	86	87	86	86
c:	83	88	90	87	85	87	85	84	84	86
d:	88	82	86	84	85	85	88	83	85	84
e:	83	86	89	84	85	85	83	86	84	84

Reducing to a Working Mean of 80:

i \ j	1	2	3	4	5	6	7	8	9	10
1	3	4	4	6	5	4	7	8	6	6
2	4	3	5	4	4	8	6	7	6	6
3	3	8	10	7	5	7	5	4	4	6
4	8	2	6	4	5	5	8	3	5	4
5	3	6	9	4	5	5	3	6	4	4

$$c\sum_i (x_{i.} - x_{..})^2 = 6{\cdot}8$$

$$r\sum_j (x_{.j} - x_{..})^2 = 25{\cdot}6$$

$$\sum_{ij}(x_{ij} - x_{i.} - x_{.j} + x_{..})^2 = 122{\cdot}4$$

$$\sum_{ij}(x_{ij} - x_{..})^2 = 154{\cdot}8$$

(These sums may be calculated by first calculating the means of rows and columns, but more conveniently by the method given in the next section.) We obtain the following analysis of variance.

Components of Variance:

	Sum of Squares	*d.f.*	*Mean Square*
Between machines . . .	6·8	4	1·70
Between operators . . .	25·6	9	2·83
Residual	122·4	36	3·40
Total	154·8	49	3·16

The F test gives for between machines mean square against residual mean square, $\dfrac{3{\cdot}40}{1{\cdot}70} = 2{\cdot}00$ (testing the greater against the lesser) and

between operators mean square against residual mean square $\frac{3 \cdot 40}{2 \cdot 83} = 1 \cdot 20$.
With the appropriate degrees of freedom, neither of these is significant and so there is no evidence that the homogeneity of machines and operators should be rejected. The 'total mean square', 3·16, is taken as estimate of the variance, i.e. $\sigma = 1 \cdot 78$.

We shall therefore expect 99% of the pellet weights to lie within

$$x_{..} \pm 2 \cdot 58\sigma = 85 \cdot 28 \pm 4 \cdot 59 = 80 \cdot 69 \text{ and } 89 \cdot 87 \text{ oz.}$$

12.3.2 Method of computation

As in the single-factor experiment the computational procedure is capable of simplification.

As before, $c\sum_i(x_{i.} - x_{..})^2 = \frac{\sum_i(\sum_j x_{ij})^2}{c} - \frac{(\sum_{ij} x_{ij})^2}{rc}$ equivalent to equation (5).

Similarly, $r\sum_j(x_{.j} - x_{..})^2 = \frac{\sum_j(\sum_i x_{ij})^2}{r} - \frac{(\sum_{ij} x_{ij})^2}{rc}$ equivalent to equation (5)

and $\sum_{ij}(x_{ij} - x_{..})^2 = \sum_{ij} x_{ij}^2 - \frac{(\sum_{ij} x_{ij})^2}{rc}$ equivalent to equation (4)

$\sum_{ij}(x_{ij} - x_{i.} - x_{.j} + x_{..})^2$ reduces to

$$\sum_{ij} x_{ij}^2 - \sum_i \frac{(\sum_j x_{ij})^2}{c} - \sum_j \frac{(\sum_i x_{ij})^2}{r} + \frac{(\sum_{ij} x_{ij})^2}{rc} \quad . \quad . \quad (19)$$

These may be stated as follows, s.s. denoting sum of squares.

$$\text{Between rows s.s.} = \frac{1}{c}\Sigma R^2 - \frac{1}{rc}A^2$$

$$\text{Between cols. s.s.} = \frac{1}{r}\Sigma C^2 - \frac{1}{rc}A^2$$

$$\text{Residual s.s.} = S - \frac{1}{c}\Sigma R^2 - \frac{1}{r}\Sigma C^2 + \frac{1}{rc}A^2$$

$$\text{Total s.s.} = S - \frac{1}{rc}A^2$$

The above example becomes in computational form:

											$\sum_j x$	$(\sum_j x)^2$	$\sum_j x^2$
	3(9)	4(16)	4(16)	6(36)	5(25)	4(16)	7(49)	8(64)	6(36)	6(36)	53	2809	303
	4(16)	3(9)	5(25)	4(16)	4(16)	8(64)	6(36)	7(49)	6(36)	6(36)	53	2809	303
	3(9)	8(64)	10(100)	7(49)	5(25)	7(49)	5(25)	4(16)	4(16)	6(36)	59	3481	389
	8(64)	2(4)	6(36)	4(16)	5(25)	5(25)	8(64)	3(9)	5(25)	4(16)	50	2500	284
	3(9)	6(36)	9(81)	4(16)	5(25)	5(25)	3(9)	6(36)	4(16)	4(16)	49	2401	269
$\sum_i x$	21	23	34	25	24	29	29	28	25	26	264		
$(\sum_i x)^2$	441	529	1156	625	576	841	841	784	625	676			
$\sum_i x^2$	107	129	258	133	116	179	183	174	129	140			1548

$$\sum_{ij} x^2 = 1548,\ \sum_i\Big(\sum_j x\Big)^2 = 14000,\ \sum_j\Big(\sum_i x\Big)^2 = 7094,\ \sum_{ij} x = 264,$$

$$\frac{\sum_i\big(\sum_j x\big)^2}{c} = 1400,\ \frac{\sum_j\big(\sum_i x\big)^2}{r} = 1418{\cdot}8,\ \frac{\big(\sum_{ij} x\big)^2}{rc} = 1393{\cdot}2.$$

$$\frac{\sum_i\big(\sum_j x\big)^2}{c} - \frac{\big(\sum_{ij} x\big)^2}{rc} = 6{\cdot}8$$

$$\frac{\sum_j\big(\sum_i x\big)^2}{r} - \frac{\big(\sum_{ij} x\big)^2}{rc} = 25{\cdot}6$$

$$\sum_{ij} x^2 - \frac{\sum_i\big(\sum_j x\big)^2}{c} - \frac{\sum_j\big(\sum_i x\big)^2}{r} + \frac{\big(\sum_{ij} x\big)^2}{rc} = 122{\cdot}4$$

and $\sum_{ij} x^2 - \big(\sum_{ij} x\big)^2/rc = 154{\cdot}8$, as already given.

Ex. iv

Four chemists each estimate the strength of five solutions of the same acid by neutralizing against a standard homogeneous alkali. The following are the number of c.c. of standard used in each of twenty tests. Estimate whether there is any difference between the chemists and between the strengths of the five solutions.

	Chemist			
Solution	*A*	*B*	*C*	*D*
a:	43	41	42	46
b:	45	41	42	46
c:	49	46	44	46
d:	45	42	42	45
e:	44	42	44	44

Reducing to a working mean of 40 the analysis becomes

					$\sum_j x_{ij}$	$\big(\sum_j x_{ij}\big)^2$	$\sum_j x_{ij}^2$
	3(9)	1(1)	2(4)	6(36)	12	144	50
	5(25)	1(1)	2(4)	6(36)	14	196	66
	9(81)	6(36)	4(16)	6(36)	25	625	169
	5(25)	2(4)	2(4)	5(25)	14	196	58
	4(16)	2(4)	4(16)	4(16)	14	196	52
$\sum_i x_{ij}$	26	12	14	27	79		
$\big(\sum_i x_{ij}\big)^2$	676	144	196	729			
$\sum_i x_{ij}^2$	156	46	44	149			395

$$\sum_{ij} x^2 = 395,\ \sum_i\Big(\sum_j x\Big)^2 = 1357,\ \sum_j\Big(\sum_i x\Big)^2 = 1745,\ \sum_{ij} x = 79$$

$$\frac{\sum_i\big(\sum_j x\big)^2}{4} = 339{\cdot}25,\qquad \frac{\sum_j\big(\sum_i x\big)^2}{5} = 349,\qquad \frac{\big(\sum_{ij} x\big)^2}{20} = 312{\cdot}05.$$

Components of Variance

	Sum of Squares	*d.f.*	*Mean Square*
Between solutions	27·20	4	6·80
Between chemists	36·95	3	12·32
Residual	18·80	12	1·57
Total	82·95	19	4·34

Testing the between solutions mean square against the residual mean square gives $F = \frac{6{\cdot}80}{1{\cdot}57} = 4{\cdot}33$ with d.f. 4 and 12 $P < 5\%$

Testing the between chemists mean square against the residual mean square gives $F = \frac{12{\cdot}32}{1{\cdot}57} = 7{\cdot}87$ with d.f. 3 and 12 $P < 1\%$

Both of these are significant and we deduce that neither the set of chemists nor the set of solutions can be regarded as constant.

12.3.3 Rejection of the null hypothesis

When the null hypothesis of §12.3.1 is untenable on examination by the F test, the assumption of constancy of means is discarded and a new hypothesis is formed:

Each element in each combination of row and column position is selected at random from a normal population (μ_{ij}, σ).

All that is now assumed besides normality is constancy of variance in the rc populations.

Let
$$x_{ij} = \mu_{ij} + y_{ij}$$
then
$$x_{i\cdot} = \mu_{i\cdot} + y_{i\cdot}$$
$$x_{\cdot j} = \mu_{\cdot j} + y_{\cdot j}$$
$$x_{\cdot\cdot} = \mu_{\cdot\cdot} + y_{\cdot\cdot}$$
where y_{ij} is normal $(0, \sigma)$.

$$E\sum_{ij}(x_{ij} - x_{i\cdot} - x_{\cdot j} + x_{\cdot\cdot})^2$$
$$= E\sum_{ij}(\mu_{ij} - \mu_{i\cdot} - \mu_{\cdot j} + \mu_{\cdot\cdot})^2 + E\sum_{ij}(y_{ij} - y_{i\cdot} - y_{\cdot j} + y_{\cdot\cdot})^2 \quad . \quad (20)$$

since y and μ are independent and $E(y_{ij} - y_{i\cdot} - y_{\cdot j} + y_{\cdot\cdot}) = 0$.

We can say for y, as for x in equation (14),

$$E\sum_{ij}(y_{ij} - y_{i\cdot} - y_{\cdot j} + y_{\cdot\cdot})^2 = (r - 1)(c - 1)\sigma^2 \quad . \quad . \quad (21)$$

From equations (20) and (21) we see that the residual

$$\sum_{ij}(x_{ij} - x_{i\cdot} - x_{\cdot j} + x_{\cdot\cdot})^2$$

will provide an unbiased estimate of σ^2 only if

$$\sum_{ij}(\mu_{ij} - \mu_{i\cdot} - \mu_{\cdot j} + \mu_{\cdot\cdot})^2 = 0$$

i.e. if $$\mu_{ij} - \mu_{i\cdot} - \mu_{\cdot j} + \mu_{\cdot\cdot} = 0, \text{ for all } i, j \quad . \quad . \quad (22)$$

This leads to making the further assumption that

$$\left.\begin{aligned} \mu_{ij} &= a_i + b_j \\ \mu_{i\cdot} &= a_i + b. \\ \mu_{\cdot j} &= a. + b_j \\ \mu_{\cdot\cdot} &= a. + b. \end{aligned}\right\} \quad . \quad . \quad . \quad . \quad (23)$$

We thus assume that the means of the rc populations have two components, a row component and a column component. Equation (22) is then satisfied and (20) and (21) now give

$$E\sum_{ij}(x_{ij} - x_{i\cdot} - x_{\cdot j} + x_{\cdot\cdot})^2 = (r - 1)(c - 1)\sigma^2 \quad . \quad . \quad (24)$$

The expectations of the other terms of equation (10) are next to be considered. We note that since $(y_{i\cdot} - y_{\cdot\cdot})$ is distributed normally $(0, \sigma^2/c)$.

$$E\sum_{i}(y_{i\cdot} - y_{\cdot\cdot})^2 = (r - 1)\sigma^2/c \quad . \quad . \quad . \quad (25)$$

Now
$$\begin{aligned} E\sum_{ij}(x_{i\cdot} - x_{\cdot\cdot})^2 &= cE\sum_{i}(x_{i\cdot} - x_{\cdot\cdot})^2 \\ &= cE\sum_{i}(a_i + b_{\cdot} + y_{i\cdot} - a_{\cdot} - b_{\cdot} - y_{\cdot\cdot})^2 \\ &= cE\sum_{i}(a_i - a_{\cdot})^2 + cE\sum_{i}(y_{i\cdot} - y_{\cdot\cdot})^2 \\ &= cE\sum_{i}(a_i - a_{\cdot})^2 + (r - 1)\sigma^2, \quad . \quad . \quad (26) \end{aligned}$$

using equation (25).

Similarly,

$$E\sum_{ij}(x_{\cdot j} - x_{\cdot\cdot})^2 = rE\sum_{j}(b - b_{\cdot})^2 + (c - 1)\sigma^2 \quad . \quad . \quad (27)$$

Finally,

$$\begin{aligned} E\sum_{ij}(x_{ij} - x_{\cdot\cdot})^2 &= E\sum_{ij}(a_i + b_j + y_{ij} - a_{\cdot} - b_{\cdot} - y_{\cdot\cdot})^2 \\ &= E\sum_{ij}(a_i - a_{\cdot})^2 + E\sum_{ij}(b_j - b_{\cdot})^2 + E\sum_{ij}(y_{ij} - y_{\cdot\cdot})^2 \\ &= cE\sum_{i}(a_i - a_{\cdot})^2 + rE\sum_{j}(b_j - b_{\cdot})^2 + (rc - 1)\sigma^2 \quad (28) \end{aligned}$$

We can regard $\sum_{i}(a_i - a_{\cdot})^2/(r - 1)$ as estimating σ_r^2, the between rows variance, and $\sum_{j}(b_j - b_{\cdot})^2/(c - 1)$ as σ_c^2, the between columns variance. Incorporating these in the R.H.S. of equations (26), (27) and (28) we arrive at a new table of components of variance.

Cause	*Sum of Squares*	*d.f.*	*Mean Square Estimates*
Between rows .	$c\sum_{i}(x_{i\cdot} - x_{\cdot\cdot})^2$	$r - 1$	$c\sigma_r^2 + \sigma^2$
Between columns	$r\sum_{j}(x_{\cdot j} - x_{\cdot\cdot})^2$	$c - 1$	$r\sigma_c^2 + \sigma^2$
Residual (row × column)	$\sum_{ij}(x_{ij} - x_{i\cdot} - x_{\cdot j} + x_{\cdot\cdot})^2$	$(r - 1)(c - 1)$	σ^2
Total .	$\sum_{ij}(x_{ij} - x_{\cdot\cdot})^2$	$rc - 1$	$\frac{c(r - 1)}{rc - 1}\sigma_r^2 + \frac{r(c - 1)}{rc - 1}\sigma_c^2 + \sigma^2$

Referring to Ex. iv, σ_r^2 will represent the between solutions variance, σ_c^2 the between chemists variance and these are given by

$$4\sigma_r^2 + \sigma^2 = 6{\cdot}80$$
$$5\sigma_c^2 + \sigma^2 = 12{\cdot}32$$
$$\sigma^2 = 1.57$$
$$\text{whence } \sigma_r^2 = 1{\cdot}31,\ \sigma_c^2 = 2{\cdot}15$$

The implication is that if a given chemist estimates the strength of a number of solutions we expect his results to display a variance $\sigma^2 + \sigma_r^2$; if a given solution is tested by a number of chemists we expect a variance $\sigma^2 + \sigma_c^2$; lastly if many chemists test many solutions, the expected variance is $\sigma^2 + \sigma_c^2 + \sigma_r^2$. The last quantity is the limit as r and c tend to infinity of our 'total' mean square estimate.

The sequence of hypotheses in analysis of variance amounts to taking first the simplest working assumption; if this proves to be untenable we try the next in simplicity. From the results of this section we see that if there is a between row or between column effect we shall not only expect the F test to give a significant result but we shall also expect the between row or between column mean square to be significantly greater than the residual mean square. If the residual mean square is significantly greater than either of the between mean squares we cannot merely reject the homogeneity of means, since heterogeneity of means by itself leads to the opposite effect of that observed. In such a case it is safest to assume that readings are not independent. Non-independence can easily arise in experiments. For example, if pellets are filled with powder until a hopper is exhausted, then if one pellet receives a greater amount of powder, others will receive less and so there will be a degree of dependence of results one on another.

We now show how to obtain, in special cases, a better estimate of the residual variance σ^2.

If one of the effects, say the between columns effect is not significant, then $\sigma_c^2 = 0$ and we have two estimates of σ^2, from the between columns mean square and from the residual mean square. As in equation (25) of Chapter VI (see also §6.7), these may be combined to give a new and improved estimate of σ^2, by forming a weighted mean of the two estimates, the weights of each being its number of d.f. This is equivalent to adding the corresponding sums of squares in the table of components of variance and dividing by the sum of the degrees of freedom, as in the following example.

Ex. v

From the following data:

5	15	7	5
4	17	6	4
6	13	9	6
8	14	8	7

the analysis gives:

Cause	*Sum of Squares*	*d.f.*	*Mean Square*
Between rows . . .	5·3	3	1·8
Between columns . . .	226·3	3	75·4
Residual	22·2	9	2·5
Total	253·8	15	16·9

The F test gives no significance for the between row mean square and a high significance ($< 0{\cdot}001$) for the between column mean square. A better estimate for the residual mean square is now obtained by combining the between row mean square with the residual mean square:

$$\text{Estimate of } \sigma^2 = \frac{\text{Between row sum of squares} + \text{Residual sum of squares}}{\text{Between row d.f.} + \text{Residual d.f.}}$$

$$= \frac{5{\cdot}3 + 22{\cdot}2}{3 + 9} = 2{\cdot}30.$$

12.4.1 Three-factor Analysis of Variance

As a final extension of the method we consider the case when a variate has three influences acting upon it. For instance, powder pellets are made by automatic filling from machines. A tray with empty pellet cases is placed under a hopper on each machine and the hopper fills the cases to a stipulated weight automatically. Hence there are here (at least) three effects to be considered—machines, hoppers, trays. We can measure the total weight on each tray and use each combination of machine, hopper and tray in an analysis of variance.

In order to generalize the problem we shall refer to the three factors as ROW (tray), COLUMN (machine), BLOCK (hopper) and we consider one reading at each of the combinations.

Let x_{ijk} refer to the observed variate (tray weight) at a particular combination of row, column and block where i refers to row, j to column and k to block.

Data Table

COLUMNS

	1 *Block*					2 *Block*					c *Block*				
Rows	1	2	3	.	b	1	2	3	.	b	1	2	3	.	b
1:	x_{111}	x_{112}	x_{113}	.	x_{11b}	x_{121}	x_{122}	x_{123}	.	x_{12b}	x_{1c1}	x_{1c2}	x_{1c3}	.	x_{1cb}
2:	x_{211}	x_{212}	x_{213}	.	x_{21b}	x_{221}	x_{222}	x_{223}	.	x_{22b}	x_{2c1}	x_{2c2}	x_{2c3}		x_{2cb}
3:	x_{311}	x_{312}	x_{313}	.	x_{31b}	x_{321}	x_{322}	x_{323}	.	x_{32b}	x_{3c1}	x_{3c2}	x_{3c3}	.	x_{3cb}
r:	x_{r11}	x_{r12}	x_{r13}	.	x_{r1b}	x_{r21}	x_{r22}	x_{r23}	.	x_{r2b}	x_{rc1}	x_{rc2}	x_{rc3}	.	x_{rcb}

Using the 'dot' notation, $x_{ij\cdot} = \dfrac{\sum_{k=1}^{b} x_{ijk}}{b}$

$$x_{i\cdot\cdot} = \frac{\sum_{j=1}^{c}\sum_{k=1}^{b} x_{ijk}}{bc}$$

$$x_{\cdots} = \frac{\sum_{i=1}^{r}\sum_{j=1}^{c}\sum_{k=1}^{b} x_{ijk}}{rcb}$$

The initial null hypothesis is

'Each observation in each cell (i.e. each combination of row, column, block) has been taken at random from a normal population (μ, σ)'.

This is the usual starting-point, assuming constancy of means and variances.

By algebra analogous to that used in §§12.2.1, 12.3.1 we express the total sum of squares as the sum of seven components which, with the degrees of freedom given in the table below, and on the null hypothesis, are independent χ^2 variates. Each mean square yields an estimate of σ^2.

The total sum of squares is the sum of the component sums of squares and the total degrees of freedom is the sum of the component degrees of freedom.

The mean squares of the three interactions can each be tested in turn against the residual. If any of them is significant, the null hypothesis has to be rejected and a fresh hypothesis stated if the negative conclusion of rejection is not sufficient.

It should be noted that when a number of comparisons of variances is made, the level of significance must be tightened. For example, if only one comparison is made, a 5% level may be regarded as significant. But such a level is not significant if it occurs once while making 20 comparisons on the F test (see §6.6.1).

12.4.2 Method of computation

The sums of squares required can be expressed in forms which do not necessitate the evaluation of the various means. The reader will perhaps find it convenient to proceed directly to Ex. vi below, referring back to equations (29) to (36) where these are used.

$$\sum_{ijk}(x_{i\cdot\cdot} - x_{\cdots})^2 = \frac{\sum_i\left(\sum_{jk} x_{ijk}\right)^2}{bc} - \frac{\left(\sum_{ijk} x_{ijk}\right)^2}{rbc} \quad . \quad . \quad . \quad . \quad . \quad (29)$$

$$\sum_{ijk}(x_{\cdot j\cdot} - x_{\cdots})^2 = \frac{\sum_j\left(\sum_{ik} x_{ijk}\right)^2}{rb} - \frac{\left(\sum_{ijk} x_{ijk}\right)^2}{rbc} \quad . \quad . \quad . \quad . \quad . \quad (30)$$

$$\sum_{ijk}(x_{\cdot\cdot k} - x_{\cdots})^2 = \frac{\sum_k\left(\sum_{ij} x_{ijk}\right)^2}{rc} - \frac{\left(\sum_{ijk} x_{ijk}\right)^2}{rbc} \quad . \quad . \quad . \quad . \quad . \quad (31)$$

Components of Variance

Cause	*Sum of Squares*	*d.f.*	*Mean Square Estimates*
Between rows . .	$\sum_{ijk}(x_{i\cdot\cdot} - x_{\cdot\cdot\cdot})^2$	$r - 1$	σ^2
Between columns .	$\sum_{ijk}(x_{\cdot j\cdot} - x_{\cdot\cdot\cdot})^2$	$c - 1$	σ^2
Between blocks . .	$\sum_{ijk}(x_{\cdot\cdot k} - x_{\cdot\cdot\cdot})$	$b - 1$	σ^2
Interaction (columns × blocks) . .	$\sum_{ijk}(x_{\cdot jk} - x_{\cdot\cdot k} - x_{\cdot j\cdot} + x_{\cdot\cdot\cdot})^2$	$(c - 1)(b - 1)$	σ^2
Interaction (rows × blocks) . .	$\sum_{ijk}(x_{i\cdot k} - x_{i\cdot\cdot} - x_{\cdot\cdot k} + x_{\cdot\cdot\cdot})^2$	$(b - 1)(r - 1)$	σ^2
Interaction (rows × columns) . .	$\sum_{ijk}(x_{ij\cdot} - x_{i\cdot\cdot} - x_{\cdot j\cdot} + x_{\cdot\cdot\cdot})^2$	$(r - 1)(c - 1)$	σ^2
Residual (rows × columns × blocks)	$\sum_{ijk}(x_{ijk} - x_{ij\cdot} - x_{i\cdot k} - x_{\cdot jk} + x_{i\cdot\cdot} + x_{\cdot j\cdot} + x_{\cdot\cdot k} - x_{\cdot\cdot\cdot})^2$	$(r - 1)(c - 1)(b - 1)$	σ^2
Total . . .	$\sum_{ijk}(x_{ijk} - x_{\cdot\cdot\cdot})^2$	$rcb - 1$	σ^2

$$\sum_{ijk}(x_{\cdot jk} - x_{\cdot\cdot k} - x_{\cdot j\cdot} + x_{\cdot\cdot\cdot})^2$$
$$= \frac{\sum_{jk}\left(\sum_{i} x_{ijk}\right)^2}{r} - \frac{\sum_{k}\left(\sum_{ij} x_{ijk}\right)^2}{cr} - \frac{\sum_{j}\left(\sum_{ik} x_{ijk}\right)^2}{rb} + \frac{\left(\sum_{ijk} x_{ijk}\right)^2}{rbc} \quad . \quad . \quad (32)$$

$$\sum_{ijk}(x_{i\cdot k} - x_{i\cdot\cdot} - x_{\cdot\cdot k} + x_{\cdot\cdot\cdot})^2$$
$$= \frac{\sum_{ik}\left(\sum_{j} x_{ijk}\right)^2}{c} - \frac{\sum_{i}\left(\sum_{jk} x_{ijk}\right)^2}{cb} - \frac{\sum_{k}\left(\sum_{ij} x_{ijk}\right)^2}{rc} + \frac{\left(\sum_{ijk} x_{ijk}\right)^2}{rbc} \quad . \quad . \quad . \quad (33)$$

$$\sum_{ijk}(x_{ij\cdot} - x_{i\cdot\cdot} - x_{\cdot j\cdot} + x_{\cdot\cdot\cdot})^2$$
$$= \frac{\sum_{ij}\left(\sum_{k} x_{ijk}\right)^2}{b} - \frac{\sum_{j}\left(\sum_{ik} x_{ijk}\right)^2}{rb} - \frac{\sum_{i}\left(\sum_{jk} x_{ijk}\right)^2}{cb} + \frac{\left(\sum_{ijk} x_{ijk}\right)^2}{rbc} \quad . \quad . \quad . \quad (34)$$

$$\sum_{ijk}(x_{ijk} - x_{ij\cdot} - x_{i\cdot k} - x_{\cdot jk} + x_{i\cdot\cdot} + x_{\cdot j\cdot} + x_{\cdot\cdot k} - x_{\cdot\cdot\cdot})^2$$
$$= \sum_{ijk} x^2_{ijk} - \frac{\sum_{jk}\left(\sum_{i} x_{ijk}\right)^2}{r} - \frac{\sum_{ik}\left(\sum_{j} x_{ijk}\right)^2}{c} - \frac{\sum_{ij}\left(\sum_{k} x_{ijk}\right)^2}{b}$$
$$+ \frac{\sum_{i}\left(\sum_{jk} x_{ijk}\right)^2}{cb} + \frac{\sum_{j}\left(\sum_{ik} x_{ijk}\right)^2}{rb} + \frac{\sum_{k}\left(\sum_{ij} x_{ijk}\right)^2}{rc} - \frac{\left(\sum_{ijk} x_{ijk}\right)^2}{rbc} \quad . \quad . \quad . \quad (35)$$

$$\sum_{ijk}(x_{ijk} - x_{\cdot\cdot\cdot})^2 = \sum_{ijk} x_{ijk}{}^2 - \frac{\left(\sum_{ijk} x_{ijk}\right)^2}{rcb} \quad . \quad . \quad . \quad . \quad . \quad . \quad (36)$$

Ex. vi

Pellet weights, one for each combination of tray, machine and hopper are as in the following table. Obtain the eight mean squares and estimate whether there are any significant effects. (Weights are in tenths of an oz. with working origin at 8 oz.)

	Machine											
	$j = 1$				$j = 2$				$j = 3$			
	Hopper				*Hopper*				*Hopper*			
Tray	$k = 1$	2	3		$k = 1$	2	3		$k = 1$	2	3		
$i = 1$	5	4	3	12	1	6	4	11	3	3	3	9	32
$i = 2$	1	5	3	9	1	4	1	6	4	2	4	10	25
$i = 3$	5	2	3	10	3	6	3	12	2	2	0	4	26
	11	11	9		5	16	8		9	7	7		
	31				29				23				

Some sub-totals have been inserted. We have

$$r = c = b = 3.$$

$$\Sigma x = \underline{83} \qquad (\Sigma x)^2/rcb = 6889/27 = \underline{255{\cdot}1}$$

$$\Sigma x^2 = \underline{319}$$

$$\sum_{i}\left(\sum_{jk} x\right)^2 = 32^2 + 25^2 + 26^2 = 2325;\ 2325/9 = \underline{258{\cdot}3}$$

$$\sum_{j}\left(\sum_{ik} x\right)^2 = 31^2 + 29^2 + 23^2 = 2331;\ 2331/9 = \underline{259{\cdot}0}$$

$$\sum_{k}\left(\sum_{ij} x\right)^2 = (11 + 5 + 9)^2 + (11 + 16 + 7)^2 + (9 + 8 + 7)^2$$
$$= 2357; \qquad 2357/9 = \underline{261{\cdot}9}$$

Note that, for example, the sum 11 + 5 + 9 is the sum when $k = 1$ for all values of i and j; 11 + 16 + 7 is the sum when $k = 2$ and so on.

$$\sum_{jk}\left(\sum_i x\right)^2 = 11^2 + 11^2 + 9^2 + 5^2 + 16^2 + 8^2 + 9^2 + 7^2 + 7^2$$
$$= 847; \qquad 847/3 = \underline{282 \cdot 3}$$

Here each $\sum_i x$ is the sum for fixed j and k.

$$\sum_{ik}\left(\sum_j x\right)^2 = 9^2 + 13^2 + 10^2 + 6^2 + 11^2 + 8^2 + 10^2 + 10^2 + 6^2$$
$$= 807 \qquad 807/3 = \underline{269 \cdot 0}$$

This is the hardest sum. We must keep i and k fixed while j varies. For instance, when $i = 1$ and $k = 1$, we have 5 + 1 + 3 = 9.

$$\sum_{ij}\left(\sum_k x\right)^2 = 12^2 + 11^2 + 9^2 + 9^2 + 6^2 + 10^2 + 10^2 + 12^2 + 4^2$$
$$= 823 \qquad 823/3 = \underline{274 \cdot 3}$$

Applying equations (29) to (36) we obtain the required sums of squares:

Between trays (rows)	258·3 − 255·1 = 3·2
Between machines (columns)	259·0 − 255·1 = 3·9
Between hoppers (blocks)	261·9 − 255·1 = 6·8
Interaction machine × hopper	282·3 − 261·9 − 259·0 + 255·1 = 16·5
Interaction hopper × tray	269·0 − 258·3 − 261·9 + 255·1 = 3·9
Interaction tray × machine	274·3 − 259·0 − 258·3 + 255·1 = 12·1
	Total, 319 − 255·1 = 63·9

The residual sum of squares may be found by subtracting from this total the sum of the preceding six results, or by equation (35). It is 17·5.

Introducing the d.f. we now have the following analysis of variance.

Components of Variance

Cause	*Sum of Squares*	*d.f.*	*Mean Square*
Between trays . . .	3·2	2	1·6
Between machines . . .	3·9	2	2·0
Between hoppers . . .	6·8	2	3·4
Interaction: Hoppers × machines . .	16·5	4	4·1
Interaction: Trays × hoppers . .	3·9	4	1·0
Interaction: Machines × trays . .	12·1	4	3·0
Residual	17·5	8	2·2
Total.	63·9	26	2·46

The procedure is now to compare the interaction mean squares, starting with the smallest, with the residual m.s. If the F-test shows no significance the sum of squares concerned, with its d.f., may be pooled with the residual sum and its d.f., to give a new residual m.s. After the interaction mean squares have been so investigated in order of magnitude, we proceed to the between mean squares, again in order of magnitude. In the present example no mean square is significantly large, so that we are left finally with 2·46 as the best estimate of variance; all the six possible causes of variability are of no significance. The reasons for the procedure adopted are discussed in the next section.

12.4.3 Rejection of the null hypothesis

When the null hypothesis, that all values are random values from a common normal parent population, proves untenable, we must adopt a new one which takes into account the possible causes of variation which we have under consideration.

We can allow for these causes by supposing that any value of x is the sum of variates corresponding to the six causes ('between' or 'interaction'), together with a variability due to all other causes and responsible for the residual variance.

Then $x_{ijk} = a_i + b_j + c_k + d_{jk} + e_{ki} + f_{ij} + y_{ijk}$ . . . (37)

If we were able to rule out the interactions as significant causes of variability we could adopt the simpler hypothesis:

$$x_{ijk} = a_i + b_j + c_k + y_{ijk}$$

We test the interactions first in the hope that this simpler hypothesis is tenable.

On the more general hypothesis (37) we can determine the expectations of the various mean squares. There is a variance corresponding to each of the terms of (37) and by algebra analogous to that of §12.3.3 we can obtain the required expectations in terms of these variances. Each mean square obtained in our analysis of a set of observed values may then be regarded as estimating the corresponding expectation, as follows.

	Estimates	
The between row m.s.	$bc\sigma_r^2 + b\sigma_{rc}^2 + c\sigma_{rb}^2 + \sigma^2$	
The between cols. m.s.	$br\sigma_c^2 + b\sigma_{rc}^2 + r\sigma_{cb}^2 + \sigma^2$	
The between blocks m.s.	$rc\sigma_b^2 + r\sigma_{bc}^2 + c\sigma_{br}^2 + \sigma^2$	
The interaction col.-block m.s.	$r\sigma_{bc}^2 + \sigma^2$	. (38)
The interaction block-row m.s.	$c\sigma_{br}^2 + \sigma^2$	
The interaction row-col. m.s.	$b\sigma_{rc}^2 + \sigma^2$	
The residual m.s.	σ^2	

Here σ_r^2 is the variance of the term a_i, that is of the component variate representing the variation between rows. Similarly σ_c^2, σ_b^2 are variances due to the between columns and between blocks effects. σ_{br}^2 or σ_{rb}^2 is the variance due to interaction between blocks and rows, and similarly for the other interaction effects. For instance, in Ex. vi if the row-block m.s. had proved significantly large compared with the residual m.s. the inference would be that the performance of a hopper is affected by the tray used with it.

In the following example, rows correspond to days of the week. The figures suggest, on mere inspection, that the between days effect is significant. It is more marked for operator B than for the other operators so that we may find an operator-day interaction. The analysis of variance is designed to test whether these and other effects could be accounted for by the operation of 'pure chance'—of random sampling, or are to be considered real, or 'significant'.

Ex. vii

The following data represent the internal dimension (measured in thousandths of an inch from an origin of 1·634 in.) of a component. The factors are days, operators, machines. Discuss whether these factors, or any combination of them, influence the product.

	Operator A			*Operator B*			*Operator C*		
	Machine			*Machine*			*Machine*		
	a	b	c	a	b	c	a	b	c
Monday . .	6	4	9	0	4	10	7	7	10
Wednesday .	18	21	21	17	16	17	13	12	14
Friday . .	25	26	37	32	29	32	22	23	23

The analysis is:

Components of Variance

Cause	*Sums of Squares*	*d f.*	*Mean Square*
Between machines . . .	76·77	2	38·38
Between days	2049·18	2	1024·59
Between operators . . .	76·73	2	38·36
Machines × days . . .	17·45	4	4·36
Days × operators . . .	130·16	4	32·54
Operators × machines . .	15·80	4	3·95
Residual	63·31	8	7·91
Total.	2429·40	26	93·44

Only one interaction, namely day-operator, is found to be significant. The corresponding variance is σ_{rc}^2. We could set $\sigma_{br}^2 = c_{bc}^2 = 0$ and pool the machine-day and operator-machine sums of squares with the residual to form a new estimate of the residual σ^2, as $\dfrac{17{\cdot}45 + 15{\cdot}80 + 63{\cdot}31}{4 + 4 + 8} = 6{\cdot}04.$

All the 'between' mean squares are found to be significantly high. The inference is that machines, days and operators all have distinct and different influences on the dimension, and in addition there is evidence that certain operators work differently according to the day of the week, but not according to the machine they use.

Ex. viii

An analysis gave the following components of variance:

Cause	*Sum of Squares*	*d.f.*	*Mean Square*
Between rows	4185·0	5	837·0
Between columns . . .	5784·0	7	826·3
Between blocks . . .	2484·0	10	248·4
Columns × blocks . . .	532·0	70	7·6
Blocks × rows . . .	2040·0	50	40·8
Rows × columns . . .	5075·0	35	145·0
Residual	2240·0	350	6·4
Total	22340·0	527	42·4

Estimate the component variances.

$$\begin{aligned} 88\sigma_r^2 + 11\sigma_{rc}^2 + 8\sigma_{rb}^2 + \sigma^2 &= 837{\cdot}0 \\ 66\sigma_c^2 + 11\sigma_{rc}^2 + 6\sigma_{cb}^2 + \sigma^2 &= 826{\cdot}3 \\ 48\sigma_b^2 + 8\sigma_{br}^2 + 6\sigma_{cb}^2 + \sigma^2 &= 248{\cdot}4 \\ 6\sigma_{cb}^2 + \sigma^2 &= 7{\cdot}6 \\ 8\sigma_{br}^2 + \sigma^2 &= 40{\cdot}8 \\ 11\sigma_{rc}^2 + \sigma^2 &= 145{\cdot}0 \\ \sigma^2 &= 6{\cdot}4 \end{aligned}$$

These give

$$\begin{aligned} \sigma^2 &= 6{\cdot}4 \\ \sigma_{rc}^2 &= 12{\cdot}6 \\ \sigma_{br}^2 &= 4{\cdot}3 \\ \sigma_{cb}^2 &= 0{\cdot}2 \\ \sigma_b^2 &= 4{\cdot}3 \\ \sigma_c^2 &= 10{\cdot}3 \\ \sigma_r^2 &= 7{\cdot}4 \end{aligned}$$

We notice that although the Column × Block effect is not significant, σ_{cb}^2 can still be estimated.

Alternatively we can treat σ_{cb}^2 as being zero, since the Block × Column mean square is not significant, and pool the old residual sum of squares with the Block × Column sum of squares, giving a new residual estimate of

$$\sigma^2 = \frac{2240 + 532}{350 + 70} = 6{\cdot}6.$$

This would give only slightly different solutions to the components.

It sometimes happens that the estimate of a component variance is found to be negative. The assumption that the variate is the sum of *independent* variates must then be abandoned, if the negative variance is significantly large.

12.4.4 Two-factor analysis with replications

In §12.3.1 the two-factor analysis was based on a single observation for each of the rc combinations of row and column. Suppose instead that b observations were made for each such combination. Then the table of values would have the appearance of a table for three-factor analysis. There is, however, the difference that in the three-factor case the first, second, etc., of any one set of b observations was to be associated with the first, second, etc., respectively of any other set; for instance, in Ex. vi the first of each set of three values was taken using the first hopper. We now suppose merely that three readings were taken with each machine, used with each tray, and the order of the three readings is immaterial. Any component of variance associated with hoppers (blocks) is now suppressed. Such components contain the suffix b in equation (38) and we may set

$$\sigma_b^2 = \sigma_{br}^2 = \sigma_{bc}^2 = 0.$$

Pooling the sums of squares which lead now to estimates of σ^2, we have the reduced analysis of variance.

Source	*d.f.*	*m.s. estimates*
Between rows . . .	$r - 1$	$bc\sigma_r^2 + b\sigma_{rc}^2 + \sigma^2$
Between columns . .	$c - 1$	$br\sigma_c^2 + b\sigma_{rc}^2 + \sigma^2$
Interaction row × column .	$(r - 1)(c - 1)$	$b\sigma_{rc}^2 + \sigma^2$
Residual	$rc(b - 1)$	σ^2

It may be noted that as compared with the earlier two-factor analysis with one observation ($b = 1$) for each combination of row and column, what was previously a residual mean square has now been resolved into two components—an interaction and a residual component. The first of these becomes the residual component when $b = 1$ and σ_{rc}^2 becomes the residual σ^2.

The computation of the sums of squares required in a two-factor analysis with replications may be indicated as follows:

Let
$$A = \frac{1}{rcb}\left(\sum_{ijk} x\right)^2$$

$$B = \sum_{ijk} x^2$$

$$C = \frac{1}{bc}\sum_i\left(\sum_{jk} x\right)^2$$

$$D = \frac{1}{rb}\sum_j\left(\sum_{ki} x\right)^2$$

$$E = \frac{1}{b}\sum_{ij}\left(\sum_k x\right)^2.$$

Then the between rows sum of squares $= C - A$
Then the between column sum of squares $= D - A$
Then the interaction sum of squares $= E - C - D + A$
Then the residual sum of squares $= B - E$
Then the total sum of squares $= B - A$

Ex. ix

Use the data of Ex. vi to represent three readings for each combination of machine and tray and tabulate the analysis of variance.

We need only

$$(\Sigma x)^2/rcb = 255{\cdot}1$$

$$\Sigma x^2 = 319{\cdot}0$$

$$\frac{1}{bc}\sum_i\Big(\sum_{jk}x\Big)^2 = 258{\cdot}3$$

$$\frac{1}{rb}\sum_j\Big(\sum_{ik}x\Big)^2 = 259{\cdot}0$$

and

$$\frac{1}{b}\sum_{ij}\Big(\sum_k x\Big)^2 = 274{\cdot}3$$

It will be noted that these are the easier sums obtainable from the sub-totals included in the data. The analysis is as follows.

Source	*Sum of Squares*	*d.f.*	*Mean Square*
Between machines	3·2	2	1·6
Between trays	3·9	2	2·0
Machine-tray interaction . .	12·1	4	3·0
Residual	44·7	18	2·5
Total	63·9	26	2·

12.5 Further developments

The methods we have considered do not exhaust the subject of Analysis of Variance, but should be sufficient to elucidate the principles involved. Perhaps the most important of the further developments are the use of Latin and Greco-Latin squares, by means of which an analysis of three or four factors may be made on the basis of a relatively small set of data. Space does not permit of their treatment here, but the reader who is particularly interested in the Analysis of Variance will find them dealt with in more complete treatments.

Suggestions for further reading:

R. A. Fisher, *Statistical Methods for Research Workers* (Oliver and Boyd).
M. G. Kendall, *Advanced Theory of Statistics*, Chapters 23, 24 (Griffin).
K. A. Brownlee, *Industrial Experimentation* (H.M.S.O.).

EXAMPLES ON CHAPTER XII

1. Five mixes of the same alloy are made and for each mix six determinations of the density are effected. The results are:

Densities

Mix A:	5·6	5·5	5·7	5·1	5·1	5·2
Mix B:	5·3	5·5	5·4	5·2	5·4	5·4
Mix C:	5·5	5·3	5·4	5·4	5·3	5·2
Mix D:	5·5	5·4	5·0	5·3	5·3	5·8
Mix E:	5·7	5·4	5·6	5·5	5·6	5·4

Is there evidence that certain mixes have higher mean density than others?

2. An electrical contractor receives from his supplier four batches of resistors. Measurement of a random sample of ten resistors from each batch yields the values given in the table below. Is there evidence of a batch-to-batch difference in mean resistance?

Resistances in ohms

Batch 1: 811, 807, 809, 810, 808, 805, 807, 806, 808, 809.
Batch 2: 815, 811, 809, 811, 808, 809, 807, 811, 807, 808.
Batch 3: 810, 814, 806, 813, 813, 812, 812, 807, 808, 808.
Batch 4: 806, 810, 807, 812, 808, 813, 811, 809, 807, 806.

3. Six machines produce steel wire. The following data give the diameters at ten positions along the wire for each machine. Examine whether the machine means can be regarded as constant. If not, estimate the variance of group means.

Case A. *Diameters in thousandths of an inch.*

Machine (a): 12, 13, 13, 14, 15, 14, 16, 15, 19, 15.
(b): 11, 12, 12, 13, 14, 15, 15, 15, 15, 18.
(c): 12, 14, 14, 14, 15, 15, 15, 16, 20, 13.
(d): 13, 13, 14, 14, 15, 15, 16, 17, 17, 18.
(e): 14, 16, 17, 18, 13, 12, 16, 14, 16, 14.
(f): 13, 19, 18, 18, 15, 17, 16, 14, 17, 14.

Case B.

Machine (a): 12, 13, 13, 16, 16, 14, 15, 15, 16, 17.
(b): 12, 14, 14, 16, 16, 18, 17, 19, 20, 18.
(c): 14, 21, 17, 14, 19, 18, 17, 17, 16, 15.
(d): 23, 27, 25, 21, 26, 24, 27, 24, 20, 21.
(e): 12, 14, 13, 16, 13, 17, 16, 15, 15, 14.
(f): 13, 18, 13, 16, 17, 15, 15, 16, 16, 17.

4. The resistance per unit length is calculated at five different points in each of four batches of wire. Examine whether the batch means can be regarded as constant and, if not, estimate the variance of batch means.

Resistance per unit length in ohms.

Batch A: 2·3, 2·3, 2·5, 2·5, 2·2.
Batch B: 3·2, 3·0, 3·2, 3·2, 3·0.
Batch C: 2·9, 2·9, 3·0, 2·9, 2·6.
Batch D: 2·7, 2·9, 3·0, 2·7, 2·6.

5. The following data represent external diameters of piston heads made from five machines by five operators. Determine whether there is any 'between machines' or 'between operators' effect and interpret the results.

Machine

	A	B	C	D	E
Operator (a):	517	512	519	521	518
(b):	515	520	520	517	529
(c):	513	515	515	521	524
(d):	522	523	521	516	527
(e):	522	525	525	525	528

6. A confectionery company is worried about the variability in weights of its '2-oz.' bars of chocolate. They assume that the cause is either due to the mix (I, II, III, IV) or the machine (A, B, C, D) and take a sample as follows:

Bar Weights (oz.)

Machine

		A	B	C	D
	I:	2·00	2·06	2·04	2·00
Mix	II:	2·02	2·08	2·04	2·06
	III:	2·00	2·04	2·04	2·04
	IV:	2·04	2·06	2·02	2·06

What conclusions can be drawn from this sample?

7. How far are the conclusions in Question 6 confirmed by the following sample of five bars from each machine × mix combination?

Bar Weights (oz.)

Mix		A	B	C	D
		2·03	2·06	2·01	2·02
		1·97	2·06	2·00	2·02
	I	2·02	2·04	2·02	2·03
		1·98	2·08	2·02	2·06
		2·05	2·07	2·00	2·04
		2·03	2·06	2·02	2·04
		2·04	2·05	2·06	2·01
	II	2·05	2·06	2·00	2·00
		1·98	2·08	2·04	2·02
		2·05	2·05	2·04	2·02
		2·02	2·04	2·01	1·99
		2·02	2·07	2·02	2·00
	III	2·00	2·09	2·06	2·02
		2·01	2·05	2·04	2·03
		2·06	2·02	2·04	2·00
		1·98	2·05	2·03	2·05
		2·02	2·09	2·03	1·98
	IV	2·00	2·05	2·00	2·07
		2·06	2·04	2·04	2·02
		2·00	2·08	2·01	2·03

Assuming that homogeneity of means can be obtained without affecting the present residual variance, at what level should the mean bar weight be set in order that not more than one bar in ten thousand shall be underweight?

8. Three operators make three hardness readings on each of four blocks of metal. Examine the data to determine whether
   (a) the mean readings of the operators are constant,
   (b) the mean hardnesses of the blocks are constant,
   (c) there is any interaction between operators and blocks.

	Block A	B	C	D
Operator				
(a):	716, 713, 714	717, 713, 713	712, 708, 711	706, 705, 706
(b):	720, 717, 719	713, 710, 711	709, 712, 713	708, 708, 708
(c):	718, 720, 718	717, 715, 716	706, 708, 709	707, 706, 706

9. What do you understand by degrees of freedom?

   An experiment is carried out by 8 operatives on 6 machines, so that each operative uses each machine once only. Complete the following analysis of variance table and comment on the results.

Source of Variation	*Sums of Squares*	*Degrees of Freedom*	*Variance*
Between machines .	822		
Between operatives .	1524		
Residual . . .			
Total . . .	4667		

(A.I.S.)

10. Five doctors each test five treatments for a certain disease, and observe the number of days each patient takes to recover. The results are as follows (figures in days):

Treatment	1	2	3	4	5
Doctor A	10	14	23	18	20
„ B	11	15	24	17	21
„ C	9	12	20	16	19
„ D	8	13	17	17	20
„ E	12	15	19	15	22

Using the method of Analysis of Variance, discuss the difference between (*a*) doctors and (*b*) treatments. (A.I.S.)

CHAPTER XIII

THE PRINCIPLE OF MAXIMUM LIKELIHOOD. PROBIT ANALYSIS

The general problem of estimation. Probit analysis.

13.1 The problem of estimation: a method of solution

One of the recurring problems in statistics is that of attempting to draw conclusions about a parent population on the basis of a sample drawn from it. In such a problem it is important to know the type of sampling employed (e.g. whether random or stratified, etc.). When random sampling has been used, the Principle of Maximum Likelihood can be adopted to estimate the parameters of the population. The following is an outline of the method of R. A. Fisher. In the simple examples considered in the first part of this chapter the maximum likelihood estimates are in general identical with those which have previously been used by intuition. However, as will be shown in the section on Probit Analysis, the principle does lend itself to the solution of an important class of problems which would be very difficult to solve by other methods.

The principle may be stated thus: on the basis of a random sample drawn from a population, the estimates of the parameters of the distribution function of the population are those values of the parameters which make the probability of the observed sample a maximum.

If, for instance, the parent population is assumed normal (μ, σ) and the sample is $x_1, x_2 \ldots x_n$, the chance P of the sample of n falling one in each of the n ranges $x_1 \pm \frac{dx_1}{2}, x_2 \pm \frac{dx_2}{2} \ldots x_n \pm \frac{dx_n}{2}$ is

$$P = \frac{1}{\sigma^n (2\pi)^{n/2}} \cdot$$

$$\exp\left\{-\frac{(x_1 - \mu)^2 + (x_2 - \mu)^2 + \ldots + (x_n - \mu)^2}{2\sigma^2}\right\} dx_1, dx_2 \ldots dx_n \quad (1)$$

Given $x_1, x_2 \ldots x_n$ the R.H.S. of equation (1) is a function of μ and σ and we choose as estimates of these parameters the values of μ and σ which maximize P. The probability P is called the likelihood of the sample.

The first case considered, however, is the case where only one parameter has to be estimated.

13.2 Estimation of a single parameter

For each value of the single parameter, θ (say), the chance of the observed sample of n, $x_1, x_2 \ldots x_n$, of the variate x can be evaluated—

$$P(x_1, x_2 \ldots x_n/\theta) = g(\theta) \quad . \quad . \quad . \quad (2)$$

(The notation on the L.H.S. of equation (2) means the probability of the set $x_1, x_2 \dots x_n$ for an assumed θ.)

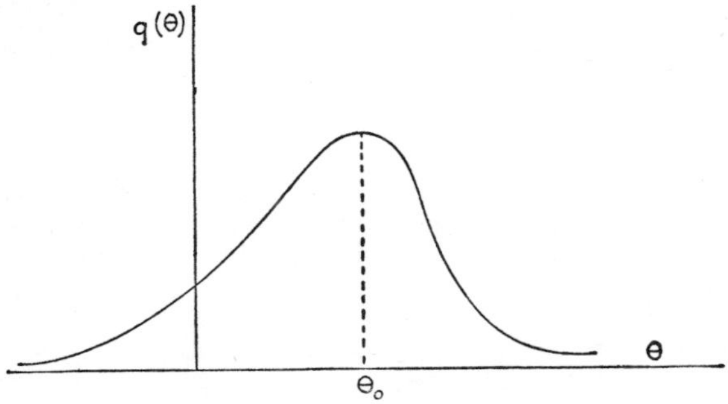

FIG. 45

The estimate of the true (population) value, $\bar{\theta}$ of θ is that value of θ, say θ_o, which maximizes $g(\theta)$, i.e. it is the solution of the two relations:

$$g'(\theta) = 0$$
$$g''(\theta) < 0$$

This value θ_o is still only an estimate of θ since the likelihood has only been maximized for the particular sample obtained.

What gives the principle its importance in the field of estimation is that for any random sample of size n (under broad assumptions which need not be stated here) the distribution of $(\theta_o - \bar{\theta})\sqrt{n}$ tends, as n increases, to a normal distribution with zero mean, i.e. the estimate is asymptotically unbiased.

If the distribution of $g(\theta)$ is not continuous, the principle still obtains in the sense that $g(\theta_o)$ is the maximum ordinate of $g(\theta)$.

Ex. i

Consider Poisson's distribution $p(x) = \dfrac{e^{-\lambda}\lambda^x}{x!}$

where x takes discrete values $0, 1, 2 \dots, r, \dots$

The probability P of the random sample $x_1, x_2 \dots x_n$ is

$$P = \frac{e^{-\lambda}\lambda^{x_1}}{x_1!} \cdot \frac{e^{-\lambda}\lambda^{x_2}}{x_2!} \cdots \frac{e^{-\lambda}\lambda^{x_n}}{x_n!}$$

Let L denote $\log P$; then

$$L = -n\lambda + (x_1 + x_2 + \dots + x_n)\log\lambda - (\log x_1! + \log x_2! + \dots + \log x_n!)$$

$$\frac{dL}{d\lambda} = -n + \frac{(x_1 + x_2 + \dots + x_n)}{\lambda}$$

But if $\dfrac{dL}{d\lambda} = 0$, $\dfrac{dP}{d\lambda} = 0$ and vice versa; hence the estimate λ_0 of $\bar{\lambda}$ the true (population) value of λ is given by

$$\lambda_o = \frac{x_1 + x_2 + \ldots + x_n}{n}$$

Thus the maximum likelihood estimate of the mean of a Poisson distribution on the evidence of a random sample is the sample mean as previously assumed (see §5.3.2).

Ex. ii

Consider the binomial distribution $p(x) = {}^nC_x p^x(1 - p)^{n-x}$ where the x's assume discrete values and n is fixed. The parameter to be estimated is the probability p of success in a single trial, or the mean np of the distribution.

Then the probability P of the sample values $x_1, x_2 \ldots x_m$ is

$$P = [{}^nC_{x_1} p^{x_1} (1 - p)^{n-x_1}] [{}^nC_{x_2} p^{x_2} (1 - p)^{n-x_2}] \ldots [{}^nC_{x_m} p^{x_m} (1 - p)^{n-x_m}]$$

$$\begin{aligned} L = \log P &= m \log n! - (\log x_1! + \ldots + \log x_m!) \\ &- \{\log (n - x_1)! + \ldots + \log (n - x_m)!\} \\ &+ (x_1 + \ldots + x_m) \log p + \{mn - (x_1 + x_2 + \ldots + x_m\} \log(1 - \mathrm{p}) \end{aligned}$$

$$\frac{dL}{dp} = \frac{x_1 + x_2 + \ldots + x_m}{p} - \frac{mn - (x_1 + x_2 + \ldots x_m)}{1 - p}$$

$$= 0 \text{ for } p \equiv p_0 = \frac{x_1 + x_2 + \ldots + x_m}{mn}$$

Hence the maximum likelihood estimate of np is the sample mean.

Ex. iii

An example of a continuous distribution with one parameter is the Pearson Type III distribution:

$$dp = \frac{x^{k-1} e^{-x/\lambda}}{\Gamma(k) \lambda^k} dx, \; x \geqslant 0,$$

where k is a known positive constant and λ is a positive parameter which is to be estimated.

(If $\nu = 2k$, $\lambda = 2$ the distribution is that of χ^2 (see §6.7).

In this case the probability density is

$$P = \frac{(x_1 . x_2 \ldots x_n)^{k-1} \exp\{- (x_1 + x_2 + \ldots + x_n)/\lambda\}}{\{\Gamma(k)\}^n \lambda^{nk}}$$

$$\begin{aligned} L &= (k - 1) \{\log x_1 + \ldots + \log x_n\} - \frac{x_1 + x_2 + \ldots + x_n}{\lambda} \\ &- n \log \{\Gamma(k)\} - nk \log \lambda \end{aligned}$$

$$\frac{dL}{d\lambda} = \frac{x_1 + x_2 + \ldots + x_n}{\lambda^2} - \frac{nk}{\lambda}$$

$$= 0 \text{ if } \lambda \equiv \lambda_0 = \frac{x_1 + x_2 + \ldots + x_n}{nk}$$

13.3 Discussion of the problem of estimation

It has been noted before that the estimate θ_0 will vary with the sample data. It is possible to determine theoretically the sampling distribution of θ_0. There are three distributions to be considered:

(a) The original distribution function, $f(x, \bar{\theta})$, of the variate, containing a single parameter $\bar{\theta}$ which has to be estimated. A random sample is drawn from this population.

(b) For any trial value of $\bar{\theta}$, say θ, the likelihood $g(\theta)$ of the sample can be determined. This is the second distribution function. The estimate θ_0 of $\bar{\theta}$ is such that,

$$\text{(i)}\ \frac{dg}{d\theta} = 0 \qquad \text{(ii)}\ \frac{d^2g}{d\theta^2} < 0$$

(The inequality (ii) should be verified by the reader in Exs. i–iii above.)

(c) There is a sampling distribution function, $h(\theta_0)$ (say), giving the probability of the estimate θ_0 being obtained. (It has been stated above that $h\left(\frac{\theta_0 - \bar{\theta}}{\sqrt{n}}\right)$ tends to a normal distribution with zero mean as $n \to \infty$ for all maximum likelihood estimates satisfying certain broad conditions.)

A method of comparing different approaches to the problem of estimation would be to compare the functions $h(\theta_0)$ which derive from each manner of estimation. For a method to be unbiased, the mean of $h(\theta_0)$ must tend to $\bar{\theta}$ as the sample size increases. What should really be of interest in comparing different estimators for a given sample size is to see the chance of the estimate θ_0 being within a given range of $\bar{\theta}$. A discussion on these lines is outside the scope of this book, but it can be noted that the principle of maximum likelihood gives a method in which

(i) θ_0 is easy to calculate;

(ii) $h(\theta_0)$ tends asymptotically to a normal distribution with mean $\bar{\theta}$.

It is not difficult to formulate other methods of estimating $\bar{\theta}$. The mean θ_1 (say) of the distribution function $g(\theta)$ could be used and this could be called 'mean likelihood' and the properties of such a method of estimation could be discovered.

Solely in order to illustrate the ideas and problems of estimation, fresh estimates for Exs. i and iii above based on 'mean likelihood' will now be obtained.

Ex. iv

For the Poisson distribution, writing θ for λ:

$$g(\theta) = \frac{e^{-n\theta}\,\theta^{(x_1+x_2+\ldots+x_n)}}{x_1!\,x_2!\ldots x_n!}, \quad \theta \geqslant 0$$

The mean value of θ is $$\theta_1 = \frac{\int_0^\infty \theta g(\theta)d\theta}{\int_0^\infty g(\theta)d\theta} \quad . \quad . \quad . \quad . \quad . \quad (3)$$

$$\equiv \frac{N}{D}$$

If $n\theta = \phi$, $$N = \int_0^{\infty} \frac{\phi^{(x_1+x_2+\ldots+x_n+1)}e^{-\phi}\,d\phi}{x_1!\,x_2!\ldots x_n!\,n^{(x_1+x_2+\ldots+x_n+2)}}$$

$$= \frac{\Gamma(x_1 + x_2 + \ldots + x_n + 1)}{x_1!\,x_2!\ldots x_n!\,n^{(x_1+x_2\ldots+x_n+1)}}$$

Similarly, $$D = \frac{\Gamma(x_1 + x_2 + \ldots + x_n)}{x_1!\,x_2!\ldots x_n!\,n^{(x_1+x_2+\ldots+x_n+1)}}$$

But $$\Gamma(k) = (k-1)\,\Gamma(k-1),$$

$$\therefore \theta_1 = \frac{N}{D} = \frac{x_1 + x_2 + \ldots + x_n + 1}{n}.$$

That is, the estimate differs from the sample mean, which was the previous estimate, although they both tend to the same limit as $n \to \infty$.

Ex. v

In the distribution function of Ex. iii, writing θ for λ:

$$g(\theta) = \frac{(x_1, x_2 \ldots x_n)^{k-1}\,exp\{-(x_1 + \ldots x_n)/\theta\}}{\{\Gamma(k)\}^n\,\theta^{nk}}$$

Using equation (3),

$$N = \frac{(x_1, x_2 \ldots x_n)^{k-1}\,(x_1 + x_2 + \ldots x_n)^{-nk+2}\Gamma(nk-2)}{\{\Gamma(k)\}^n}$$

Hence $\theta_1 = \dfrac{x_1 + x_2 + \ldots + x_n}{nk-2}$, which differs from the previous estimator λ_0, although they both tend to the same limit as $n \to \infty$.

13.4 Estimation of two parameters

Suppose now that there are two independent parameters θ, ϕ in the distribution (such a case is the normal distribution with mean μ and variance σ^2). The likelihood of any random sample can be calculated for any combination of θ and ϕ—

$$P(x_1, x_2 \ldots x_n/\theta, \phi) = g(\theta, \phi)$$

If θ, ϕ are discrete the solution by maximum likelihood is that pair (θ, ϕ) for which $g(\theta, \phi)$ is greatest. If θ, ϕ are continuous the equations to solve are

$$\frac{\partial g}{\partial \theta} = 0, \quad \frac{\partial g}{\partial \phi} = 0$$

Ex. vi

Consider the estimation of the mean and variance of a normal distribution by a random sample $(x_1, x_2 \ldots x_n)$.

Here, $$dp = \frac{1}{\sigma\sqrt{2\pi}} \exp\left\{-\frac{(x-\mu)^2}{2\sigma^2}\right\} dx$$

The likelihood, P, of the sample is

$$\frac{1}{\sigma^n(2\pi)^{n/2}} \exp\left\{-\frac{(x_1-\mu)^2 + (x_2-\mu)^2 + \ldots + (x_n-\mu)^2}{2\sigma^2}\right\}.$$

$$L \equiv \log P = -\frac{n}{2}\log(2\pi) - n\log\sigma - \frac{(x_1-\mu)^2 + (x_2-\mu)^2 + \ldots + (x_n-\mu)^2}{2\sigma^2}$$

$$\frac{\partial L}{\partial \mu} = \frac{(x_1-\mu) + (x_2-\mu) + \ldots + (x_n-\mu)}{\sigma^2}$$

$$= 0 \qquad \text{if } \mu \equiv \mu_0 = \frac{x_1 + x_2 + \ldots + x_n}{n} = \bar{x}$$

Hence the maximum likelihood estimator of μ is $\bar{x}$ and this estimator is normally distributed $\left(\mu, \frac{\sigma}{\sqrt{n}}\right)$.

Also, $\dfrac{\partial L}{\partial \sigma} = -\dfrac{n}{\sigma} + \dfrac{(x_1-\mu)^2 + (x_2-\mu)^2 + \ldots + (x_n-\mu)^2}{\sigma^3}$

Substituting $\mu = \bar{x}$ and putting $\dfrac{\partial L}{\partial \sigma} = 0$,

$$\sigma_0^2 = \frac{(x_1-\bar{x})^2 + (x_2-\bar{x})^2 + \ldots + (x_n-\bar{x})^2}{n}$$

$$= s^2, \text{ the sample variance.}$$

This is not an unbiased estimator of σ^2 (see §6.2.2), but it tends to be one as $n \to \infty$.

Ex. vii

Consider the bivariate population of x, y, such that when x is fixed y is normal (μ_x, σ), the mean varying with x but the variance remaining constant for all x.

Assume also that μ_x is a linear function of x (this is the problem of linear regression of Chapter XI).

Let $\mu_x = a + bx$. Then the probability of the pair (x, y), given x, is

$$\frac{1}{\sigma\sqrt{2\pi}} \exp\left\{-\frac{(y-\mu_x)^2}{2\sigma^2}\right\} = \frac{1}{\sigma\sqrt{2\pi}} \exp\left\{-\frac{(y-a-bx)^2}{2\sigma^2}\right\}$$

The likelihood of the set of pairs (x_1, y_1), (x_2, y_2) . . . (x_n, y_n), is

$$P = \frac{1}{\sigma^n(2\pi)^{n/2}} \cdot \exp\left\{\frac{(y_1-a-bx_1)^2 + (y_2-a-bx_2)^2 + \ldots + (y_n-a-bx_n)^2}{2\sigma^2}\right\}$$

$$L = \log P = \frac{n}{2}\log(2\pi) - n\log\sigma - \frac{1}{2\sigma^2}\Sigma(y-a-bx)^2$$

$$\frac{\partial L}{\partial a} = \frac{1}{\sigma^2}\Sigma(y \quad a - bx); \quad \frac{\partial L}{\partial b} = \frac{1}{\sigma^2}\Sigma x(y-a-bx)$$

$$\frac{\partial L}{\partial a} = 0, \frac{\partial L}{\partial b} = 0 \text{ if}$$

$$\Sigma y = na + b\Sigma x$$

$$\Sigma xy = a\Sigma x + b\Sigma x^2$$

These are the familiar 'least squares' equations of §10.2.1.

The problems so far considered are straightforward and have been presented to illustrate the working of the process. In the sections that follow, the principle of maximum likelihood is used to tackle a rather more difficult but quite usual type of problem.

13.5 Probit analysis

An interesting application of Maximum Likelihood Estimation is in the field of probit analysis. This type of problem can be introduced by an example.

Consider a batch of pellets each of which will explode when a sufficiently large impulse is imparted to it and is such that, in addition, the application of an impulse even if it fails to explode the pellet will affect its sensitivity for subsequent impulses. Hence we have a set of pellets, where each will have a (different) minimum impulse sufficient to explode it, and where each pellet can be tested only once with a preassigned impulse. The result of the experiment will be an explosion if the impulse is greater than that particular pellet's 'minimum effective impulse' and there will be no reaction if the impulse is less than that particular pellet's 'minimum effective impulse'. How can one estimate an impulse sufficient to explode (say) half the pellets or any given proportion of them?

This problem also arises frequently in toxicology. Given a batch of mice and a poison, what dose of poison when injected would be sufficient to kill a given proportion of the mice? Here the object being tested is a mouse, the stimulus is a dose of poison, the minimum stimulus is the minimum lethal dose and the result is either death or no death.

In general terms, given a unit with an unknown minimum effective stimulus (x), how can we obtain percentiles of the distribution of x?

The method of attacking this problem which is discussed here is to select a set of stimuli $x_1, x_2 \ldots x_r$ and to test random samples of the units, of sizes $n_1, n_2 \ldots n_r$ respectively, and to note the proportions $p_1, p_2 \ldots p_r$ respectively which respond to the stimulus. In the case of the pellets above, if the stimulus be that of a 1 lb. ball dropping a given distance on to the pellet, then the experiment could be arranged by dropping the ball 8 in. on to each of ten pellets, 9 in. on to each of ten pellets, up to (say) 12 in. on to each of ten pellets. Suppose that the numbers exploding are 1, 2, 5, 8, 9 (respectively).

Hence:

Stimulus (x_i)	*Sample size* (n_i)	*Number responding*	*Proportion responding* (p_i)
$x_1 = 8$	$n_1 = 10$	1	$p_1 = 0{\cdot}1$
$x_2 = 9$	$n_2 = 10$	2	$p_2 = 0{\cdot}2$
$x_3 = 10$	$n_3 = 10$	5	$p_3 = 0{\cdot}5$
$x_4 = 11$	$n_4 = 10$	8	$p_4 = 0{\cdot}8$
$x_5 = 12$	$n_5 = 10$	9	$p_5 = 0{\cdot}9$

These results can be plotted:

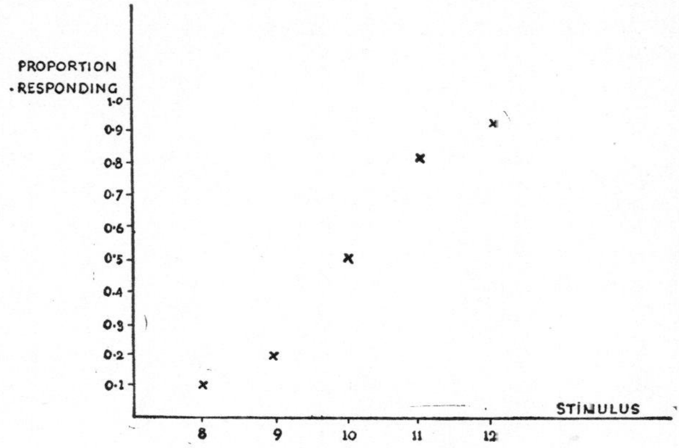

FIG. 46

In general, apart from sampling errors, one would expect the plot of proportion responding to increase with the impulse. Our problem is to find the 'true' or population curve relating proportion responding to the impulse.

The first stage is to relate the population curve of proportion responding against impulse to the distribution of minimum effective impulse, x.

Let this distribution be $g(x)\,dx$ (Fig. 47).

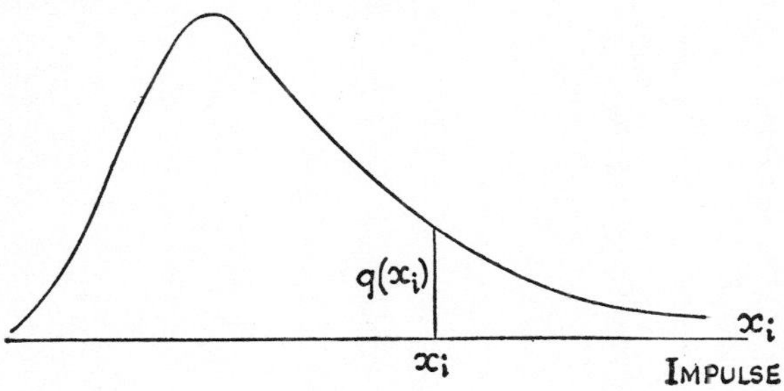

FIG. 47

For any impulse x_i all units with minimum effective impulse less than x_i will respond. The proportion of the population responding to impulse x_i is

$$p_i = \int_{-\infty}^{x_i} g(x)\,dx$$

The 'true' or population curve for proportion responding against impulse will be the cumulative $\int_{-\infty}^{x_i} g(x)\,dx$ against x_i (Fig. 48).

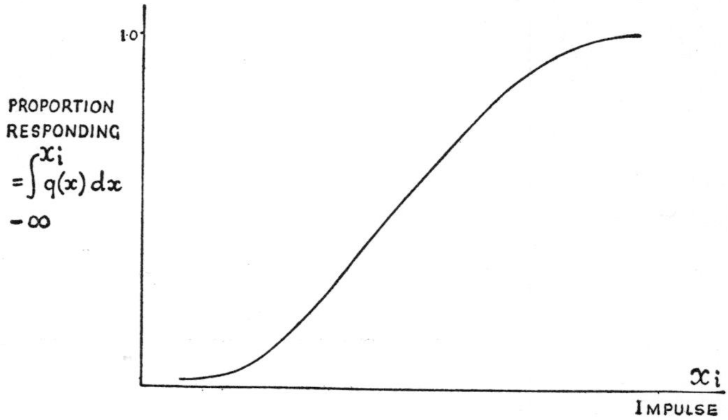

FIG. 48

$$\text{Proportion responding} = \int_{-\infty}^{x_i} g(x)\,dx$$

It is convenient for the solution of the problem to select a scale of impulse x, for which the distribution $g(x)\,dx$ is normal. In some cases the units in which x is initially measured are such that the minimum effective impulse is normally distributed. In other cases the logarithm of the impulse, $\log x$, is such that $g(\log x)\,d(\log x)$ is normal (Fig. 49).

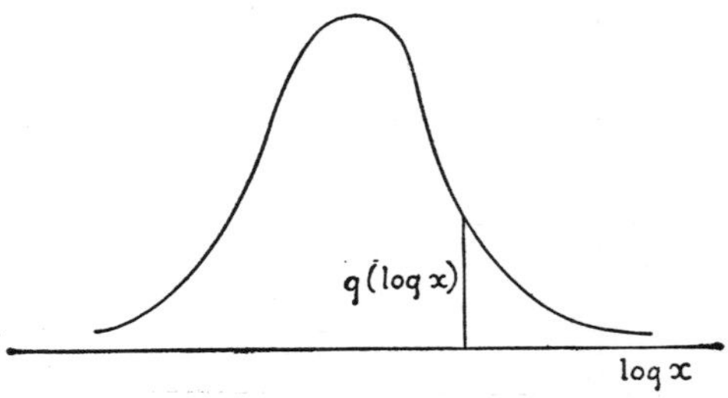

FIG. 49

In other cases $x^k (k \leqslant 1)$ suffices, i.e. $g(x^k)\,d(x^k)$ is a normal distribution function. In conformity with the literature of the subject, if x is a measure of the impulse such that $g(x)\,dx$ is normal, x is called the dosage and the untransformed (original) impulse is called the dose. The cumulative of such a normal $g(x)\,dx$ is called a normal sigmoid curve.

If $g(x) = \dfrac{1}{\sqrt{2\pi}\sigma} \exp\left\{-\dfrac{(x-\mu)^2}{2\sigma^2}\right\}$, the dosage for which a 50%

response can be expected is $x = \mu$ and the dosage for which an $R\%$ response can be expected is $\mu + t\sigma$, where t is a function of R.

The problem is to fit a normal sigmoid curve to the data. This is not done directly. The data are transformed to a form for which a straight line fit can be expected. The straight line is fitted by more or less usual methods and then the line is transformed back into the normal sigmoid curve.

The first step then is to transform, by means of what is known as the probit transformation, the normal sigmoid curve into a straight line.

The probit corresponding to a proportion P is defined as the abscissa, t, such that

$$P = \frac{1}{\sqrt{2\pi}}\int_{-\infty}^{t-5} e^{-x^2/2}\,dx$$

The use of the 5 is purely arbitrary and is merely to ensure that t is always positive and thus make tabulation more simple.

Table 7 gives the probits corresponding to proportions of the normal curve. It will be seen that the abscissae are 5 more than the corresponding abscissae of the normal curve (see Table 1).

Consider the data above. Assuming that the stimuli x_i are dosages, the transformation consists of replacing the proportions p_i by the corresponding probits, viz.:

Dosage	p	*Probit* $= y$
8	0·1	3·718
9	0·2	4·158
10	0·5	5·000
11	0·8	5·842
12	0·9	6·282

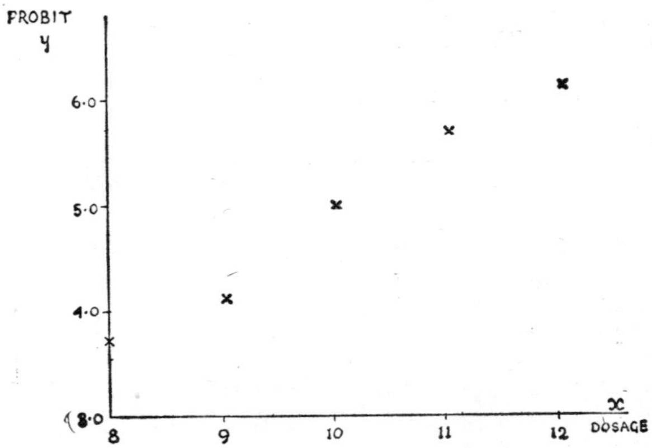

FIG. 50

What has been done effectively is to plot in Fig. 50, for each dosage, the normal equivalent deviation corresponding to the proportionate response.

If $g(x) = \frac{1}{\sigma\sqrt{2\pi}} \exp\left\{-\frac{(x-\mu)^2}{2\sigma^2}\right\}$ and if the response is p at dosage x, it can be expected that, apart from sampling causes,

$$\frac{x-\mu}{\sigma} = y - 5$$

where

$$\frac{1}{\sqrt{2\pi}}\int_{-\infty}^{y-5} e^{-x^2/2}\,dx = p$$

i.e.

$$x = \sigma y + (\mu - 5\sigma)$$

or

$$y = \frac{x}{\sigma} - \frac{(\mu - 5\sigma)}{\sigma}$$

If a rough answer only is needed, a straight-line fit by eye or by least squares will give a line of the form $y = ax + b$, where a estimates $\frac{1}{\sigma}$ and b estimates $\frac{5\sigma - \mu}{\sigma}$. The normal sigmoid curve is thus estimated and a solution has been obtained.

For example, the least squares fit to the points in the figure is

$$y = 0{\cdot}681x - 1{\cdot}81 \quad . \quad . \quad . \quad . \quad (4)$$

and the estimate of σ is $\frac{1}{0{\cdot}681} = 1{\cdot}47$ and the estimate of μ is $1{\cdot}47\,(1{\cdot}81 + 5) = 10{\cdot}00$.

For dosages 8, 9, 10, 11, 12 the corresponding expected probits Y (say) are 3·65, 4·33, 5·00, 5·69, 6·37, giving the following expected responses:

Dosage	*Expected Probit, Y*	*Expected Proportion, P*
8	3·65	0·09
9	4·33	0·25
10	5·00	0·50
11	5·69	0·75
12	6·37	0·92

Alternatively the expected dosage to produce any desired response can be calculated. For proportions 0·01, 0·05, 0·10, 0·15 respective probits are 2·674, 3·355, 3·718, 3·964 and equation (4) gives dosages 6·59, 7·58, 8·13, 8·48 respectively.

A flaw in the transformation above is that for responses of 0 or 1, i.e. complete failure or complete success, the probits are minus or plus infinity. No straight-line fit can take into account such ordinates unless some modification of method is used. For a quick rough answer such complete responses must be neglected and this throwing away of information is clearly unsatisfactory. The following sections will deal with this problem and will also develop methods of greater accuracy.

13.6 A more accurate method

It may well be that the methods so far outlined are not of sufficient precision. In that case an appeal to the principle of maximum likelihood yields a method of greater accuracy.

Suppose $L(x)$ is a function of x and two parameters α, β, and it is required to estimate α and β, so that $\partial L/\partial\alpha$ and $\partial L/\partial\beta$ vanish.

Let a and b be approximations or guesses to α and β respectively.

* Corrections to these values are $\delta\alpha$, $\delta\beta$ given by:

$$\left.\begin{aligned}\frac{\partial L}{\partial a} + \delta\alpha\,\frac{\partial^2 L}{\partial a^2} + \delta\beta\,\frac{\partial^2 L}{\partial a\,\partial b} &= 0\\ \frac{\partial L}{\partial b} + \delta\alpha\,\frac{\partial^2 L}{\partial a\,\partial b} + \delta\beta\,\frac{\partial^2 L}{\partial b^2} &= 0\end{aligned}\right\} \quad . \quad . \quad (5)$$

where the notation implies that the partial derivatives take their values at $\alpha = a$, $\beta = b$.

Now consider the present case where in a batch of n there are r responses and where the expected value of r/n is P and $Q = 1 - P$.

Then the chance of r responses out of n is $\dfrac{n!}{r!(n-r)!}P^r Q^{n-r}$.

If C is the chance of the set of results obtained for the different stimuli,

$$L \equiv \log C = (\text{constant}) + \Sigma r \log P + \Sigma(n-r)\log Q$$

$$\frac{\partial L}{\partial a} = \Sigma\frac{r}{P}\frac{\partial P}{\partial a} + \Sigma\frac{n-r}{Q}\frac{\partial Q}{\partial a}$$

$$= \Sigma\frac{n(p-P)}{PQ}\frac{\partial P}{\partial a}, \text{ where } np = r.$$

Similarly $\dfrac{\partial L}{\partial b} = \Sigma\dfrac{n(p-P)}{PQ}\dfrac{\partial P}{\partial b}$.

In equations (5) above, $\dfrac{\partial^2 L}{\partial a^2}, \dfrac{\partial^2 L}{\partial a \partial b}, \dfrac{\partial^2 L}{\partial b^2}$ may be replaced by their expectations, obtained by differentiating and putting $p = P$, viz.:

$$\frac{\partial^2 L}{\partial a^2} = \frac{\partial}{\partial a}\left(\Sigma\frac{n(p-P)}{PQ}\frac{\partial P}{\partial a}\right)$$

$$= \Sigma\frac{n(p-P)}{PQ}\frac{\partial^2 P}{\partial a^2} - \Sigma\frac{n}{PQ}\left(\frac{\partial P}{\partial a}\right)^2 - \Sigma\frac{n(p-P)}{P^2Q}\left(\frac{\partial P}{\partial a}\right)^2$$

$$- \Sigma\frac{n(p-P)}{PQ^2}\left(\frac{\partial P}{\partial a}\right)\left(\frac{\partial Q}{\partial a}\right)$$

Setting $p = P$ for the expectation, $\frac{\partial^2 L}{\partial a^2} = -\Sigma\frac{n}{PQ}\left(\frac{\partial P}{\partial a}\right)^2$

Hence equations (5) can be rewritten:

$$\left.\begin{aligned} \delta a\Sigma\frac{n}{PQ}\left(\frac{\partial P}{\partial a}\right)^2 + \delta b\Sigma\frac{n}{PQ}\frac{\partial P}{\partial a}\frac{\partial P}{\partial b} &= \Sigma\frac{n(p-P)}{PQ}\frac{\partial P}{\partial a} \\ \delta a\Sigma\frac{n}{PQ}\frac{\partial P}{\partial a}\frac{\partial P}{\partial b} + \delta b\Sigma\frac{n}{PQ}\left(\frac{\partial P}{\partial b}\right)^2 &= \Sigma\frac{n(p-P)}{PQ}\frac{\partial P}{\partial b} \end{aligned}\right\} \quad . \quad (6)$$

It is assumed that $P = \frac{1}{\sigma\sqrt{2\pi}}\int_{-\infty}^{x} e^{-\frac{(x-\mu)^2}{2\sigma^2}}\,dx$

and that $\quad \mathrm{Y} = \alpha + \beta x$, where $P = \frac{1}{\sqrt{2\pi}}\int_{-\infty}^{Y-5} e^{-\frac{1}{2}t^2}\,dt$

But $\quad \frac{\partial P}{\partial Y} = \frac{1}{\sqrt{2\pi}}e^{-\frac{1}{2}(Y-5)^2} \equiv Z$ (say)

and $\quad \frac{\partial P}{\partial \alpha} = \frac{\partial P}{\partial Y}\cdot\frac{dY}{d\alpha} = Z$

$$\frac{\partial P}{\partial \beta} = \frac{\partial P}{\partial Y}\cdot\frac{dY}{d\beta} = Zx$$

*

Thus if a and b are first approximations to α and β, better approximations are given by $a + \delta a$, $b + \delta b$ where

$$\left.\begin{aligned} \delta a\Sigma\frac{nZ^2}{PQ} + \delta b\Sigma\frac{nZ^2}{PQ}x &= \Sigma\frac{nZ^2}{PQ}\left(\frac{p-P}{Z}\right) \\ \delta a\Sigma\frac{nZ^2}{PQ}x + \delta b\Sigma\frac{nZ^2}{PQ}x^2 &= \Sigma\frac{nZ^2}{PQ}\left(\frac{p-P}{Z}\right)x \end{aligned}\right\} \quad . \quad . \quad (7)$$

Equations (7) are those which we would obtain for the linear regression of $(p - P)/Z$ on x, the weight for each point $[x, (p - P)/Z]$ being taken as nZ^2/PQ. The line obtained is

$$\frac{p-P}{Z} = \delta a + \delta b.x.$$

Which, with the first approximation

$$Y = a + bx$$

yields $Y + \frac{p - P}{Z} = (a + \delta a) + (b + \delta b)x.$

Hence we fit, by least squares, the best straight line through the points

$$\left(x,\ Y + \frac{p - P}{Z}\right),$$

each having weight nZ^2/PQ. Writing y for $Y + (p - P)/Z$, the new regression line is $y = a^1 + b^1x$ and the process may be repeated, if necessary, as in Ex. ix below, to obtain a still better fit.

The computational work has been simplified by Fisher and Yates (*Statistical Tables for Biological, Agricultural and Medical Research*, Oliver and Boyd), who have tabulated $w = \frac{Z^2}{PQ}$, $Y - \frac{P}{Z}$, $Y + \frac{Q}{Z}$, $\frac{1}{Z}$. These are reproduced in Table 8. The above may be summarized by saying:

(i) Make an estimate $Y = a + bx$ . . . . (8)
at the 'true' regression $Y = \alpha + \beta x$.

(ii) Evaluate $Z = \frac{1}{\sqrt{2\pi}} \exp\left\{-\frac{(Y-5)^2}{2}\right\}$,

$$P = \frac{1}{\sqrt{2\pi}} \int_{-\infty}^{Y-5} e^{-t^2/2} dt;$$

$Q = 1 - P$, where Y is given by equation (8) for each x,

and $w = \frac{Z^2}{PQ}$.

(iii) A better regression is then obtained by the least squares fit to the points $\left(x, Y + \frac{p - P}{Z}\right)$ with weighting coefficient at each point nw.

13.7.1 The case of complete failure or complete response

It is not possible to use points where $p = 0$ or 1, in the first approximation since y is infinite. Once the first approximation has been made we have an expected response, P, at the points where the data give $p = 0$ or 1, where

$$P = \frac{1}{\sqrt{2\pi}} \int_{-\infty}^{Y-5} e^{-t^2/2} dt,\ Y \text{ being given by the fitted line.}$$

Now $$\delta p = Z\delta y$$

i.e. $$Q - \frac{s}{n} = Z\delta y, \text{ where } r + s = n.$$

The average value of the deviation of y from its expected value when s out of n do not respond to the stimulus is

$$\frac{1}{Z}\left(Q - \frac{s}{n}\right)$$

When $$s = 0, \text{ this is } \frac{Q}{Z},$$

Hence the average value of y, for complete response, is $Y + \frac{Q}{Z}$, and similarly the average value of y for zero response is $Y - \frac{P}{Z}$. These are the ordinates to be used, with weights $\frac{nZ^2}{PQ}$, together with the other points, to get a further approximation to the regression line.

Ex. viii

Detonators are tested by dropping them on to a metal plate. The heights of drop are at one-inch intervals and twenty are tested at each height. Estimate from the following data the heights at which 1% can be expected to explode and the proportion that can be expected to explode at 14 in. (The heights can be regarded as dosages.)

Height of Drop (in.)	*Number Exploding*	*Proportion (p)*
10	0	0
11	2	0·10
12	3	0·15
13	7	0·35
14	13	0·65
15	15	0·75
16	17	0·85
17	20	1·00

The first stage is to transform the proportions to probits, ignoring the complete response or complete failure dosages, using Table 7.

Height (x)	*Proportion (p)*	*Probit (y)*
11	0·10	3·72
12	0·15	3·96
13	0·35	4·61
14	0·65	5·39
15	0·75	5·67
16	0·85	6·04

The first approximation can be taken by the least squares fit to the pairs of x, y in the above table.

The least squares fit is $y = 0{\cdot}51x - 1{\cdot}85$ . . . . . . (9)

Let Y, P, Z be the values corresponding to equation (9) at points where $x = 10, 11, 12 \ldots 17$.

Then, by Table 8:

p	x	Y	$Y - P/Z$	$1/Z$	p/Z	$Y - P/Z + p/Z$	w
0	10	3·2	2·74	12·67	0	2·74	0·18
0·10	11	3·8	3·21	5·15	0·52	3·73	0·37
0·15	12	4·3	3·53	3·20	0·48	4·01	0·53
0·35	13	4·8	3·72	2·56	0·90	4·62	0·63
0·65	14	5·3	3·68	2·62	1·70	5·38	0·62
0·75	15	5·8	3·08	3·45	2·59	5·67	0·50
0·85	16	6·3	1·03	5·83	4·96	5·99	0·34
1·00	17	6·8	7·26†	‡	‡	7·26	0·18

† This is $Y + Q/Z$.
‡ Not needed for complete response or complete failure.

The least squares fit to the pairs (x, $Y - P/Z + p/Z$) with weights w (‖) is:

$$y = 0{\cdot}55\,x - 2{\cdot}56 \quad . \quad . \quad . \quad . \quad (10)$$

This gives numbers expected at $x = 10, 11, 12 \ldots 17$ and a χ^2 evaluation as follows:

x:	10	11	12	13	14	15	16	17
y:	2·9	3·4	4·0	4·5	5·1	5·6	6·2	6·7
proportion expected $= p_e$:	0·02	0·06	0·16	0·31	0·54	0·71	0·85	0·96
number expected $= n_e$:	0·4	1·2	3·2	6·2	10·8	14·2	17·0	19·2
number obtained $= n_o$:	0	2	3	7	13	15	17	20
$(n_e - n_o)$:	0·4	0·8	0·2	0·8	2·2	0·8	0	0·8
$(n_e - n_o)^2$:	0·16	0·64	0·04	0·64	4·84	0·64	0	0·64
$\dfrac{(n_e - n_o)^2}{n_e}$:	0·40	0·53	0·12	0·10	0·45	0·04	0	0·03

$$\chi^2 = \Sigma(n_e - n_o)^2/n_e = 1{\cdot}67 \text{ with 7 d.f.}$$

This lies on the 97% level of significance, and so we can take

$y = 0{\cdot}55x - 2{\cdot}56$ as the probit-dosage relation.

When $x = 14$, $y = 5{\cdot}14$, and $p = 0{\cdot}56$
When $p = 0{\cdot}01$, $y = 2{\cdot}67$, and $x = 9{\cdot}5$ in.

Ex. ix

The proportions of mosquitoes killed by an insecticide in varying dosages are given in the following data. Obtain the population relationship and the minimum dosage sufficient to kill 99·9% of all mosquitoes.

Dosage x	*Number Tested*	*Number Killed*
4%	20	1
6%	10	4
7%	30	15
10%	40	30
12%	30	28
15%	20	20

For the first approximation, a least squares fit must be made to:

p	x	y
0·05	4	3·36
0·40	6	4·75
0·50	7	5·00
0·75	10	5·67
0·93	12	6·48

$$\text{viz.: } y = 0{\cdot}35x + 2{\cdot}3 \quad . \quad . \quad . \quad . \quad (11)$$

‖ See note at end of this section, §13.7.2

Using Table 8, and letting Y, P, Z be the values corresponding to $x = 4, 6, 7, 10, 12$ in equation 11.

p	x	Y	$Y - P/Z$	$1/Z$	p/Z	$Y - P/Z + p/Z$	w	wn
0·05	4	3·7	3·14	5·84	0·29	3·43	0·34	6·8
0·40	6	4·4	3·58	3·00	1·20	4·78	0·56	5·6
0·50	7	4·8	3·72	2·56	1·28	5·00	0·63	18·9
0·75	10	5·8	3·08	3·45	2·59	5·67	0·50	20·0
0·93	12	6·5	– 0·71	7·72	7·18	6·47	0·27	8·1
1·00	15	7·6	† 7·94	‡	‡	† 7·94	0·04	0·8

† This is $Y + Q/Z$.
‡ Not needed for complete response.

The least squares fit to the pairs $(x, Y - P/Z + p/Z)$ with weights w (||) is:

$$y = 0{\cdot}27x + 3{\cdot}3 \quad . \quad . \quad . \quad . \quad (12)$$

This gives expected numbers responding (corresponding to the unequal numbers) and a χ^2 evaluation as follows:

x:	4	6	7	10	12	15
p_e:	0·28	0·46	0·58	0·84	0·93	0·99
n_e:	5·6	4·6	16·6	33·2	27·9	19·8
n_o:	1·0	4·0	15·0	30·0	28·0	20·0
$n_e - n_o$:	4·6	0·6	1·6	3·2	0·1	0·2
$(n_e - n_o)^2$:	21·16	0·36	2·56	10·24	0·01	0·04
$\frac{(n_e - n_o)^2}{n_e}$:	3·78	0·08	0·15	0·31	0·00	0·00

$\chi^2 = 4{\cdot}32$ on 5 d.f. lying on the 50% level.
A further fit is therefore necessary.
Using $y = 0{\cdot}27x + 3{\cdot}3$ and repeating the process as above:

p	x	Y	$Y - P/Z$	$1/Z$	p/Z	$Y - P/Z + p/Z$	w	nw
0·05	4	3·4	2·91	9·01	0·45	3·36	0·24	4·8
0·40	6	4·8	3·72	2·62	1·05	4·77	0·62	6·2
0·50	7	6·0	2·52	4·13	2·06	4·58	0·44	13·2
0·75	10	5·7	3·27	3·20	2·40	5·67	0·53	21·2
0·93	12	6·5	– 0·71	7·72	7·18	6·48	0·27	5·4
1·00	15	7·9	† 8·21	‡	‡	† 8·21	0·02	0·6

† This is $Y + Q/Z$.
‡ Not needed for complete response.
The least squares fit to the pairs $(x, Y - P/Z + p/Z)$ with weights nw (||) is:

$$y = 0{\cdot}36x + 2{\cdot}1$$

For the χ^2 evaluation:

x:	4	6	7	10	12	15
p_e:	0·07	0·23	0·35	0·73	0·92	0·99
n_e:	1·4	2·3	10·5	29·2	27·6	19·8
n_o:	1	4	15	30	28	20
$(n_e - n_o)$:	0·4	1·7	4·5	0·8	0·4	0·2
$(n_e - n_o)^2$:	0·16	2·89	20·25	0·64	0·16	0·04
$\frac{(n_e - n_o)^2}{n_e}$:	0·11	1·26	1·93	0·02	0·01	0·00

$\chi^2 = 3{\cdot}33$ on 5 d.f. lying on the 70% level. We may thus take

$$y = 0{\cdot}36x + 2{\cdot}1$$

as the probit-dosage relationship.

|| See note at end of this section, §13.7.2.

For 99·9% response, $y = 8{\cdot}09$, and so the dosage for such a response is 16·6%.

As stated in §13.5, in the probit-dosage relation, $y = ax + b$, a estimates $\frac{1}{\sigma}$ and b estimates $\frac{5\sigma - \mu}{\sigma}$.

In this example, therefore, we may take $\sigma = 2{\cdot}78$, $\mu = 8{\cdot}06$ and the proportion responding, p, — dosage, x, relationship is:

$$p = \frac{1}{2{\cdot}78\sqrt{2\pi}} \int_{-\infty}^{x} \exp\left\{ - \frac{(t - 8{\cdot}06)^2}{2(2{\cdot}78)^2} \right\} dt$$

$$= 0{\cdot}143 \int_{-\infty}^{x} \exp\left\{ - 0{\cdot}0647\,(t - 8{\cdot}06)^2 \right\} dt$$

13.7.2 Note on fitting straight lines to weighted points

Let the points (x_i, y_i) have weights w_i.

To fit $y = ax + b$ we minimize $S \equiv \Sigma w_i\,(y_i - ax_i - b)^2$ by solving:

$$\frac{\partial s}{\partial a} = 0 = \Sigma w_i\, x_i\, y_i - a\Sigma w_i\, x_i^2 - b\Sigma w_i\, x_i$$

$$\frac{\partial s}{\partial b} = 0 = \Sigma w_i\, y_i - a\Sigma w_i\, x_i - b\Sigma w_i.$$

In the case of probit analysis, if for a given (x_i, y_i) there were n_i observations and the weighting coefficient was k_i, then we take $w_i = n_i\, k_i$.

Suggestions for further reading:

D. J. Finney, *Probit Analysis* (Cambridge University Press).
R. A. Fisher and F. Yates, *Statistical Tables for Biological, Agricultural and Medical Research,* Preface to 3rd Edition (Oliver and Boyd).
J. O. Irwin, *Supp., J.R.S.S.,* IV, I, 1937.

EXAMPLES ON CHAPTER XIII

1. A one-pound weight is dropped on to igniters and the following data result. Estimate the population height (dosage)—response relationship.

Height (in.)	*Number Tested*	*Number Responding*
15	5	1
16	5	2
17	5	3
18	5	3
19	5	4

2. The proportion of cells killed by a reagent at different concentrations is shown. Assuming that the logarithm of the concentration is a dosage, estimate the response to a concentration of 3·6. (The number of cells at each test may be regarded as constant.)

Concentration:	1·3	2·1	3·0	3·9	5·2
Proportionate response:	0·07	0·26	0·41	0 72	0·93

3. The numbers of rats killed by concentrations (dosages) of poison is shown. What dosage is sufficient to kill 60% of rats?

Dosage:	4·2	4·7	5·1	5·7	6·1	6·9
Number tested:	10	10	10	10	10	10
Number killed:	0	2	3	5	7	10

4. The proportion of valves failing to work in radio sets subjected to standard drop tests (dosages) is shown. Assuming that each set has five valves and will function if four valves are good, find the chance of a set failing owing to valve breakage after dropping a height of 20 ft.

Height of drop (ft.):	10	12	16	22	26
Number of valves tested:	100	80	95	45	80
Number of valves failing:	2	3	18	35	70

5. A weed killer is tried out in different concentrations. What concentration (dosage) is sufficient to kill 95% of the particular weed plants considered in the following data?

Concentration:	0·10	0·20	0·30	0·40	0·50
Number tested:	10	20	20	10	10
Number killed:	0	7	10	6	10

6. The proportions of igniters responding to an electrical impulse (dosage) are given below. What proportion can be expected to respond to an impulse of 12?

Impulse:	9	11	15	16	17	20
Number tested:	10	10	8	12	6	10
Number responding:	0	2	3	7	4	10

TABLES

Table 1

THE NORMAL PROBABILITY INTEGRAL

$$\frac{1}{\sigma\sqrt{2\pi}}\int_{\mu}^{x}\exp\left\{-\frac{(x-\mu)^2}{2\sigma^2}\right\}dx = \frac{1}{\sqrt{2\pi}}\int_{0}^{\frac{x-\mu}{\sigma}}\epsilon^{-t^2/2}dt.$$

$\frac{x-\mu}{\sigma}$	0	1	2	3	4	5	6	7	8	9	1	2	3	4	5	6	7	8	9
0	0000	0040	0080	0120	0160	0199	0239	0279	0319	0359	4	8	12	16	20	24	28	32	36
·1	0398	0438	0478	0517	0557	0596	0636	0675	0714	0753	4	8	12	16	20	24	28	32	36
·2	0793	0832	0871	0909	0948	0987	1026	1064	1103	1141	4	8	12	15	19	23	27	31	35
·3	1179	1217	1255	1293	1331	1368	1406	1443	1480	1517	4	8	11	15	18	23	26	30	34
·4	1555	1591	1628	1664	1700	1736	1772	1808	1844	1879	4	7	11	14	18	22	25	29	32
·5	1915	1950	1985	2019	2045	2088	2123	2157	2190	2224	3	7	10	14	17	21	24	27	31
·6	2257	2291	2324	2357	2389	2422	2454	2486	2517	2549	3	6	10	13	16	19	23	26	29
·7	2580	2611	2642	2673	2703	2734	2764	2794	2823	2852	3	6	9	12	15	18	21	24	27
·8	2881	2910	2939	2967	2995	3023	3051	3078	3106	3133	3	6	8	11	14	17	19	22	25
·9	3159	3186	3212	3238	3264	3289	3315	3340	3365	3389	3	5	8	10	13	15	18	20	23
1·0	3413	3438	3461	3485	3508	3531	3554	3577	3599	3621	2	5	7	9	11	14	16	18	21
1·1	3643	3665	3686	3708	3729	3749	3770	3790	3810	3830	2	4	6	8	10	12	14	16	19
1·2	3849	3869	3888	3907	3925	3944	3962	3980	3997	4015	2	4	5	7	9	11	13	15	16
1·3	4032	4049	4066	4082	4099	4115	4131	4147	4162	4177	2	3	5	6	8	10	11	13	14
1·4	4192	4207	4222	4236	4251	4265	4279	4292	4306	4319	1	3	4	6	7	8	10	11	13
1·5	4332	4345	4357	4370	4382	4394	4406	4418	4429	4441	1	2	4	5	6	7	8	10	11
1·6	4452	4463	4474	4484	4495	4505	4515	4525	4535	4545	1	2	3	4	5	6	7	8	9
1·7	4554	4564	4573	4582	4591	4599	4608	4616	4625	4633	1	2	3	3	4	5	6	7	8
1·8	4641	4649	4656	4664	4671	4678	4686	4693	4699	4706	1	1	2	3	4	4	5	6	6
1·9	4713	4719	4726	4732	4738	4744	4750	4756	4761	4767	1	1	2	2	3	4	4	5	5
2·0	4772	4778	4783	4788	4793	4798	4803	4808	4812	4817	0	1	1	2	2	3	3	4	4
2·1	4821	4826	4830	4834	4838	4842	4846	4850	4854	4857									
2·2	4861	4865	4868	4871	4875	4878	4881	4884	4887	4890									
2·3	4893	4896	4898	4901	4904	4906	4909	4911	4913	4916									
2·4	4918	4920	4922	4925	4927	4929	4931	4932	4934	4936									
2·5	4938	4940	4941	4943	4946	4947	4948	4949	4951	4952									
2·6	4953	4955	4956	4957	4959	4960	4961	4962	4963	4964									
2·7	4965	4966	4967	4968	4969	4970	4971	4972	4973	4974									
2·8	4974	4975	4976	4977	4977	4978	4979	4979	4980	4981									
2·9	4981	4982	4982	4983	4984	4984	4985	4985	4986	4986									
3	4987	4990	4993	4995	4997	4998	4998	4999	4999	4999									

P	*One tail*	*Both tails*
·1	1·282	1·645
·05	1·645	1·960
·01	2·326	2·576
·001	3·090	3·291
·0001	3·719	3·891
·00001	4·265	4·417

These values are half those given by P(x) on page 62.

Table 2

THE χ^2– DISTRIBUTION

Degree of freedom ν	χ^2 as a function of ν and p (= 100P)													
	p = 99	98	95	90	80	70	50	30	20	10	05	02	01	00·1
1	0·000	0·001	0·004	0·016	0·064	0·148	0·455	1·074	1·642	2·706	3·841	5·412	6·635	10·827
2	0·020	0·040	0·103	0·211	0·446	0·713	1·386	2·408	3·219	4·605	5·991	7·824	9·210	13·815
3	0·115	0·185	0·352	0·584	1·005	1·424	2·366	3·665	4·642	6·251	7·815	9·837	11·341	16·268
4	0·297	0·429	0·711	1·064	1·649	2·195	3·357	4·878	5·989	7·779	9·488	11·668	13·277	18·465
5	0·554	0·752	1·145	1·610	2·343	3·000	4·351	6·064	7·289	9·236	11·070	13·388	15·086	20·517
6	0·872	1·134	1·635	2·204	3·070	3·828	5·348	7·231	8·558	10·645	12·592	15·033	16·812	22·457
7	1·239	1·564	2·167	2·833	3·822	4·671	6·346	8·383	9·803	12·017	14·067	16·622	18·475	24·322
8	1·646	2·032	2·733	3·490	4·594	5·527	7·344	9·524	11·030	13·362	15·507	18·168	20·090	26·125
9	2·088	2·532	3·325	4·168	5·380	6·393	8·343	10·656	12·242	14·684	16·919	19·679	21·666	27·877
10	2·558	3·059	3·940	4·865	6·179	7·267	9·342	11·781	13·442	15·987	18·307	21·161	23·209	29·588
11	3·053	3·609	4·575	5·578	6·989	8·148	10·341	12·899	14·631	17·275	19·675	22·618	24·725	31·264
12	3·571	4·178	5·226	6·304	7·807	9·034	11·340	14·011	15·812	18·549	21·026	24·054	26·217	32·909
13	4·107	4·765	5·892	7·042	8·634	9·926	12·340	15·119	16·985	19·812	22·362	25·472	27·688	34·528
14	4·660	5·368	6·571	7·790	9·467	10·821	13 339	16·222	18·151	21·064	23·685	26·873	29·141	36·123
15	5·229	5·985	7·261	8·547	10·307	11·721	14·339	17·322	19·311	22·307	24·996	28·259	30·578	37·697
16	5·812	6·614	7·962	9·312	11·152	12·624	15·338	18·418	20·465	23·542	26·296	29·633	32·000	39·252
17	6·408	7·255	8·672	10·085	12·002	13·531	16·338	19·511	21·615	24·769	27·587	30·995	33·409	40·790
18	7·015	7·906	9·390	10·865	12·857	14·440	17·338	20·601	22·760	25·989	28·869	32·346	34·805	42·312
19	7·633	8·567	10·117	11·651	13·716	15·352	18·338	21·689	23·900	27·204	30·144	33·687	36·191	43·820
20	8·260	9·237	10·851	12·443	14·578	16·266	19·337	22·775	25·038	28·412	31·410	35·020	37·566	45·315
21	8·897	9·915	11·591	13·240	15·445	17·182	20·337	23·858	26·171	29·615	32·671	36·343	38·932	46·797
22	9·542	10·600	12·338	14·041	16·314	18·101	21·337	24·939	27·301	30·813	33·924	37·659	40·289	48·268
23	10·196	11·293	13·091	14·848	17·187	19·021	22·337	26·018	28·429	32·007	35·172	38·968	41·638	49·728
24	10·856	11·992	13·848	15·659	18·062	19·943	23·337	27·096	29·553	33·196	36·415	40·270	42·980	51·179
25	11·524	12·697	14·611	16·473	18·940	20·867	24·337	28·172	30·675	34·382	37·652	41·566	44·314	52·620
26	12·198	13·409	15·379	17·292	19·820	21·792	25·336	29·246	31·795	35·563	38·885	42·856	45·642	54·052
27	12·879	14·125	16·151	18·114	20·703	22·719	26·336	30·319	32·912	36·741	40·113	44·140	46·963	55·476
28	13·565	14·847	16·928	18·939	21·588	23·647	27·336	31·391	34·027	37·916	41·337	45·419	48·278	56·893
29	14·256	15·574	17·708	19·768	22·475	24·577	28·336	32·461	35·139	39·087	42·557	46·693	49·588	58·302
30	14·953	16·306	18·493	20·599	23·364	25·508	29·336	33·530	36·250	40·256	43·773	47·962	50·892	59·703

Table 2 is reprinted from Table IV of Fisher and Yates, *Statistical Tables for Biological, Agricultural and Medical Research* published by Oliver and Boyd, Ltd., Edinburgh, by permission of the authors and publishers.

Table 3

TABLE OF *t*

Degrees of freedom ν	*t* as a function of ν and p (= 100P)												
	p = 90	80	70	60	50	40	30	20	10	05	02	01	00·1
1	0·158	0·325	0·510	0·727	1·000	1·376	1·963	3·078	6·314	12·706	31·821	63·657	636·619
2	0·142	0·289	0·445	0·617	0·816	1·061	1·386	1·886	2·920	4·303	6·965	9·925	31·598
3	0·137	0·277	0·424	0·584	0·765	0·978	1·250	1·638	2·353	3·182	4·541	5·841	12·941
4	0·134	0·271	0·414	0·569	0·741	0·941	1·190	1·533	2·132	2·776	3·747	4·604	8·610
5	0·132	0·267	0·408	0·559	0·727	0·920	1·156	1·476	2·015	2·571	3·365	4·032	6·859
6	0·131	0·265	0·404	0·553	0·718	0·906	1·134	1·440	1·943	2·447	3·143	3·707	5·959
7	0·130	0·263	0·402	0·549	0·711	0·896	1·119	1·415	1·895	2·365	2·998	3·499	5·405
8	0·130	0·262	0·399	0·546	0·706	0·889	1·108	1·397	1·860	2·306	2·896	3·355	5·041
9	0·129	0·261	0·398	0·543	0·703	0·883	1·100	1·383	1·833	2·262	2·821	3·250	4·781
10	0·129	0·260	0·397	0·542	0·700	0·879	1·093	1·372	1·812	2·228	2·764	3·169	4·587
11	0·129	0·260	0·396	0·540	0·697	0·876	1·088	1·363	1·796	2·201	2·718	3·106	4·437
12	0·128	0·259	0·395	0·539	0·695	0·873	1·083	1·356	1·782	2·179	2·681	3·055	4·318
13	0·128	0·259	0·394	0·538	0·694	0·870	1·079	1·350	1·771	2·160	2·650	3·012	4·221
14	0·128	0·258	0·393	0·537	0·692	0·868	1·076	1·345	1·761	2·145	2·624	2·977	4·140
15	0·128	0·258	0·393	0·536	0·691	0·866	1·074	1·341	1·753	2·131	2·602	2·947	4·073
16	0·128	0·258	0·392	0·535	0·690	0·865	1·071	1·337	1·746	2·120	2·583	2·921	4·015
17	0·128	0·257	0·392	0·534	0·689	0·863	1·069	1·333	1·740	2·110	2·567	2·898	3·965
18	0·127	0·257	0·392	0·534	0·688	0·862	1·067	1·330	1·734	2·101	2·552	2·878	3·922
19	0·127	0·257	0·391	0·533	0·688	0·861	1·066	1·328	1·729	2·093	2·539	2·861	3·883
20	0·127	0·257	0·391	0·533	0·687	0·860	1·064	1·325	1·725	2·086	2·528	2·845	3·850
21	0·127	0·257	0·391	0·532	0·686	0·859	1·063	1·323	1·721	2·080	2·518	2·831	3·819
22	0·127	0·256	0·390	0·532	0·686	0·858	1·061	1·321	1·717	2·074	2·508	2·819	3·792
23	0·127	0·256	0·390	0·532	0·685	0·858	1·060	1·319	1·714	2·069	2·500	2·807	3·767
24	0·127	0·256	0·390	0·531	0·685	0·857	1·059	1·318	1·711	2·064	2·492	2·797	3·745
25	0·127	0·256	0·390	0·531	0·684	0·856	1·058	1·316	1·708	2·060	2·485	2·787	3·725
26	0·127	0·256	0·390	0·531	0·684	0·856	1·058	1·315	1·706	2·056	2·479	2·779	3·707
27	0·127	0·256	0·389	0·531	0·684	0·855	1·057	1·314	1·703	2·052	2·473	2·771	3·690
28	0·127	0·256	0·389	0·530	0·683	0·855	1·056	1·313	1·701	2·048	2·467	2·763	3·674
29	0·127	0·256	0·389	0·530	0·683	0·854	1·055	1·311	1·699	2·045	2·462	2·756	3·659
30	0·127	0·256	0·389	0·530	0·683	0·854	1·055	1·310	1·697	2·042	2·457	2·750	3·646
40	0·126	0·255	0·388	0·529	0·681	0·851	1·050	1·303	1·684	2·021	2·423	2·704	3·551
60	0·126	0·254	0·387	0·527	0·679	0·848	1·046	1·296	1·671	2·000	2·390	2·660	3·460
120	0·126	0·254	0·386	0·526	0·677	0·845	1·041	1·289	1·658	1·980	2·358	2·617	3·373
∞	0·126	0·253	0·385	0·524	0·674	0·842	1·036	1·282	1·645	1·960	2·326	2·576	3·291

Table 3 is reprinted from Table III of Fisher and Yates, *Statistical Tables for Biological, Agricultural and Medical Research*, published by Oliver and Boyd, Ltd., Edinburgh, by permission of the authors and publishers.

Table 4

THE VARIANCE RATIO (1)

p	ν_2 \ ν_1	1	2	3	4	5	6	8	10	15	20	30	Inf.
10	1	39·9	49·5	53·6	55·8	57·2	58·2	59·4	60·2	61·2	61·7	62·3	63·3
5		161	199	216	225	230	234	239	242	246	248	250	254
1		4050	5000	5400	5630	5760	5860	5980	6060	6160	6210	6360	6370
·1		405000	500000	540000	563000	576000	586000	598000	606000	616000	621000	627000	637000
10	2	8·53	9·00	9·16	9·24	9·29	9·33	9·37	9·39	9·42	9·44	9·46	9·49
5		18·5	19·0	19·2	19·2	19·3	19·3	19·4	19·4	19·4	19·4	19·5	19·5
1		98·5	99·0	99·2	99·3	99·3	99·4	99·3	99·4	99·4	99·5	99·5	99·5
·1		999	999	999	999	999	999	999	999	999	1000	1000	1000
10	3	5·54	5·46	5·39	5·34	5·31	5·28	5·25	5·23	5·20	5·18	5·16	5·13
5		10·1	9·55	9·28	9·12	9·01	8·94	8·84	8·78	8·70	8·66	8·62	8·53
1		34·1	30·8	29·5	28·7	28·2	27·9	27·5	27·2	26·9	26·7	26·5	26·1
·1		167	148	141	137	135	133	131	128	127	125	124	123
10	4	4·54	4·32	4·19	4·11	4·05	4·01	3·95	3·92	3·87	3·84	3·82	3·76
5		7·72	6·94	6·59	6·39	6·26	6·16	6·04	5·96	6·86	5·80	5·75	5·63
1		21·2	18·0	16·7	16·0	15·5	15·2	14·8	14·5	14·2	14·0	13·8	13·5
·1		74·1	61·3	56·2	53·4	51·7	50·5	49·0	48·2	46·9	46·2	45·6	44·1
10	5	4·06	3·78	3·62	3·52	3·45	3·40	3·34	3·30	3·24	3·21	3·17	3·10
5		6·61	5·79	5·41	5·19	5·05	4·95	4·82	4·74	4·62	4·56	4·50	4·36
1		16·3	13·3	12·1	11·4	11·0	10·7	10·3	10·1	9·72	9·55	9·38	9·00
·1		47·0	36·6	33·2	31·1	29·8	28·8	27·6	26·9	25·8	25·4	24·9	23·8

$p = 100P$

THE VARIANCE RATIO (2)

p	ν_2 \ ν_1	1	2	3	4	5	6	8	10	15	20	30	Inf.
10	6	3·78	3·46	3·29	3·18	3·11	3·05	2·98	2·94	2·87	2·84	2·80	2·72
5		5·99	5·14	4·76	4·53	4·39	4·28	4·15	4·06	3·94	3·87	3·81	3·67
1		13·7	10·9	9·78	9·15	8·75	8·47	8·10	7·86	7·56	7·39	7·23	6·88
·1		35·5	27·0	23·7	21·9	20·8	20·0	19·0	18·4	17·6	17·2	16·7	15·8
10	7	3·59	3·26	3·07	2·96	2·88	2·83	2·75	2·70	2·63	2·59	2·56	2·47
5		5·59	4·74	4·35	4·12	3·97	3·87	3·73	3·63	3·52	3·44	3·38	3·23
1		12·2	9·55	8·45	7·85	7·46	7·19	6·84	6·62	6·31	6·15	5·99	5·65
·1		29·2	21·7	18·8	17·2	16·2	15·5	14·6	14·2	13·3	13·0	12·5	11·7
10	8	3·46	3·11	2·92	2·81	2·73	2·67	2·59	2·54	2·46	2·42	2·38	2·29
5		5·32	4·46	4·07	3·84	3·69	3·58	3·44	3·34	3·22	3·14	3·08	2·93
1		11·3	8·65	7·59	7·01	6·63	6·37	6·03	5·82	5·52	5·36	5·20	4·86
·1		25·4	18·5	15·8	14·4	13·5	12·9	12·0	11·6	10·8	10·5	10·1	9·34
10	9	3·36	3·01	2·81	2·69	2·61	2·55	2·47	2·42	2·34	2·30	2·25	2·16
5		5·12	4·26	3·86	3·63	3·48	3·37	3·23	3·13	3·01	2·93	2·86	2·71
1		10·6	8·02	6·99	6·42	6·06	5·80	5·47	5·26	4·96	4·80	4·65	4·31
·1		22·9	16·4	13·9	12·6	11·7	11·1	10·4	10·0	9·23	8·88	8·55	7·81
10	10	3·28	2·92	2·73	2·61	2·52	2·46	2·38	2·32	2·24	2·20	2·15	2·06
5		4·96	4·10	3·71	3·48	3·33	3·22	3·07	2·97	2·84	2·77	2·70	2·54
1		10·	7·56	6·55	5·99	5·64	5·39	5·06	4·85	4·56	4·41	4·25	3·91
·1		21·0	14·9	12·6	11·3	10·5	9·92	9·20	8·80	8·13	7·80	7·48	6·76

$p = 100P$

THE VARIANCE RATIO (3)

p	ν_2 \ ν_1	1	2	3	4	5	6	8	10	15	20	30	Inf.
10	12	3·18	2·81	2·61	2·48	2·39	2·33	2·24	2·19	2·10	2·06	2·01	1·90
5		4·75	3·88	3·49	3·26	3·11	3·00	2·85	2·76	2·62	2·54	2·47	2·30
1		9·33	6·93	5·95	5·41	5·06	4·82	4·50	4·30	4·01	3·86	3·70	3·36
·1		18·6	13·0	10·8	9·63	8·89	8·38	7·71	7·27	6·65	6·38	6·10	5·42
10	14	3·10	2·73	2·52	2·39	2·31	2·24	2·15	2·10	2·00	1·96	1·90	1·80
5		4·60	3·74	3·34	3·11	2·96	2·85	2·70	2·60	2·46	2·39	2·31	2·13
1		8·86	6·51	5·56	5·03	4·69	4·46	4·14	3·94	3·66	3·50	3·34	3·00
·1		17·1	11·8	9·73	8·62	7·92	7·43	6·80	6·40	5·81	5·58	5·27	4·60
10	16	3·05	2·67	2·46	2·33	2·24	2·18	2·09	2·03	1·93	1·89	1·83	1·72
5		4·49	3·63	3·24	3·01	2·85	2·74	2·59	2·49	2·36	2·28	2·20	2·01
1		8·53	6·23	5·29	4·77	4·44	4·20	3·89	3·69	3·41	3·25	3·10	2·75
·1		16·1	11·0	9·00	7·94	7·27	6·81	6·19	5·81	5·23	5·02	4·71	4·06
10	18	3·01	2·62	2·42	2·29	2·20	2·13	2·04	1·98	1·88	1·84	1·78	1·66
5		4·41	3·55	3·16	2·93	2·77	2·66	2·51	2·41	2·27	2·19	2·11	1·92
1		8·28	6·01	5·09	4·58	4·25	4·01	3·71	3·51	3·23	3·07	2·91	2·57
·1		15·4	10·4	8·49	7·46	6·81	6·35	5·76	5·37	4·83	4·58	4·30	3·67
10	20	2·97	2·59	2·38	2·25	2·16	2·09	2·0	1·94	1·84	1·79	1·74	1·61
5		4·35	3·49	3·10	2·87	2·71	2·60	2·45	2·35	2·20	2·12	2·04	1·84
1		8·10	5·85	4·94	4·43	4·10	3·87	3·56	3·37	3·09	2·92	2·78	2·42
·1		14·8	9·95	8·10	7·10	6·46	6·02	5·44	5·06	4·52	4·26	4·02	3·38

$p = 100P$

THE VARIANCE RATIO (4)

p	ν_2 \ ν_1	1	2	3	4	5	6	8	10	15	20	30	Inf.
10	25	2·92	2·53	2·32	2·18	2·09	2·02	1·93	1·87	1·77	1·70	1·66	1·52
5		4·24	3·38	2·99	2·76	2·60	2·49	2·34	2·24	2·09	2·00	1·92	1·71
1		7·77	5·57	4·68	4·18	3·86	3·63	3·32	3·13	2·85	2·70	2·54	2·17
·1		13·9	9·22	7·45	6·49	5·88	5·46	4·91	4·54	4·04	3·79	3·53	2·89
10	30	2·88	2·49	2·28	2·14	2·05	1·98	1·88	1·82	1·72	1·67	1·61	1·46
5		4·17	3·32	2·92	2·69	2·53	2·42	2·27	2·16	2·01	1·93	1·84	1·62
1		7·56	5·39	4·51	4·02	3·70	3·47	3·17	2·98	2·70	2·55	2·39	2·01
·1		13·3	8·77	7·05	6·12	5·53	5·12	4·58	4·25	3·76	3·50	3·24	2·59
10	50	2·83	2·42	2·21	2·07	1·98	1·90	1·80	1·73	1·63	1·56	1·50	1·33
5		4·03	3·18	2·79	2·56	2·40	2·29	2·13	2·02	1·87	1·78	1·69	1·44
1		7·17	5·06	4·20	3·72	3·41	3·18	2·88	2·70	2·42	2·26	2·10	1·68
·1		12·2	7·94	6·32	5·46	4·90	4·49	3·98	3·66	3·19	2·95	2·69	2·03
10	100	2·76	2·36	2·14	2·02	1·92	1·83	1·73	1·66	1·56	1·49	1·42	1·21
5		3·94	3·09	2·70	2·46	2·30	2·19	2·03	1·92	1·77	1·68	1·57	1·28
1		6·90	4·82	3·98	3·51	3·20	2·99	2·69	2·51	2·25	2·06	1·89	1·43
·1		11·8	4·40	5·88	5·02	4·48	4·09	3·61	3·28	2·81	2·57	2·31	1·62
10	Inf.	2·71	2·30	2·08	1·94	1·85	1·77	1·67	1·60	1·49	1·42	1·34	1·00
5		3·84	2·99	2·60	2·37	2·21	2·09	1·94	1·83	1·66	1·57	1·46	1·00
1		6·64	4·60	3·78	3·32	3·02	2·80	2·51	2·32	2·03	1·87	1·69	1·00
·1		10·8	6·91	5·42	4·62	4·10	3·74	3·27	2·96	2·50	2·26	2·01	1·00

$p = 100P$

Most of Table 4 is reprinted from Table V of Fisher and Yates, *Statistical Tables for Biological, Agricultural and Medical Research*, published by Oliver and Boyd, Ltd., Edinburgh, by permission of the authors and publishers.

Table 5

TABLE OF r

d.f.	0·10	0·05	0·02	0·01	0·001
1	·988	·997	·999	1·000	1·000
2	·900	·950	·980	·990	·999
3	·805	·878	·934	·959	·991
4	·729	·811	·882	·917	·974
5	·669	·754	·833	·874	·951
6	·621	·707	·789	·834	·925
7	·582	·666	·750	·798	·898
8	·549	·632	·716	·765	·872
9	·521	·602	·685	·735	·847
10	·497	·576	·658	·708	·823
11	·476	·553	·634	·684	·801
12	·457	·532	·612	·661	·780
13	·441	·514	·592	·641	·760
14	·426	·497	·574	·623	·742
15	·412	·482	·558	·606	·725
16	·400	·468	·543	·590	·708
17	·389	·456	·528	·575	·693
18	·378	·444	·516	·561	·679
19	·369	·433	·503	·549	·665
20	·360	·423	·492	·537	·652
25	·323	·381	·445	·487	·597
30	·296	·349	·409	·449	·554
35	·275	·325	·381	·418	·519
40	·257	·304	·358	·393	·490
45	·243	·287	·338	·372	·465
50	·231	·273	·322	·354	·443
60	·211	·250	·295	·325	·408
70	·195	·232	·274	·302	·380
80	·183	·217	·256	·283	·357
90	·173	·205	·242	·267	·337
100	·164	·195	·230	·254	·321

Table 6

RANDOM NUMBERS

03 47 43 73 86	36 96 47 36 61	46 98 63 71 62	33 26 16 80 45
97 74 24 67 62	42 81 14 57 20	42 53 32 37 32	27 07 36 07 51
16 76 62 27 66	56 50 26 71 07	32 90 79 78 53	13 55 38 58 59
12 56 85 99 26	96 96 68 27 31	05 03 72 93 15	57 12 10 14 21
55 59 56 35 64	38 54 82 46 22	31 62 43 09 90	06 18 44 32 53
16 22 77 94 39	49 54 43 54 82	17 37 93 23 78	87 35 20 96 43
84 41 17 53 31	57 24 55 06 88	77 04 74 47 67	21 76 33 50 25
63 01 63 78 59	16 95 55 67 19	98 10 50 71 75	12 86 73 58 07
33 21 12 34 29	78 64 56 07 82	52 42 07 44 38	15 51 00 13 42
57 60 86 32 44	09 47 27 96 54	49 17 46 09 62	90 52 84 77 27
18 18 07 92 46	44 17 16 58 09	79 83 86 19 62	06 76 50 03 10
26 62 38 97 75	84 16 07 44 99	83 11 46 32 24	20 14 85 88 45
23 42 40 64 74	82 97 77 77 81	07 45 32 14 08	32 98 94 07 72
52 36 28 19 95	50 92 26 11 97	00 56 76 31 38	80 22 02 53 53
37 85 94 35 12	83 39 50 08 30	42 34 07 96 88	54 42 06 87 98
70 29 17 12 13	40 33 20 38 26	13 89 51 03 74	17 76 37 13 04
56 62 18 37 35	96 83 50 87 75	97 12 25 93 47	70 33 24 03 54
99 49 57 22 77	88 42 95 45 72	16 64 36 16 00	04 43 18 66 79
16 08 15 04 72	33 27 14 34 09	45 59 34 68 49	12 72 07 34 45
31 16 93 32 43	50 27 89 87 19	20 15 37 00 49	52 85 66 60 44
68 34 30 13 70	55 74 30 77 40	44 22 78 84 26	04 33 46 09 52
74 57 25 65 76	59 29 97 68 60	71 91 38 67 54	13 58 18 24 76
27 42 37 86 53	48 55 90 65 72	96 57 69 36 10	96 46 92 42 45
00 39 68 29 61	66 37 32 20 30	77 84 57 03 29	10 45 65 04 26
29 94 98 94 24	68 49 69 10 82	53 75 91 93 30	34 25 20 57 27
16 90 82 66 59	83 62 64 11 12	67 19 00 71 74	60 47 21 29 68
11 27 94 75 06	06 09 19 74 66	02 94 37 34 02	76 70 90 30 86
35 24 10 16 20	33 32 51 26 38	79 78 45 04 91	16 92 53 56 16
38 23 16 86 38	42 38 97 01 50	87 75 66 81 41	40 01 74 91 62
31 96 25 91 47	96 44 33 49 13	34 86 82 53 91	00 52 43 48 85

Table 6 is abridged from Table XXXIII of Fisher and Yates, *Statistical Tables for Biological, Agricultural and Medical Research*, published by Oliver and Boyd, Ltd., Edinburgh, by permission of the authors and publishers.

Table 7

RELATION BETWEEN PERCENTAGES AND PROBITS

Percentage	0	1	2	3	4	5	6	7	8	9
0	—	2·67	2·95	3·12	3·25	3·36	3·45	3·52	3·59	3·66
10	3·72	3·77	3·82	3·87	3·92	3·96	4·01	4·05	4·08	4·12
20	4·16	4·19	4·23	4·26	4·29	4·33	4·36	4·39	4·42	4·45
30	4·48	4·50	4·53	4·56	4·59	4·61	4·64	4·67	4·69	4·72
40	4·75	4·77	4·80	4·82	4·85	4·87	4·90	4·92	4·95	4·97
50	5·00	5·03	5·05	5·07	5·10	5·13	5·15	5·18	5·20	5·23
60	5·25	5·28	5·31	5·33	5·36	5·39	5·41	5·44	5·47	5·50
70	5·52	5·55	5·58	5·61	5·64	5·67	5·71	5·74	5·77	5·81
80	5·84	5·88	5·92	5·95	5·99	6·04	6·08	6·13	6·18	6·23
90	6·28	6·34	6·41	6·48	6·55	6·64	6·75	6·88	7·05	7·33

For example, the probit corresponding to **25 %** is **4·33**.

Table 7 is abridged from Table IX of Fisher and Yates, *Statistical Tables for Biological, Agricultural and Medical Research*, published by Oliver and Boyd, Ltd., Edinburgh, by permission of the authors and publishers.

Table 8.

Weighting Coefficients and Probit Values to

(Adapted from

Expected Probit Y	*Minimum Working Probit* $Y - P/Z$	*Range* $1/Z$	*Maximum Working Probit* $Y + Q/Z$	*Weighting Coefficient* Z^2/PQ
1·1	0·8579	5034	5035	·00082
1·2	0·9522	3425	3426	·00118
1·3	1·0462	2354	2355	·00167
1·4	1·1400	1634	1635	·00235
1·5	1·2335	1146	1147	·00327
1·6	1·3266	811·5	812·8	·00451
1·7	1·4194	580·5	581·9	·00614
1·8	1·5118	419·4	420·9	·00828
1·9	1·6038	306·1	307·7	·01104
2·0	1·6954	225·6	227·3	·01457
2·1	1·7866	168·00	169·79	·01903
2·2	1·8772	126·34	128·22	·02459
2·3	1·9673	95·96	97·93	·03143
2·4	2·0568	73·62	75·68	·03977
2·5	2·1457	57·05	59·20	·04979
2·6	2·2340	44·654	46·888	·06169
2·7	2·3214	35·302	37·623	·07563
2·8	2·4081	28·189	30·597	·09179
2·9	2·4938	22·736	25·230	·11026
3·0	2·5786	18·522	21·101	·13112
3·1	2·6624	15·240	17·902	·15436
3·2	2·7449	12·666	15·411	·17994
3·3	2·8261	10·633	13·459	·20774
3·4	2·9060	9·015	11·921	·23753
3·5	2·9842	7·721	10·705	·26907
3·6	3·0606	6·6788	9·7394	·30199
3·7	3·1351	5·8354	8·9705	·33589
3·8	3·2074	5·1497	8·3571	·37031
3·9	3·2773	4·5903	7·8676	·40474
4·0	3·3443	4·1327	7·4770	·43863
4·1	3·4083	3·7582	7·1665	·47144
4·2	3·4687	3·4519	6·9206	·50260
4·3	3·5251	3·2025	6·7276	·53159
4·4	3·5770	3·0010	6·5780	·55788
4·5	3·6236	2·8404	6·4640	·58099
4·6	3·6643	2·7154	6·3797	·60052
4·7	3·6982	2·6220	6·3202	·61609
4·8	3·7241	2·5573	6·2814	·62741
4·9	3·7407	2·5192	6·2599	·63431
5·0	3·7467	2·5066	6·2533	·63662

Table 8 is reprinted from Table XI of Fisher and Yates, *Statistical Tables for Biological, Agricultural and Medical*

PROBITS

be used in Adjustments of Special Accuracy

Bliss, 1935)

Expected Probit Y	*Minimum Working Probit* $Y - P/Z$	*Range* $1/Z$	*Maximum Working Probit* $Y + Q/Z$	*Weighting Coefficient* Z^2/PQ
5·1	3·7401	2·5192	6·2593	·63431
5·2	3·7186	2·5573	6·2759	·62742
5·3	3·6798	2·6220	6·3018	·61609
5·4	3·6203	2·7154	6·3357	·60052
5·5	3·5360	2·8404	6·3764	·58099
5·6	3·4220	3·0010	6·4230	·55788
5·7	3·2724	3·2025	6·4749	·53159
5·8	3·0794	3·4519	6·5313	·50260
5·9	2·8335	3·7582	6·5917	·47144
6·0	2·5230	4·1327	6·6557	·43863
6·1	2·1324	4·5903	6·7227	·40474
6·2	1·6429	5·1497	6·7926	·37031
6·3	1·0295	5·8354	6·8649	·33589
6·4	0·2606	6·6788	6·9394	·30199
6·5	− 0·705	7·721	7·0158	·26907
6·6	− 1·921	9·015	7·0940	·23753
6·7	− 3·459	10·633	7·1739	·20774
6·8	− 5·411	12·666	7·2551	·17994
6·9	− 7·902	15·240	7·3376	·15436
7·0	− 11·101	18·522	7·4214	·13112
7·1	− 15·230	22·736	7·5062	·11026
7·2	− 20·597	28·189	7·5919	·09179
7·3	− 27·623	35·302	7·6786	·07564
7·4	− 36·888	44·654	7·7661	·06168
7·5	− 49·20	57·05	7·8543	·04979
7·6	− 65·68	73·62	7·9432	·03977
7·7	− 87·93	95·96	8·0327	·03143
7·8	− 118·22	126·34	8·1228	·02458
7·9	− 159·79	168·00	8·2134	·01903
8·0	− 217·3	225·6	8·3046	·01457
8·1	− 297·7	306·1	8·3962	·01104
8·2	− 410·9	419·4	8·4882	·00828
8·3	− 571·9	580·5	8·5806	·00614
8·4	− 802·8	811·5	8·6734	·00451
8·5	− 1137	1146	8·7666	·00327
8·6	− 1625	1634	8·8600	·00235
8·7	− 2345	2354	8·9538	·00167
8·8	− 3416	3425	9·0478	·00118
8·9	− 5025	5034	9·1421	·00082

Research, published by Oliver and Boyd, Ltd., Edinburgh, by permission of the authors and publishers.

Chap. II, p. 24.

1. 3·87, 3·69.
2. 5·844, 0·161.
3. 0·158, 5·843.
4. 1·5 thou., 0·05 (slight positive skewness).
5.

	σ	*m.d.*	v	*Range*	*Inter-quartile range*
A .	21	17	2	120	27·2
B .	22	21·4	2·1	60	45·6

6. 46·46, 46·92. Effect of grouping: the results would be equal if the digits 0, 1 . . . 9 occurred with equal frequency.
7. (a) square root of $\frac{\omega}{2\pi}\int_0^{2\pi/\omega} I^2 dt$

   (b) square root of $\frac{1}{A}\int y^2 dA$

   (c) square root of $\frac{1}{N}\Sigma(x - \bar{x})^2$. All are r.m.s. values.
8. There are two peaks, centred at ages 2 and 37, corresponding to persons born in 1945 and 1910 respectively. A full discussion is impossible without knowledge of expected death rates. The most significant feature is the high proportion of old persons which may be expected as the persons near the second peak age become older. The relatively small numbers of the succeeding generation reflect the falling birth rate between the wars.
9. Mean = 1·756, median = mode = 1.
10. Mean = 1002·45 lb.wt., s.d. = 9·97 lb.wt.
11. Mean = 82·08, s.d. = 3·66.
12. $\frac{1}{2}\sqrt{33} = 2{\cdot}87$.

Chap. III, p. 45.

1. (i) 1/13; (ii) 1/270725.
2. (i) 44/4165; (ii) $4 \times {}^{16}C_{13}/{}^{52}C_{13}$; (iii) $16/{}^{52}C_{13}$; (iv) $4/{}^{52}C_{13}$.
3. ${}^{32}C_3/{}^{51}C_3$.
4. $2/(n - 1)$.
5. 33 to 2, 3/7.
6. 14/15.
7. 13!/(7!3!3!), 8!/(3!2!3!).
8. 7/8.
9. 0·98.

10. Probability of 4, 4 is 1/36, of 3, 5 or 5, 3 is 1/18, of 2, 6 or 6, 2 is 1/18.
11. (a) 0·285, (b) 0·608.
12. 323.
13. 0·88.
14. 2·393, 1·472; 2·4, 1·486.
15. $r/m = \log(0\cdot9)/m \log(1 - 1/m)$.
16. 0·0264; 2, 0·302; $10^{-7}(1\cdot02)$.
17. $\beta_1 = (q - p)^2/npq$.
18. (i) 3/50; (ii) 3/25.
19. (i) 1/8; (ii) 4/25.
20. (i) 3/4; (ii) 1/4.
21. 0·652.
22. 0·0152.
23. 23/144.
24. 1/36, 0·0277.

Chap. IV, p. 67.

1. (i) 0·6065, 0·3032, 0·0758, 0·0126, 0·0016, 0·0002;
(ii) 0·3679, 0·3679, 0·1839, 0·0613, 0·0153, 0·0031;
(iii) 0·0498, 0·1494, 0·2241, 0·2241, 0·1680, 0·1008.
2. 0·6065, 0·0903.
3. Mean = 1·87, variance = 1·76.
4. (i) 0·1755; (ii) 0·0821.
5. 0·9513.
6. m = 0·128, s.d. = 0·366.
20941, 2680, 171·5, 7·3, 0·2, 0·0.
7. 0·2231, 0·1913; 0·3904, 0·1872.
8. 9.
9. 5·6, 23·2, 48·4, 67·3, 70·0, 58·3, 40·5, 24·1, 12·6, 5·8, 2·5, 0·9.
10. (a) (i) Normal; (ii) Poisson; 0·3679.
(b) (i) Normal; (ii) Poisson; 0·6065.
12. 0·0062, 0·468.
13. 0·9994.
14. m = 699·7 hours, σ = 19·05 hours; 0·9955.
15. 0, $1/a^2$, $2/a^3$, $9/a^4$.
16. 69·5 in. approx.
17. 0·1074 oz.
18. 50 ± 6·62; 100 ± 9·36 ohms.
19. 0·0228, 0·001 approx. (q^n = 0·0009886, Poisson gives 0·00107).
20. 10·038 oz.

Chap. V, p. 93.

1. Linear interpolation gives 0·144. Draw graph of P against χ^2 for better value.
3. Expected frequencies 14·2, 26·5, 24·8, 15·4, 11·1 (four or more defects). $\chi^2 = 1{\cdot}47$, $\nu = 3$, $0{\cdot}5 < P < 0{\cdot}7$, good fit.
4. $\chi^2 = 12{\cdot}9$, $\nu = 9$, $0{\cdot}1 < P < 0{\cdot}2$; fair fit.
5. Expected frequencies 1·1, 2·5, 6·0, 11·7, 17·1, 19·9, 17·8, 12·6, 7·0, 3·0, 1·4. Pooling the first three and the last three frequencies, $\chi^2 = 1{\cdot}2$, $\nu = 4$; good fit.
6. Pooling the first and the last three frequencies, $\chi^2 = 3{\cdot}0$, $\nu = 4$; good fit.
7. $\chi^2 = 0{\cdot}025$, $\nu = 6$. Very good fit (too good to be true!).
8. $\chi^2 = 3{\cdot}76$, $\nu = 4$, $P = 0{\cdot}5$ nearly; hypothesis confirmed.
9. Using Yates' correction, $\chi^2 = 3{\cdot}06$, $\nu = 1$, $P > 0{\cdot}05$. Using χ as normal deviate with $\sigma = 1$, $P = 0{\cdot}08$. The hypothesis is not contradicted.
12. $\chi^2 = 4{\cdot}22$, $\nu = 2$, $P > 0{\cdot}1$. The theory is fairly well supported.
13. $\chi^2 = 9{\cdot}87$, $\nu = 1$, $P < 0{\cdot}01$. Difference significant.
14. Sections 1 and 2.
15. $\chi^2 = 2{\cdot}33$, $\nu = 2$, $0{\cdot}3 < P < 0{\cdot}5$; not significant.
    $\chi^2 = 7$, $\nu = 2$, $0{\cdot}02 < P < 0{\cdot}05$; significant.
16. The degree of association is very slight; not significant. Given the actual weight and sp. gr. of each pellet, the degree of association could be expressed by the correlation coefficient. (See Chap. XI.)

Chap. VI, p. 121.

1. For mean of differences, $t = 2{\cdot}2$, $\nu = 9$, P just $> 0{\cdot}05$. Some likelihood that improvement is real.
   For difference of means, $t = 0{\cdot}82$, $\nu = 18$. Increase not significant.
2. A: $t = 3{\cdot}58$, $P < 0{\cdot}01$; significant (reads high).
   B: $t = 3{\cdot}36$, $P < 0{\cdot}01$; significant (reads low).
   $F = 2{\cdot}29$, $P > 0{\cdot}1$; not significant.
3. $t = 2{\cdot}39$, $\nu = 10$, $P < 0{\cdot}05$; significant. Experiment more conclusive using method of matched pairs.
4. $t = 4{\cdot}29$, $\nu = 13$, $P < 0{\cdot}001$; highly significant. Control group were not of comparable quality; their high scores in first test might leave little room for improvement.
5. $t = 0{\cdot}94$, $\nu = 14$; not significant.
6. A: $t = 1{\cdot}67$, $\nu = 7$, $P > 0{\cdot}1$; not significant.
   B: Obviously not significant offset.
7. $t = 2{\cdot}325$, $\nu = 5$, $0{\cdot}05 < P < 0{\cdot}1$; doubtful.

8. $F = 2{\cdot}06$, $\nu_1 = \nu_2 = 19$; not significant at 0·05 level.
9. $F = 1{\cdot}26$; not significant.
   $t = 2{\cdot}4$, $\nu = 18$, $0{\cdot}05 > P > 0{\cdot}02$. A reads significantly higher.
10. Largest $F = 5$, nearly, and, as a single ratio, would be significant at the 0·01 level. But Bartlett's test gives P slightly less than 0·1, so that there is not conclusive evidence of varying consistency.
11. $\chi^2 = 3{\cdot}71$, $\nu = 3$; $0{\cdot}2 < P < 0{\cdot}3$; not significantly different.
12. $t = 0{\cdot}64$, $\nu = 10$; no significant difference.
    $t_1 = 1{\cdot}82$, $\nu = 5$; $t_2 = 1{\cdot}9$, $\nu = 5$. Neither mean is significantly lower than 1000.

Chap. VII, p. 137.

1. Mean: 0·5096, 0·5057, 0·4923, 0·4884 in.
   Range: 0·030, 0·008 in.
2. Mean: 0·0593, 0·0510, 0·0500, 0·0417 in.
   Range: 0·0025, 0·0007 in.
3. 8.
4. Ordinary mean: 1012·0, 1007·5, 992·5, 988·0. Range 34, 9.
   Modified mean: 1092·5, 1088·0, 912·0, 907·5.
5. Mean at sample 15. Fresh ordinary mean: 1030·6, 1026·1, 1011·1, 1006·6.
6. Mean and range at sample 22. Modified control invalid.
   Fresh ordinary mean: 1066·2, 1057·4, 1027·8, 1019·0. Range 66, 13.
7. 2.
8. At sample 14. Fresh limit $c = 3$.
9. 0·1152 in. and 0·1128 in.
10. 10·012 in. and 9·988 in.

Chap. VIII, p. 155.

1. (i) A; (ii) B.
2. 0·215.
3. (a) 0·904, (b) 0·106; 10·2%.
4. 132 (Poisson), 130 (Binomial).
5. Allow 3 defectives in sample of 102, 0·04.
6. Allow 8 defectives in sample of 175.
7. 0·953, 24·96.
8. (a) Accept; (b) Reject.
10. $s = 0{\cdot}025$, $a = 1{\cdot}33$, $P = 0{\cdot}729$.
11. $s = 0{\cdot}011$, $a = 0{\cdot}000574$, $m = m_2$.

Chap. IX, p. 168.

1. 1570·509, 0·022 ft.
2. No; p.e. = 0·41; max. residual = 1·2.
3. $\sqrt{(e_1^2 + e_2^2)}$.
4. 17, 13·5, 14·9.
5. 0·2037 cm., 0·0003 cm.
   $t = 2{\cdot}6$, $\nu = 18$; significant at 0·02 level.
6. 85 sq. ft., 118 sq. ft.
7. 0·255.
8. 0·34 ft.
9. $\{(W_1 - W_2)^2e^2 + (W - W_2)^2e_1^2 + (W - W_1)^2e_2^2\}^{\frac{1}{2}}/(W - W_2)^2$, where e_j, e_1, e_2 are the p.e.'s of the weights.
11. 3/4.
12. The variances of the three results are in the ratios $\Sigma l^2/4$; $\Sigma l^2/8$; $\Sigma l^2/2$. The weights are inversely proportional to the variances. 102·478 ft.
13. $a\sqrt{n}/\sqrt{3}$, 1·35a.
14. 1,875.

Chap. X, p. 192.

1. $y = 0{\cdot}38 + 0{\cdot}38x$.
2. (i) $y = 51 - 0{\cdot}125x$; (ii) $x = 407 - 7{\cdot}96y$.
3. $y = 54{\cdot}86 + 0{\cdot}500x$.
4. $R = 143{\cdot}9 + 0{\cdot}826v + 0{\cdot}466v^2$.
   156·8, 179·1, 210·6, 251·5, 301·7, 361·2.
5. $- 0{\cdot}024 + 1{\cdot}196x - 0{\cdot}338x^2$.
6. $0{\cdot}959 + 0{\cdot}383x - 1{\cdot}095x^2$;
   $y = b(0{\cdot}959a^2 + 0{\cdot}383ax - 1{\cdot}095x^2)/a^2$.
7. 99·9980, 99·99755, 99·9976, 299·9931, 0·00067 ft.
8. Corrected figures for seconds are 08·4, 46·7, 22·1, 15·4.
9. $[wxy]/[wx^2]$.
10. $p = 5{\cdot}82$, $q = 9{\cdot}7275$, $r = 4{\cdot}5775$ chains.
12. Corrected figures for seconds are 29·6, 15·15, 20·6.
13. Corrected figures for seconds are 18·2, 39·3, 02·5.
14. Corrected figures for seconds are 26·7, 32·7, 09·4, 47·8, 03·4.
15. B, 54·06; C, 175·68; D, 53·02.
16. $a = 0{\cdot}992$, $b = 0{\cdot}236$.
17. $m = 1{\cdot}98$, $n = 3{\cdot}08$.
18. $A = 3{\cdot}03$, $k = - 0{\cdot}141$.
19. $A = 3{\cdot}0025$, $k = - 0{\cdot}139$.

Chap. XI, p. 217.

1. $y = 0{\cdot}055x - 0{\cdot}74$; 0·07.
2. — 0·74.
4. 0·8485, 0·7071 ; 0·2582, 0·7303, 0·5477.
5. — 0·59.
6. 0·45; $\chi^2 = 4{\cdot}7$, $\nu = 1$, $0{\cdot}02 < P < 0{\cdot}05$; evidence of association.
8. Sig. (nearly 0·001), sig. (0·05), sig. (0·015 approx.).
10. $a = \sigma_1/\sigma_2$, $b = -\sigma_1/\sigma_2$.
11. 1 in 250,000, 1 in 10.
12. — 0·68.
13. 5/11.
14. 0·78.
15. $y = 3{\cdot}70 + 0{\cdot}214x$; the variance of y.
16. $r = 0{\cdot}51$, η_{xy}(estimate) $= 0{\cdot}54$, $\eta_{yx} = 0{\cdot}57$.
17. $r = 0{\cdot}76$ approx.
18. 0·49.
19. $\varrho_1/\sqrt{(1 - \varrho_2^2)}$; — 0·45.

Chap. XII, p. 249.

1. No ($F = 1{\cdot}0$).
2. Yes, < 5% significance ($F = 3{\cdot}9$).
3. (A) Yes ($F = 1{\cdot}25$).
   (B) No, < 0·1% significance ($F = 25$) $\sigma_\mu^2 = 10{\cdot}8$.
4. No, < 0·1 significance ($F = 25$) $\sigma_\mu^2 = 9{\cdot}6$.
5. Evidence of operator and machine effects ($F = 4{\cdot}9$ and 3·8 respectively).
6. Evidence that machine means are different ($F = 4{\cdot}3$) $\sigma_\mu =$ 0·035.
7. Between machine means effect ($F = 7{\cdot}5$). $\sigma_\mu^2 = 0{\cdot}031$; 2·20.
8. (a) Yes ($F = 5{\cdot}4$), (b) Yes ($F = 66{\cdot}2$), (c) Yes ($F = 5{\cdot}2$).
9. Between machines; barely significant ($F = 2{\cdot}48$).
   Between operatives; significant ($F = 3{\cdot}28$).
10. (a) Between doctors; just significant ($F = 3{\cdot}4$).
    (b) Between treatments; highly significant ($F = 50$).

Chap. XIII, p. 269.

1. $y = 0{\cdot}39x - 1{\cdot}57$ (least squares), $y = 0{\cdot}42x - 2{\cdot}05$ (first approximation).
2. 0·69.
3. 5·85.
4. 0·88.
5. 0·74.
6. 0·26.

INDEX